LIBRARY, R.A.E., WESTCOTT

REGULATIONS FOR BORROWERS

1.   Books are issued on loan for a period of 1 month
and must be returned to the Library promptly.

2.   Before books are taken from the Library receipts
for them must be filled in, signed, and handed to a
member of the Library Staff.   Receipts for books
received through the internal post must be signed
and returned to the Library immediately.

3.   Readers are responsible for books which they
have borrowed, and are required to replace any such
books which they lose.   In their own interest they
are advised not to pass on to other readers books
they have borrowed.

4.   To enable the Library Staff to deal with urgent
requests for books, borrowers who expect to be absent
for more than a week are requested either to arrange
for borrowed books to be made available to the P.A.
or Clerk to the Section, or to return them to the
Library for safekeeping during the period of absence.

MECHANICS AND ACOUSTICS

BLACKIE & SON LIMITED
66 Chandos Place, LONDON
17 Stanhope Street, GLASGOW

BLACKIE & SON (INDIA) LIMITED
103/5 Fort Street, BOMBAY

BLACKIE & SON (CANADA) LIMITED
TORONTO

PHYSICAL PRINCIPLES

OF

MECHANICS
AND ACOUSTICS

BY

R. W. POHL

Professor of Physics in the University of Göttingen

Authorized Translation by

WINIFRED M. DEANS, M.A., B.Sc.

Late of Newnham College

BLACKIE & SON LIMITED

LONDON AND GLASGOW

First published 1932
Reprinted 1947

Printed in Great Britain by Blackie & Son, Ltd., Glasgow

FROM THE PREFACE TO THE FIRST GERMAN EDITION

This book contains the first part of my course of lectures on experimental physics. The second part (*Physical Principles of Electricity and Magnetism*) has already appeared. A concluding volume on Heat and Light is to follow.

I have tried to make the book as simple as possible, so that it may be useful not only to students and teachers of physics but to others interested in the subject.

The book differs considerably from the ordinary textbooks in subject-matter. A great deal has been left out, and not only such matters as the division of the metre into 1000 millimetres, the air pump, and the external appearance of a gramophone; more extensive omissions were unavoidable in addition. Only thus has it been possible to find room for more important topics, such as the indispensable idea of angular momentum as a vector and the general principles of wave propagation.

Fundamental experiments are brought into prominence throughout; they are intended to make the ideas concerned as clear as possible to the student and to indicate the orders of magnitude involved, quantitative details being left in the background.

Some of the experiments require a good deal of space. In the Göttingen lecture-room a smooth parquet floor 12 metres by 5 is available. That troublesome obstacle in old lecture-rooms, the large fixed lecture-table, was got rid of years ago. Instead, small tables are set up as required. These, however, are not fixed to the floor, any more than the furniture in a living-room is. Owing to these handy tables the experimental arrangements gain considerably in clarity and convenience. Most of the tables can be turned round and raised or lowered as required. Troublesome perspective effects due to one piece of apparatus blocking the view of another are thus avoided. The apparatus in actual use at any moment can be made to stand out so that it is easily seen by each member of the audience.

The pieces of apparatus required are simple and not numerous. Many of them are described here for the first time. Like the ordinary lecture apparatus they are obtainable from Spindler and Hoyer, G.m.b.H., Göttingen.*

Most of the figures are based on photographs, almost all by my chief mechanic, Herr Sperber, to whom I am very much obliged for his continual help. Many of the figures have again been made into silhouettes. This form of illustration is well adapted for printing and usually gives an idea of the size of the apparatus used. Finally, a silhouette indicates whether an experiment is suited to a large lecture-room, for it is then particularly important that the outlines should be clear and uninterrupted by subsidiary material, such as clamp-stands and the like.

*[In Great Britain from Messrs. Baird and Tatlock (London), Ltd., Hatton Garden, E.C. 1, and, for the rotation experiments, from Messrs. Nibbs and Scott, Titchfield, Hants.]

PREFACE TO THE ENGLISH EDITION

This English translation is based essentially on the text of the second German edition, but corrections and additions have been made at some points, and the notation for the various physical quantities has been altered to accord with English usage throughout. A collection of examples (compiled in part by Dr. O. Stasiw) has also been added.

R. W. POHL.

GÖTTINGEN, *June, 1932.*

CONTENTS

MECHANICS

CHAPTER I

INTRODUCTION: MEASUREMENT OF LENGTH AND TIME

CHAPTER II

METHODS OF REPRESENTING MOTION (KINEMATICS)

CHAPTER III

THE FUNDAMENTAL THEOREMS OF DYNAMICS

vii

viii CONTENTS

CONTENTS

CHAPTER VII

THE ROTATION OF RIGID BODIES

CHAPTER VIII

ACCELERATED SYSTEMS OF REFERENCE

CHAPTER IX

LIQUIDS AND GASES AT REST

CHAPTER X

THE MOTION OF LIQUIDS AND GASES

ACOUSTICS

CHAPTER XI

VIBRATIONS

CHAPTER XII

WAVES AND RADIATION

xii

MECHANICS

CHAPTER I

Introduction: Measurement of Length and Time

1. Introduction.

According to tradition the subject of physics is divided into Mechanics, Sound, Heat, Light, and Electricity. This classification of physical phenomena is based on our direct sense-impressions. In the present state of physical knowledge, however, we find that it possesses decided weaknesses. Frequently it takes no account of important but not immediately obvious relationships. This may be illustrated by an example from Heat. Suppose we are warming our hands at a hot steam-pipe. We may do this in two ways, either by actually touching the pipe, or by merely bringing our hands *near* the pipe without touching it. In the light of modern knowledge the phenomena concerned in the two cases are entirely different. In the case of *direct contact* the skin is excited by *mechanical* vibrations of the substance of the pipe, of very high frequency. These *mechanical* vibrations are essentially the same as those, of a lower frequency,* which affect another of our sense organs, i.e. the ear. To put it briefly, the heat of the pipe, like that of all other material bodies, consists in mechanical vibrations which are no longer audible; the heat of the pipe forms a special case of an *acoustical* problem. The second case, where the hand is merely brought *near* the hot pipe, is quite different. Here the skin is excited by radiation which is emitted by the pipe and traverses the air of the room. We are now concerned with *electric* waves which are essentially the same as those which are used in wireless telegraphy or which constitute visible light and X-rays. They differ from these only as regards one numerical quantity, namely, the frequency. The problem of the heat received by radiation accordingly belongs to Electricity or to Light.

Examples of this kind may be multiplied indefinitely. Nevertheless,

---

* *Frequency* means the number of vibrations per second.

little is to be gained by discarding the traditional subdivision of physical phenomena. On penetrating more deeply into the study of physics we soon come to realize both the advantages and the disadvantages of the old classification, and also the direction in which physics is tending to advance, namely, towards an increasing and often truly astonishing unification of phenomena which seem to be essentially different, together with a deliberate exclusion of all anthropomorphic features. The person observing or describing natural phenomena retreats farther and farther into the background. Yet even to-day, in spite of all the advances that have been made in this direction, more of our personal subjective impressions still remain concealed in our physical thinking than the beginner would suspect.

By way of introduction we shall describe some of the difficulties (now overcome) which the observer meets with as a result of the behaviour of our principal sense organ, the eye.

(a) *The Colours of Shadows*.—Fig. 1 shows a white wall W, an incandescent gas-light, and an electric lamp. P is any opaque body,

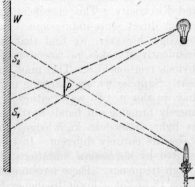

say a piece of cardboard. To begin with, let only the electric lamp be switched on, the gaslight being screened off, or not lighted. The white wall is then illuminated all over except for the region of shadow S_1. Light from the electric lamp does not reach this region. We mark the region in some way, e.g. by sticking on bits of paper. Then nothing *physical* takes place in the marked region when the electric lamp is switched on. We next switch off the electric lamp and light the gas.

Fig. 1.—The colours of shadows

The wall again shines white, *including* the marked region S_1, the black shadow thrown by the cardboard being now at S_2. Now for the actual experiment. The electric light is switched on while the gas is still burning. This makes not the slightest physical or objective difference to the region S_1. Yet what we see is completely altered. At S_1, we see a vivid *olive-green* shadow, in strong contrast to the shadow at S_2, which now appears reddish-brown. But the light thrown by S_1 on the retina of our eye still comes from the gas-light only, just as before. All that has happened is that the region S_1 is now enclosed in a bright *border* due to the light from the electric lamp, and it is this border that is alone responsible for the very striking change in the colour of S_1.

There is nothing significant in the choice of the two lamps: if we replace the electric lamp by an arc lamp we obtain beautiful yellow and lilac shadows.

The experiment is a very instructive one for the beginner. *Colour* is not a subject for physics, but for psychology or physiology. Neglect of this has led to an immense amount of unprofitable work.

(b) *The Apparent Shape of the Sky.*—In the open we see the sky as a flattened bell, in section somewhat as in fig. 2. This is an every-day observation, largely independent of the weather and of the time of day. A number of observers are asked in turn to indicate with their arms or walking-sticks the point of the sky (P) which they think is equidistant from the zenith (Z) and the horizon (H), i.e. they

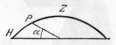

Fig. 2.—Shape of the sky for an observer in the open

are to attempt to halve the arc joining zenith and horizon. It is found that all the observers are remarkably unanimous in raising their arms or walking-sticks to make an angle (α) of only 20° or 30° with the horizontal, never an angle of 45°; no one sees the sky as a hemisphere.

The same observers are then placed with their backs to a high tower, e.g. that of a wireless transmitting station. The experiment now gives quite a different result, the arms or walking-sticks being raised to about 50° from the horizontal. To an eye guided by a vertical line the sky appears like a high vault some-what as in fig. 3. Thus the appearance of the sky is essentially altered by the inclusion of the vertical line. The whole phenomenon again belongs not to physics, but to psychology.

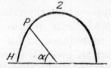

Fig. 3.—Shape of the sky for an observer at the foot of a high tower.

(c) *Mach's Bands.*—Suppose we have a piece of paper with a white strip on the left and a black strip on the right, joined by a continuous succession of all tones of grey. Let it be illuminated by day or artificial light. The quantity of light reaching our eye from this striped paper should be dis-tributed over the breadth of the paper as shown in fig. 4.

This may be realized experimentally in many different ways. As an example we may take the circular disc with a white star on a black ground, shown in fig. 5. The disc is set in rapid rotation by some kind of motor. The points of the star then appear

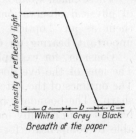

Fig. 4.—How Mach's bands arise

thoroughly blurred, and we obtain the required distribution of the reflected light in the form of concentric rings. The circle within the inner points of the star corresponds to the region *a* in fig. 4 and the ring outside the outer points of the star to the region *c*. Between them lies the intermediate region *b*, in which the quantity of light decreases continuously as the radius increases, since the breadth of the white points steadily diminishes.

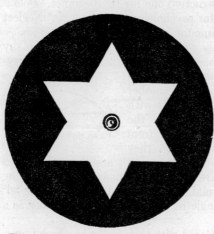

Observation, however, yields the surprising result reproduced in fig. 6 (Plate I). The inner bright circle is surrounded by a white border, while the black ring is bounded on the inner side by an even blacker border. According to the forcible evidence of our eyes the white border appears to send the most light to the eye and the black border the least. Any unsuspecting observer would suppose that the reflection of the light takes place to the greatest and least extents in these rings. Or to put it another way: the photograph is reproduced in fig. 6 (Plate I) by a half-tone block. In this reproduction process an increasing degree of blackness is obtained by an increase in the size of small isolated equidistant black dots. Any unsuspecting observer would expect to find the smallest dots in the region of the white border and the largest in the region of the black border. But it is not so. Examination of both sides of the grey ring with a magnifying glass shows that the size of the dots varies continuously throughout as we pass into the white or black zone.

Fig. 5.—Rapid rotation of this disc gives rise to the appearance shown in fig. 6 (Plate I)

These Mach's bands have led to much trouble in the carrying out of physical observations. But we must not ill-humouredly dismiss them as an "optical illusion", for the phenomenon has a most important bearing on our eyesight.

Consider, for example, the reading of black print on white paper. The lens of the eye does not by any means depict objects perfectly. The outlines of the letters on the retina at the back of the eye are not sharp. The boundary between the dark letters and the light paper is blurred as in a badly focussed photograph. But the retina or brain is able to compensate for this defect by means of Mach's bands. Figuratively speaking, the eye draws a white line at the boundary of the light paper in the image of the print and a black line at the boundary of the dark letters. We thus obtain the impression of a sharp outline despite the blurring of the retinal image.

(d) *The Spiral Illusion.*—Anyone looking at fig. 7 sees a system of concentric spirals. But the figure really consists of concentric circles, as may immediately be verified by tracing them out with the point of a pencil. This example shows very clearly how careful we should be in the interpretation of what our eyes tell us.

So much for our examples of optical effects which may arise in the observation of physical phenomena. As we mentioned above, these effects seldom raise difficulties for practised observers nowadays; nevertheless, they are a warning to be careful. How many other

Fig. 7.—The spiral illusion

subjective effects not yet recognized by us may still lie concealed in our so-called knowledge of nature! Most suspicious of all are the very general fundamental ideas, such as Space, Time, Weight, &c., which we have been led to form as a result of the experience of generations. No doubt physics has still to get rid of many prejudices and misinterpretations in this direction. We have taken the opportunity of giving at least a passing mention of these problems here. We shall consider them somewhat more closely in Chapter VIII (p. 140), but a thorough discussion of them is beyond the scope of this book.

2. The Measurement of Length: Direct Methods.

All the results of physics depend in the last instance on observation and experiment. On this point all are agreed. Experiment and observation have no doubt led to new results, often of great importance, even when they have only been carried out *qualitatively*. Yet experiment and observation do not attain their full value unless they are made to yield quantitative results. Hence measurement plays a very important part in physics. The technique of physical measurement is highly developed, the number of methods used is large, and the subject has an extensive literature devoted specially to it.

Among all the variety of physical measurements, measurements of length and of time are particularly frequent, either alone or in combination with the measurement of other quantities. Thus it is convenient to begin with the measurement of length and of time, that is, with an explanation of the fundamental ideas on which these measurements are based, not of technical details connected with the actual performance of the measurements.

All direct measurements * *of length depend on the application (repeated if necessary) of a standard to the object being measured.* At first sight this seems an extremely trivial statement. Yet the fact expressed in it has only very recently been realized (Einstein, 1905). Unless it is consistently applied some of the most famous discoveries in physics defy all attempts at interpretation.

The actual process of measurement, i.e. in this case the repeated application of the standard, is not enough. We must also define our unit.

All ways of defining physical units are perfectly arbitrary. The most important requirement is always that international agreement should extend as far as possible. It is also desirable that a unit should be easy to reproduce and that the numerical values occurring in the most common measurements in technology and everyday life should be convenient.

In electricity the two fundamental units ampere and volt have been universally adopted. Everyone who has to do with electrical quantities, no matter where, carries out his measurements and calculations in terms of the ampere and the volt. It is only a small coterie of physicists that persists in the use of two older systems of units which have very arbitrarily been called " absolute " units.† The reverse is true of the units of length. Generally speaking there is a perfectly hopeless muddle of all manner of units of length, except for the praiseworthy exception of physical literature, the overwhelming majority of physical measurements being based on one and the same unit of length, the Paris standard metre.‡

This unit depends on a standard of length which is kept in the

---

* [Ger. *echte Längenmessung*.]

† In these units, for example, the pressure at which an electric lamp works is not 220 volts, but either 0·73 gm.$^{\frac{1}{2}}$ cm.$^{\frac{3}{2}}$ sec.$^{-1}$ or 2·2.10^{10} gm.$^{\frac{1}{2}}$ cm.$^{\frac{3}{2}}$ sec.$^{-2}$. See also p. 70.

‡ The failure of the metre to win general adoption is, of course, due to the awkward length of this unit. For the needs of everyday life the metre is too large and its thousandth part, the millimetre, is too small. The graduation of ordinary rulers is too rough for tenths of a millimetre to be estimated. A practically useful unit of about the length of the ell ($\frac{7}{10}$ yard) or foot, divided into 100 parts, would without doubt have been adopted internationally for practical work. The inventors of the metre unit can scarcely have been accustomed to the use of plane or file or even of a die-stock.

Bureau des Poids et Mesures in Paris, and which consists of a metal bar of an alloy containing 90 per cent of platinum and 10 per cent of iridium. The bar has a peculiar x-shaped section as shown in fig. 8. Two marks are engraved on the surface marked N. The metre is defined as the distance between them (at a temperature of 0° C.). The x-shaped section causes the distance between the two marks to be independent of unavoidable sagging of the bar ("neutral zones"). Thirty-one copies of this standard metre have been made and distributed by lot to the nations which signed the international metre convention.

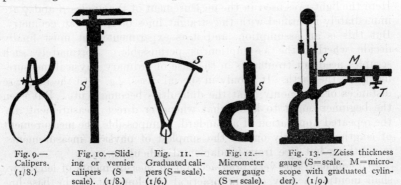

Fig. 8. — Section of the Paris standard metre: height about 2 cm.

In spite of all imaginable care taken in the handling of the standard metre and of its copies, a gradual change in the distance between the marks defining the metre has certainly to be reckoned with. All bars of metal change slightly in length in the course of decades or centuries, for their microcrystalline structure alters in course of time. Hence physics has for a long time sought to guard against unpleasant discoveries of this kind by comparing the Paris standard metre with the wave-length of a definite red line in the spectrum of cadmium vapour ($\lambda = 0.6438\ \mu$).* In the year 1913 the distance between the marks on the standard bar corresponded to 1,553,164.13 of these wavelengths (at normal atmospheric pressure and a temperature of 15° C.). According to our present knowledge this is the most secure way of preserving the metre unit for posterity.

Fig. 9.— Calipers. (1/8.)　Fig. 10.—Sliding or vernier calipers (S = scale). (1/8.)　Fig. 11. — Graduated calipers (S=scale). (1/6.)　Fig. 12.— Micrometer screw gauge (S = scale). (1/5.)　Fig. 13.— Zeiss thickness gauge (S=scale. M=microscope with graduated cylinder). (1/9.)

Practical measurements of length are carried out by means of foot- and metre-rules and a variety of measuring instruments, the most important of which are well known in ordinary life. Typical forms of useful instruments are shown in figs. 9–13. Details, such as the reading-off of the indications by means of a lens or vernier, are discussed in elementary textbooks on practical physics.

* [μ is the symbol for a *micron*, i.e. 10^{-3} millimetre.]

3. Direct Measurement of Length using a Microscope.

Direct methods of measuring length are also applicable to objects of microscopic dimensions. As an example we shall measure the diameter of a hair in front of a large audience. An image of the hair is projected on to a screen by means of a simple microscope. The breadth of the hair in the image is marked off by two arrows (fig. 14 *a*, Plate I). The hair is removed and its place taken by a small scale engraved on glass (objective micrometer), say a millimetre divided into 100 parts. The field of view is then as shown in fig. 14 *b*. We read off the distance between the points of the arrows (4 scale divisions). That is, the thickness of the hair is $4 \cdot 10^{-2}$ mm. or 40μ.

4. Indirect Methods of Measuring very great Lengths (Base-line methods, Stereophotogrammetry).

Very great lengths are often incapable of being measured directly, e.g. the distance between two mountain peaks or the distance of a heavenly body from the earth. We then have to use an indirect method, e.g. the well-known base-line method illustrated in fig. 15. BC, the length of the base-line, is obtained by direct measurement if possible, and the angles β, γ are measured. The required distance x may then be ascertained graphically or by calculation from the known values of the angles and the length of the base-line.

This method, which will be familiar to the student from his school mathematics, is not exempt from the need of careful consideration. Here the light-rays used in the measurement of the angles β and γ are immediately identified with the straight lines of Euclidean geometry. But this is an assumption, and it is experiment that must finally decide whether the assumption is permissible. Fortunately such scruples need not trouble us in the case of ordinary physical measurements on the earth. It is only in special cases, e.g. with the immense distances of astronomy, that the difficulties become acute. But even the beginner ought to know that whenever direct measurement, i.e. the repeated application of a standard, is impossible, the measurement of length, apparently one of the simplest of physical measurements, is anything but free from fundamental difficulties.

To conclude this very brief account of measurements of length we shall mention a very elegant practical development of the base-line method, namely the so-called *stereophotogrammetry*. In practice it is very useful in surveying, especially in mountainous country. In physics it is used, among other purposes, for determining complicated paths in space, e.g. those of lightning flashes.

In fig. 15 the angles β, γ were determined by means of some instrument for measuring angles (e.g. a telescope on a divided circle). In stereophotogrammetry the two angle-measuring instruments at the ends of the base-line are replaced by two cameras, the lenses of which

PLATE I

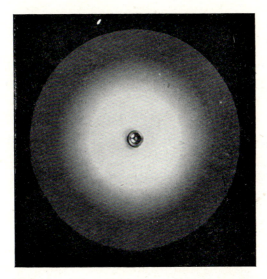

Chap. I, Fig. 6.—Mach's bands at the borders of white and grey
and of grey and black

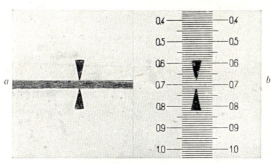

Chap. I, Fig. 14 *a*, *b*.—Measurement of a length under the
microscope

Chap. I, Fig. 17.—Stereoscope with floating mark: photographs showing branched lightning

are indicated by I and II in fig. 16. B and C, the images of the same
object A, are displaced through the distances BL, CR respectively
relative to the centres of the plates. As we
easily see from the geometry of the figure,
if we know BL or CR and the total distance
BC, we can calculate x, the required distance
of the object A. A table of values may be
drawn up for a given base-line I–II and a
given distance between the lenses and the
plates (f).

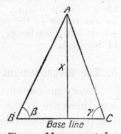

Fig. 15.—Measurement of a
length by means of a base-line

So far there is nothing particularly note-
worthy about the method. Now, however, there
emerges a serious difficulty. To pick out the
corresponding pictures B and C of the individual elements of the path
would take up a good deal of time and would often be impossible, e.g.
in the case of the tortuous path of a lightning-flash. The difficulty
may be got over in the following way. The photographs are combined
by means of a stereoscope in the familiar way to form *one* field of view
apparently extending in space. Fig. 17 (Plate I) shows the two
photographs mounted in the stereoscope.
And here is where a distinctive artifice
comes in, namely the use of a "floating
mark".

The floating mark consists of two
similar pointers 1 and 2, which can be
moved together over the surface of the
photographs both vertically and horizontally. The amounts of their displacements
are read off on the scales S_1 and S_2. The
distance between the two pointers can
also be altered by a known amount (by
means of S_3 and the graduated shaft).

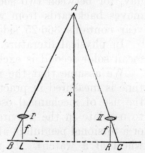

Fig. 16.—Measurement of a length
by stereophotogrammetry

Looking into the stereoscope, we see the two pointers as *one*,
floating freely in the field of view. If we alter the distance between
the two pointers (S_3) the floating mark appears to move towards or
away from us in space. Using the three degrees of freedom (S_1, S_2,
S_3) we can bring the mark to any point in the field of view, i.e. on to
a mountain top, on to a particular point of a tortuous flash of lightning,
&c. The experiment is an extremely striking one. From the scale
readings the three lengths fixing the point in depth, breadth, and
height (its three co-ordinates) can readily be obtained by means of a
table.

An external detail sometimes makes the experiment difficult to understand.
According to fig. 16, we should expect the mark to move *towards* us as the distance between the pointers was *increased*; and yet the mark actually moves

away from us. There is a simple reason for this. In every stereoscope the left-hand and right-hand photographs must be interchanged. In the photographs inserted in the stereoscope the *nearer* objects have the *smaller* distances between their separate images. It is only thus that the photographs can be reproduced on our retinæ in the same position as if we were looking at the objects themselves, say a landscape, *without* a stereoscope, i.e. as if I and II were the lenses of our eyes.

5. The Measurement of Time: Direct Methods.

Every measurement of time is based on uniformly repeated motions, and these are in the last instance caused by a uniform rotation. Here the term " uniformly " is at first defined merely in a subjective way, for the strict definition " equal angles in equal times " involves a previous measurement of time.

The *sidereal day* is used as a time unit. The sidereal day is defined as the time between two successive passages of the same fixed star across the meridian of the place of observation.

The sidereal day is subdivided into $24 \times 60 \times 60 = 86,400$ sidereal seconds. The mean solar second is obtained by multiplying the sidereal second by $366 \cdot 25/365 \cdot 25$. The solar day is longer than the sidereal day, for between two successive passages over the meridian the sun moves backwards from west to east relative to the fixed stars. A year contains $366 \cdot 25$ sidereal days, but only $365 \cdot 25$ solar days.

In physical literature, as in technology and in ordinary life, the mean solar second is exclusively used.

We assume that the reader is familiar with the clocks by which time is measured in practice. Here regularity of going is ensured by the use of mechanical oscillations. A suspended pendulum is made to oscillate in the gravitational field of the earth (pendulum clock), or a torsional pendulum attached to a coiled elastic spring (the balance wheel of a watch). It remains to show that the oscillations of a pendulum may be reduced to uniform rotation.

To put it briefly, *the oscillation of a pendulum is like motion in a circle looked at sideways.* If we look along the plane of the circular path we merely see the body moving to and fro, the successive stages of the motion being exactly the same as those of the motion of a pendulum. This is brought out particularly clearly by the method of photographic registration, which records successive stages in time as successive stages in space and depicts the course of the motion by tracing out a curve.

This curve may be photographed by means of the apparatus shown in fig. 18. The image of a slit S is thrown on the screen P by the lens L. The source of light illuminating the slit (an electric arc) is not shown in the figure. During the exposure the lens L is made to slide uniformly along a rail in the direction of the arrow, so that the image of the slit moves across the screen P. The screen is coated with a layer of phosphorescent crystals, which continue to shine for some time after they have been illuminated for a brief interval.

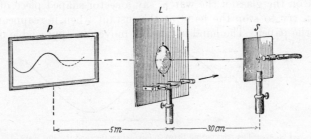

Fig. 18.—The relationship between circular motion and the sine wave. In front of the slit S a rod (pointing towards the observer) is fixed to the periphery of a rotating cylinder driven by a flexible shaft. (1/10).

In front of the slit S we successively place (1) a metal rod describing a circular path (fig. 18) and (2) a wire fastened laterally to a gravity pendulum (the pendulum of the metronome in fig. 19).

In both cases the same curve is traced out, deep black upon a bright green phosphorescent background, namely, the graph of the simplest wave form, the sine wave.

This fundamental relationship between circular motion, the motion of a pendulum, and the sine wave plays an important part in various branches of physics. From the formal mathematical point of view the relationship follows from the sketch of fig. 20, which is easy to understand. Yet in view of the great importance of this relationship the impressive experiment just described does not seem superfluous. It may also be regarded as a simple example of the analysis of motions by photographic registration.

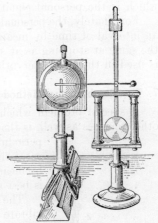

Fig. 19.—A metal rod fixed to the pendulum of a metronome and placed in front of a slit. This apparatus takes the place of S in fig. 18. (1/8.)

6. Modern Clocks: the Personal Equation.

The constructional details of modern clocks do not concern us. Nowadays very convenient pocket stop-watches are made, which enable us to read off $\frac{1}{50}$ or even $\frac{1}{100}$ second directly. Fig. 21 (Plate II) shows one of these stop-watches. The hand makes one revolution every second. (On setting the stop-watch agoing, one is invariably surprised to find how long a second lasts!)

A watch like this may be used to measure a quantity which is often important, the so-called "personal equation". We put some sort of mark on the glass of the watch, say a sector-shaped piece of paper. We then try to stop the hand at the instant when it reappears from behind the paper. The hand invariably moves an appreciable distance

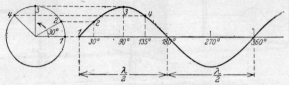

Fig. 20.—The relationship between circular motion and the sine wave

beyond the paper, usually about $\frac{1}{10}$ second. This interval of time is called the "personal equation". Its meaning is easily seen. The optical signal has to be transmitted from the eye to the brain. The brain has to notify the muscles of the fingers by way of the spinal cord. The two processes together occupy a finite interval of time, which is the personal equation of the observer.

Fortunately, the personal equation is not involved when we measure an interval of time by means of the stop-watch. For it is practically the same at stopping as at starting, provided the same organ of sense is used in the two observations.

7. The Stroboscopic Method of Measuring Time.

A special problem which often arises, not only in physics but in other sciences as well, is that of the measurement of a very short but periodically recurring interval of time. In such cases the *stroboscopic* method of measuring time by means of a rotating disc is used. The method is best explained by means of an example.

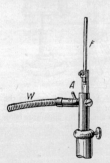

Fig. 22.—A leaf-spring (F) for measuring an interval of time stroboscopically. A picture of the spring in vibration is given in fig. 69a, Plate IX.

The leaf-spring shown in fig. 22 is made to vibrate backwards and forwards about 50 times per second (fig. 69 a, Plate IX). (In practice this is effected by an eccentric at A, which is set in rapid rotation by the flexible shaft W. For more details see the article on *Forced Vibrations* (Chapter XI, § 11, p. 253).) This spring, an image of which is projected on to the wall, say, is illuminated by intermittent light, in fact, by a uniform succession of isolated flashes of light. This illumination is most readily obtained by means of a rotating disc with e.g. 10 slit-like openings, placed in a suitable position in the path of the light.

Beginning with the disc rotating rapidly, we slow it down gradually and attempt to find the rate of occurrence of successive illuminations for which each successive flash of light finds the spring at the same (arbitrary) point in its path. We then see the spring remaining still at this point (and of course *only* at this point); if n is the number of flashes of light per second, the required period of vibration of the spring is $1/n$ second. If the interval of time between two flashes of light is somewhat greater or somewhat smaller than the period of vibration of the spring, the spring will be successively illuminated not at the same point, but at closely adjacent points of its path. Hence the image of the spring will slowly advance in one direction or the other; the spring will appear to execute greatly retarded oscillations. The appearance of these slow oscillations and their gradual dying-away to a complete standstill make the stroboscopic method of measuring time very simple to apply.

In the use of the stroboscopic method certain details have to be borne in mind. Although these are really implicit in what we have said above, we shall state them again explicitly to make sure that they are not overlooked. We must always begin with a high number of revolutions. No attention is to be paid to cases where the spring stands still at *more than one* point of its path, for in these cases we have integral multiples of n. A standstill in the *equilibrium* position of the spring, and in that position only, occurs for the first time when there are $2n$ flashes of light per second. A standstill at *any* point of the path occurs for the first time when there are n flashes of light per second. As the number of revolutions of the slotted disc diminishes, further standstills occur for all integral fractions of n or $2n$, but of course with decreasing distinctness. The student should carry out the experiment himself as a practical exercise.

8. Fundamental Difficulties in our present Methods of Measuring Time.

Instead of our present-day direct methods of measuring time, which depend on rotation, indirect methods were formerly used (e.g. water-clocks and hour-glasses).* Nowadays these survive only in the degenerate form of the egg-glass. In ancient times a great deal of trouble was taken to ensure the accuracy of water-clocks; attempts were made to increase the uniformity with which the water flowed out by the use of orifices constructed with special care, e.g. precious stones with a hole bored through them; automatic whistles warned the owner of the clock when to replenish the water, and so on. We are only too ready to laugh at these attempts of our predecessors;

---

* Suppose that the water flowing from a water-clock in which the level of water in the containing vessel is kept constant is caught in a cylindrical vessel marked with equidistant heights. This is not a direct method of measuring time. We must previously ascertain the laws according to which the lower layers of the water are compressed and the walls of the vessel forced outwards as the height of the column of water increases. Both these difficulties may be avoided if the collecting vessel is made to tip up and empty itself periodically ("relaxation oscillations", p. 228). This, however, brings us back to uniformly recurring motion. An electrical variant of the water-clock is mentioned in a companion volume to this (*Physical Principles of Electricity and Magnetism*, Blackie and Son, Ltd., 1930, p. 40, § 9).

but in reality we ought to be more modest, for even our modern methods of measuring time are far from perfect. At bottom our unit of time is no better established than our unit of length is by a standard metre bar which must inevitably perish in the course of thousands of years. This is shown by the following experiment. In fig. 23 we see a man sitting on a stool free to rotate. He is set in rotation by a push. Every time he raises his arms away from his body his rate of rotation decreases, and every time he lowers them his rate of rotation increases (for further details see p. 141). The same is true of the rotation of our earth about its axis. Any considerable displacement of mass in the direction of the radius, e.g. the building-up of a mountain range or a

Fig. 23.—Alterations in the shape of a body give rise to changes in its rate of rotation

shrinking of the earth as a whole, affects the time of rotation of the earth on its axis, and hence the length of the sidereal day. Recent technical advances have led to the construction of clocks which apparently rotate more uniformly than the earth does (cf. Chapter VI, § 5, p. 109).

For other and much more fundamental reasons, however, attempts to introduce another unit of time in place of the sidereal day have not been wanting. Such an attempt is made by Einstein's theory of relativity, in which the unit of time is defined as the interval required for light to travel along the length of the Paris standard metre and back, a source of light being supposed to be placed at one end-point of the bar and a reflecting mirror at the other.

Fortunately we do not require to pay any attention to these last difficulties in our ordinary measurements of time in physics. The beginner, however, ought to realize that deep problems are involved even in matters apparently so simple as the measurement of time.

CHAPTER II

Methods of Representing Motion (Kinematics)

1. Definition of Motion: Systems of Reference.

Motion is change of position with time, relative to a fixed rigid body ("system of reference"). The last few words are absolutely essential, as we see from an example taken at random. To a cyclist looking at his feet they appear to describe circles, whereas to an observer standing on the pavement they present quite a different aspect, appearing to describe a wavy line, the cycloid sketched in fig. 1.

The fixed rigid body to which we shall in future refer motion is the earth or the floor of the lecture-room. Here we are deliberately leaving the daily rotation of the earth out of account. In the first instance

Fig. 1.—Path of the pedal of a bicycle as it appears to an observer at rest

we shall ignore the fact that we are really carrying out physical investigations on a huge merry-go-round, and shall maintain the fiction that the earth is rigid and undeformable.

Later on we shall occasionally change our point of observation or system of reference. In a good many connexions we shall take the rotation of the earth into account, and on occasion deformation of the earth too. But we shall state this quite explicitly in every such case. Otherwise fearful confusion is likely to result, especially in the case of rotatory motion.

In order to represent or describe any motion we require measurements of length and time. Such measurements enable us to define the two ideas of *velocity* and *acceleration*. We shall begin with these.

2. Definition of Velocity: Example of the Measurement of Velocity.

Suppose a body moves through a distance Δs in the interval of time Δt: then the *velocity* is defined as the quotient

$$v = \frac{\Delta s}{\Delta t}. \qquad \ldots \ldots \ldots \quad (1)$$

(In words: Velocity equals Increment of Path divided by Increment of Time.)

15

Here the distance Δs is to be measured in such a way that the quotient remains the same no matter how small Δs is made ("passage to the limit"); otherwise we are really measuring the time-average of the velocity over a comparatively large region. Mathematically this means that we must replace Δ, the symbol for "increment", by the symbol d: the velocity is given by

$$v = \frac{ds}{dt}, \quad \ldots \ldots \ldots \quad (1a)$$

i.e. is equal to the differential coefficient of the path with respect to time.

Experimentally the condition means that it is often necessary to measure very small intervals of time. A good example is the measurement of the velocity of a bullet. A bullet has its maximum velocity when it leaves the barrel (muzzle velocity). The velocity then decreases slowly but steadily along the path of flight, owing to the resistance of the air.

Suppose that we have to measure the muzzle velocity of a pistol bullet. A suitable apparatus is shown in fig. 2 (Plate II). The increment of path Δs is bounded by two thin cardboard discs, the distance between which we may take e.g. as 22·5 cm. The measurement of time is here related to uniform motion, the fundamental basis of all measurement of time, in a very straightforward way. The times are recorded automatically by a kind of chronograph. In order to do this, the cardboard discs, which are on a common shaft, are set in rapid uniform rotation by means of an electric motor. The number of revolutions per second (also called the *frequency*) is read off on a counter such as is used in ordinary practice, and is found to be e.g. 50.

The bullet first pierces the left-hand disc, the shot-hole being our first time-indication. While the bullet is traversing the 22·5 cm. to the second disc, the "clock" or "chronograph" goes ahead. The shot-hole or time-indication on the second disc is displaced through a certain angle relative to that on the first. On stopping the disc we find that this angle is about 18°, or the corresponding arc about $\frac{1}{20}$ of the circumference.

If the experiment is shown by projection on a screen the angular displacement may be made visible if we put a stiff wire through the two shot-holes.

The time of flight Δt, then, is $\frac{1}{50} \cdot \frac{1}{20} = \frac{1}{1000}$ second. The velocity v is

$$\frac{0·225}{1/1000} \frac{\text{metre}}{\text{second}} = 225 \text{ m./sec.}$$

The experiment is then repeated with a smaller trajectory (15 cm.). The final result is the same. Thus the first trajectory was sufficiently small, so that it really did give the required muzzle velocity and not a smaller average value over a longer trajectory.

It is only in the case of motions in which the velocity is constant or uniform that the magnitudes of Δs (measured length) and of Δt (measured time) may be selected solely on grounds of practical convenience. We then briefly write $v = s/t$.

The numerical value of a velocity depends only on the units used in a particular case. We can express the velocity of the bullet equally well as $2 \cdot 25 . 10^4$ cm./sec. or as 810 km./hour. In all cases, however, we have to measure the length of a path and the duration of a time and form the quotient of their numerical values. This is expressed in physical literature by a statement which at first sight looks peculiar: Velocity has the dimensions $[lt^{-1}]$. This merely means that, independently of the units of measurement used in each case (but *not* of the system of units (p. 70)), the measurement of a velocity requires that the numerical value of a length l should be divided by the numerical value of a time t. The advantages of such dimensional expressions will become evident later on.

In everyday life we are content to indicate a particular velocity by stating its numerical value, say in ft./sec. In physics, however, this numerical value is only *one* of the quantities defining a velocity;* the *direction* must be given also. In physics velocity is always a directed quantity or *vector*, symbolized by an arrow. This is very clearly brought out in the " parallelogram of velo-cities ", or combination of two velocities to form a resultant velocity, with which even the man in the street is familiar. In fig. 3 the large velocity v_1 (e.g. the velocity of an

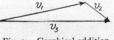

Fig. 3.—Graphical addition of velocities

aeroplane †) and v_2, the small velocity in another direction (e.g. the wind velocity), are combined to form the resultant velocity v_3 (the velocity with which the aeroplane actually advances).

3. Definition of Acceleration: the two Limiting Cases.

Motions in which the velocity is constant are rare. In general, both the magnitude and the direction of the velocity differ at different points of the path.

In fig. 4 the arrow v_1 denotes the velocity of a body at the *beginning* of an interval of time Δt. Suppose that *during* this interval of time the body receives an additional velocity Δv in any direction, represented by

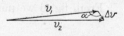

Fig. 4.—Illustrating the general definition of acceleration

* [The term *speed* is often used for the *magnitude* of a velocity irrespective of its direction.]

† [I.e. the velocity it would have in still air.]

the second short arrow. At the *end* of the interval of time Δt the body has the velocity v_2. In fig. 4 it is graphically obtained as the directed line v_2.

Then the *acceleration* is in general defined as the quotient

$$f = \frac{\Delta v}{\Delta t}. \qquad \ldots \ldots \ldots \quad (2)$$

(In words: Acceleration equals Increment of Velocity divided by Increment of Time.)

Here the increment of time Δt is to be chosen in such a way that the quotient remains the same no matter how small Δt is made (passage to the limit). Mathematically this again means that we must replace the symbol Δ by the symbol d; that is,

$$f = \frac{dv}{dt} = \frac{d^2s}{dt^2}. \qquad \ldots \ldots \ldots \quad (2a)$$

Acceleration, like velocity, is a vector. The *direction* of this vector coincides with that of the additional velocity Δv (fig. 4).

In fig. 4 the angle a between the additional velocity Δv and the initial velocity v_1 is arbitrary. We proceed to consider two limiting cases:

(1) $a = 0$ or $180°$ (fig. 5, *a*, *b*). The additional velocity lies along the line of the original velocity. The magnitude, but not the direction,

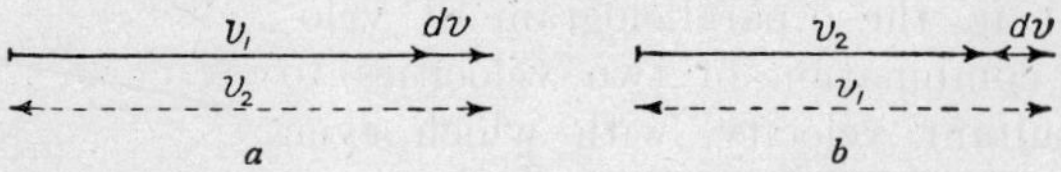

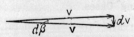

Fig. 5*a*, *b*.—To illustrate the definition of tangential acceleration

of the velocity is altered. In this case we call the acceleration dv/dt the *acceleration in the direction of motion* (f) or *tangential acceleration* (f_t).

(2) $a = 90°$ (fig. 6). The additional velocity is at right angles to the original velocity v. The magnitude of the velocity is unchanged, but its direction is altered by the small angle $d\beta$ in the interval of time dt. In this case we call dv/dt the *acceleration at right angles to the motion* or *normal acceleration* (f_n).

Fig. 6.—To illustrate the definition of normal acceleration

From fig. 6 we immediately obtain the relationship

$$dv = v\,d\beta$$

or

$$\frac{dv}{dt} = v\,\frac{d\beta}{dt}.$$

Here the quotient $d\beta/dt$ is called the *angular velocity* (ω); that is,

$$f_n = v\omega. \qquad \ldots \ldots \ldots \ldots \quad (3)$$

It will be seen from the above definitions that the term *acceleration* is used in physics in a sense quite different from that of ordinary speech.

Changes of direction are entirely ignored in the everyday use of the words velocity (or speed) and acceleration. Suppose an express train travels over 20 metres of rail per second, round curves as well as on the straight. Then in ordinary language we say: " The train travels the whole distance at the constant speed of 20 metres per second." In physics, on the contrary, we say: " It is only on the straight track that the train moves with constant velocity; on the curves it is ' accelerated '."

In most cases of motion accelerations in and at right angles to the direction of motion exist simultaneously, the velocity varying along the path of motion both in magnitude and direction. Nevertheless, we shall meanwhile confine ourselves to the limiting cases where the acceleration is wholly in the direction of motion (motion in a straight line) or wholly at right angles to the direction of motion (motion in a circle).

4. Motion in a Straight Line.

The tangential acceleration merely alters the magnitude of the velocity and not its direction, so that the motion is in a straight line.

In theory it is easy to measure accelerations along the line of motion. We find the velocities v_1 and v_2 for two successive intervals of time of length Δt, calculate Δv (which is equal to $v_2 - v_1$ and may be either positive or negative), and form the quotient $\Delta v / \Delta t$.

As we already know, Δt is to be chosen in such a way that the result of experiment remains the same no matter how far Δt is diminished. In practice this condition usually involves the measurement of very small intervals of time. There is no difficulty in this provided we can use some " self-registering " method, i.e. a method in which the course of the motion is in the first instance recorded automatically and the records are subsequently examined at leisure. It is convenient to use a high-speed cinematograph camera, but much simpler apparatus, e.g. a clock made to mark intervals of time on paper, will suffice. This marking process, of course, must not affect the motion of the body.

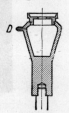

As a practical example, suppose that we have to ascertain the acceleration of a wooden bar falling freely. A suitable form of apparatus, which may be adapted for many other acceleration experiments, is shown in fig. 7 (Plate II).

Fig. 8.—The ink-sprayer in fig. 7 (Plate II); half natural size.

The essential feature is a fine jet of ink rotating in a horizontal plane. The jet emerges from the spout D on the side of an ink-pot fixed to the vertically placed shaft of an electric motor. The number of revolutions (50 per second) is ascertained from a practical counting instrument. In this apparatus it is again easy to see how the measurement of time is reduced to uniform rotation.

A piece of white paper is wrapped round the bar, which is suspended at a and released at a suitable instant by means of a wire. The bar then falls to the ground, passing through the rotating jet of ink on the way. Fig. 9 shows the result, a well-defined succession of isolated marks corresponding to times $\frac{1}{50}$ sec. apart.

The fall of the body continues as the jet of ink moves past it, hence the curvature of the marks.

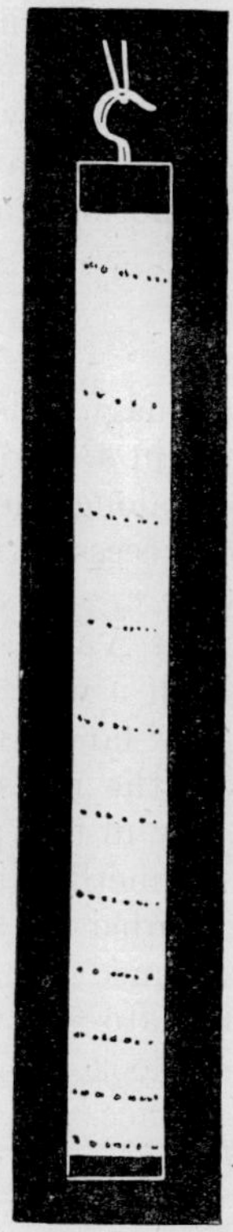

Velocity $v = \frac{\Delta s}{\Delta t}$.	Increase in Velocity (Δu) in each 1/50 sec.	Acceleration f.
cm./sec.	cm./sec.	m./sec.^{2}
285·50		
	22·50	11·25
263·00		
	17·50	8·75
245·50		
	18·00	9·00
227·50		
	21·25	10·63
206·25		
	21·25	10·63
185·00		
	18·50	9·25
166·50		
	19·00	9·50
147·50		
	18·00	9·00
129·50		
110·00	19·50	9·75
Mean:	19·50 cm./sec.	9 8 m./sec.^{2}

Fig. 9.—Falling body with marks at equal intervals of time and their interpretation (subject to the usual experimental errors). This experiment is largely meant to show that the measurement of a second differential coefficient is always a ticklish undertaking.

That the motion is accelerated is obvious at a glance. The distance between successive marks, i.e. the path Δs described in successive intervals of time Δt ($\frac{1}{50}$ sec.), increases continually. The values calculated for the velocity v (ds/dt) are given alongside. The velocity increases by the same amount, namely $\Delta v = 19\cdot5$ cm./sec. in each

PLATE II

Chap. I, Fig. 21.—Pocket stop-watch divided to show hundredths of a second; each revolution of the hand means one second

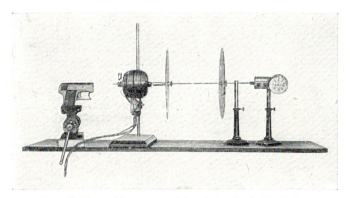

Chap. II, Fig. 2.—Measurement of the velocity of a pistol bullet by means of a simple " chronograph "

Chap. II, Fig. 7.—Measurement of the acceleration of a freely falling body

$\frac{1}{50}$ sec., if we neglect the unavoidable errors in the individual values. Thus a body falling freely is one of the uncommon cases of a *constant* or *uniform acceleration*. For the numerical value of this constant acceleration we obtain 9·8 m./sec.²

The same numerical value is obtained if we repeat the experiment with a body made of any other substance, e.g. with a brass tube instead of the wooden bar. *The constant acceleration is the same for all bodies falling freely.* It is almost always denoted by the letter g (i.e. $g = 9\cdot8$ m./sec.²) and is called the "acceleration due to gravity".*

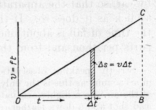

Fig. 10.—Velocity (v) and path (s) with constant acceleration in the direction of motion.

Here this is an experimental fact which we have obtained in the bygoing. Its great importance will become apparent later on.

Acceleration has the dimensions [cm./sec.²] or in general $[lt^{-2}]$.

Our experiment formed an example of the special case of *constant* acceleration in the line of motion, which is of considerable importance.

Constant acceleration means equal increases in velocity (Δv) in equal intervals of time (Δt). As shown in fig. 10, the velocity v then increases as a linear function of the time t. In each interval of time Δt the body traverses the element of path Δs. Hence $\Delta s = v\Delta t$, where v is the average value of the velocity in the interval of time Δt in question. This element of path is indicated by the shaded area in fig. 10. The whole triangular area OBC is the sum of all the elements of path described in time t. Thus when the acceleration is constant the path s described in time t is given by

Fig. 11
String and
balls

$$s = \tfrac{1}{2}ft^2, \qquad \ldots \ldots \quad (4)$$

i.e. the distance traversed increases proportionally to the square of the time of fall. A satisfactory experimental verification of this may be obtained from the apparatus of fig. 9.

Among other lecture experiments for verifying equation (4) we may mention the following. The apparatus used is a piece of thin string carrying a number of lead balls, suspended vertically with the lowest ball almost touching the ground (fig. 11). The distances between this ball and the others are in the ratios

---

* The numerical value is true for the neighbourhood of the surface of the earth and may for most purposes be regarded as constant. More accurate observations show that g varies slightly with the geographical latitude of the place of observation (Chap. VIII, § 6, p. 147). It is also affected by local peculiarities in the rocks (e.g. a subterranean ore deposit), and (though only very slightly) by the height about sea-level of the place of observation.

of the squares of the whole numbers. If the upper end of the string is let go the balls strike the floor one after the other and the successive impacts are separated by equal intervals of time.

Further, equation (4) provides us with a convenient method for determining g, the acceleration due to the earth's gravitation. The heights fallen through are taken to be several metres; the time is measured by means of a modern stop-watch, the timing being done by ear, so that the apparatus for releasing the falling body must make a click as it does so. If the height fallen through is 5 metres, i.e. if the time of fall is about one second, we obtain a value correct to a few parts per thousand from the mean of even a few observations.

Strictly speaking, observations on a body falling freely should be conducted in a vacuum, as this is the only way of getting rid of errors due to the resistance of the air. In a highly evacuated glass tube all bodies do fall equally quickly, a lead shot and a down feather reaching the bottom at the same moment, whereas in ordinary air the feather, as is well known, takes much longer than the shot. Experiments on the fall of heavy bodies with a relatively small surface area, however, are only slightly affected by the resistance of the air.

In the above experiments to verify equation (4) the acceleration used has always been that due to gravity. This is convenient but in no way essential, the origin of the constant acceleration in the line of motion being perfectly arbitrary; it may, for example, be electrical instead of mechanical.

If the body already has the initial velocity u before the acceleration sets in, equation (4) is replaced by

$$s = ut + \tfrac{1}{2}ft^2. \qquad \ldots \ldots \ldots \quad (4a)$$

5. Motion in a Circle.

The normal acceleration f_n merely alters the direction of the velocity and not its magnitude. Suppose that the normal acceleration is constant and that no other acceleration exists. Then the direction of v alters by equal angles $d\beta$ in equal intervals of time dt. The path is a circle described with constant angular velocity ω $(=d\beta/dt)$.

For closed revolutions we make the following general definitions. The *period* τ is the time of one revolution (in seconds); the *frequency* n $(=1/\tau)$ is the number of revolutions per second.

Hence it follows that for a *circular path* described with constant angular velocity:

The linear velocity $v = 2r\pi/\tau$ (arc described per second),

The angular velocity $\omega = 2\pi/\tau$ (angle described per second),

$$v = \omega r, \qquad \ldots \ldots \ldots \ldots \quad (5a)$$

$\omega = 2\pi/\tau = 2\pi n$, the number of revolutions in 2π seconds.

The student should make a point of remembering these definitions and relationships, as they are continually turning up in all branches of physics.

Equations (3) and (5a) together give

$$f_n = \omega^2 r = v^2/r. \qquad \ldots \ldots \quad (6)$$

This is the normal acceleration that must exist in order that a body may describe a circular path of radius r *with constant angular velocity* ω *or constant linear velocity* v.

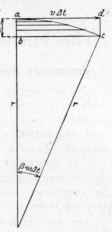

The meaning of the constant normal acceleration required if the body is to move in a circular path may be illustrated as follows (fig. 12):

Let a body describe the arc *ac* in time Δt, and let us consider its path as the combination of two successive stages, namely (1) a path *ad* $(= v \Delta t)$ *perpendicular* to the radius, described with constant velocity v, (2) a path $s \, (= \tfrac{1}{2} f_r (\Delta t)^2)$ described with accelerated motion in the direction of the radius. We see from the horizontal lines marking equal intervals of time that the motion along s is accelerated, so that equation (4) is applicable (fig. 12).

Fig. 12.—To illustrate normal acceleration

N.B.—∠ *acb* = β/2

A numerical example may be found helpful. During the time-interval of 1 sec. the moon advances by 1 km. in the direction *ad*, i.e. *perpendicular* to the radius of the path, moving slightly " farther away " from the earth. Simultaneously it " approaches " the earth with accelerated motion *along* the radius of the path by an amount s equal to $\tfrac{1}{2} f_n (1)^2 = 1 \cdot 35$ mm. The radius remains constant and the path is a circle. The normal acceleration of the moon is $2 \cdot 70$ mm./sec.^{2}.

CHAPTER III

The Fundamental Theorems of Dynamics

1. Preliminary Remarks.

In last section we discussed certain phenomena from the kinematical point of view. In kinematics the various motions occurring in nature are described in terms of geometrical ideas taken along with the notion of time. Here we lose touch with the experimental method. Even in the case of motion in a circle we no longer made use of illustrative experiments. We shall therefore delay further developments of kinematics for a while and turn to dynamics instead. In dynamics (" the science of force ") we seek to explain motions by relating them to " forces ", taking account of the particular nature of the moving bodies. The characteristic ideas of kinematics are " velocity " and " acceleration ", whereas dynamics is characterized by the additional ideas of " force " and " mass ".

In everyday language the words *force* and *mass* have had their meanings corrupted so that they are now ambiguous. This makes it difficult for the beginner to use the words in the specialized sense in which they are employed in physics. Hence we must begin with an explanation of these technical expressions, *force* and *mass*.

2. Definition of Force: Examples of Forces.

In physics and engineering, the word *force* means *that which is capable of deforming a suitably supported solid body*. As examples of forces we may mention the force exerted by muscle, the weight of a body, elastic force, friction, and forces of electrical and magnetic origin.

Every solid body is deformed by a force, no matter how feeble. This is usually overlooked by the beginner. An absolutely rigid body is a pure fiction. We shall give plain proof of this.

For this purpose we use the apparatus shown in fig. 1, the table shown being a strong oak table with a thick oaken frame Z. On the table are placed two mirrors which serve to reflect a ray of light as shown, throwing an image of the source of light (an illuminated slit S) on the wall. Any bending of the top of the table tilts the mirrors

in the directions shown by the small arrows. The great sensitivity
of the apparatus is due to the length of the beam of light (about 20 m.)
used as a pointer.

The muscular force of the little finger applied at A causes a deviation
of the beam of light which is visible at a distance. So does the force
with which the earth attracts a kilogram weight placed at A; thus
the solid table is deformed to an easily measurable extent by this

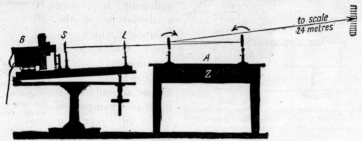

Fig. 1.—Optical proof of the deformation of a table-top by a small force,
e.g. by the pressure of a finger at A. (1/30.)

force which we call weight. The application of a compressed spiral
spring, i.e. of an elastic force, has the same effect.

Muscular forces, weights, and elastic forces are familiar to every-
body. On the other hand, beginners are often confused about the
force which we call " friction ".

*In order that the force which we call friction may arise it is necessary
that one body should move along another or slide on it.* To produce
friction we place a wooden bar H on the table-top shown in fig. 1,
and stroke it downwards with the hand quite gently
(fig. 2). Again the beam of light clearly indicates a
deformation of the table. The friction arising during
the motion has the same effect as if the muscular
force of our hand were applied to the bar at *b*.

This is an example of " external " friction. The
two bodies sliding past each other are *in contact*
along their boundary surfaces. But such contact of
the moving bodies is in no way essential for the
occurrence of a frictional force; the bodies may be
separated by relatively wide stretches of air. We may
then speak of " internal " friction, for it is the internal
layers of air in the intervening space that give rise to
the frictional force. For more details see p. 167.

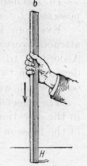

Fig. 2.—External
friction

Internal friction may also be exhibited experimentally by the
deformation of a solid body. The apparatus of fig. 1, however, is
not sensitive enough. We must replace the thick table-top by a fine
metal spring, such as is shown at F in fig. 3. To this is rigidly connected

the metal plate H, which corresponds to the wooden bar H in fig. 2. A smooth metal disc M in rapid rotation passes behind the plate H at a distance of about 1 mm.; it corresponds to the hand in fig. 2. The pointer Z indicates the deformation of the spring. The pointer and coiled spring together form a kind of balance.

If the metal disc makes about 30 revolutions per second, the internal friction is sufficient to cause a deviation of the balance which is easily visible at a distance.

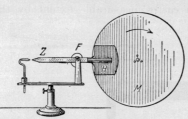

Fig. 3.—Internal friction. There is a distance of several millimetres between the rotating plate M and the plate H. (1/12.)

Fig. 4.—A bar magnet not free to rotate remains floating above a horseshoe magnet

Nowadays it is almost superfluous to mention examples of forces of electric or magnetic origin. For example, there is the child's toy in which paper dolls dance about under a disc of cellon, or the experiment shown in fig. 4, in which a thick cylindrical bar magnet NS floats freely above the poles of a horseshoe magnet, the whole being enclosed in a glass case. Glass partitions prevent rotatory motions of the bar magnet, and the forces of magnetic origin and the downward pull of the weight of the bar are "in equilibrium" when one magnet is at a definite height above the other.

3. Mass.

In mechanics the word mass is taken to imply two properties possessed by all bodies, namely "weight" and "inertia".

Everyone knows the meaning of the word "heavy". The earth attracts every body with a *force* which we call the *weight* of the body. According to everyday experience, this *force* which we call weight depends on two things, (1) a quantity characteristic of the particular *body*, which we call *mass* (m), (2) the earth itself.

The effect of the earth is quite obvious. Every body is attracted in the direction of the centre of the earth. To express this interaction of the mass m and the earth, we may briefly write

$$\text{Weight} = m \cdot F \text{ (Earth)}; \quad \ldots \ldots \quad (1)$$

or, in words: the weight of a body depends on its mass m, and also, in some way as yet unknown to us in detail, on the earth.

Everyone will agree that there is an unsatisfactory amount of arbitrariness in this definition of the word *mass*. Justification of the assumption is to be sought for solely in the subsequent results. We make no attempt to gloss over this.

For the word " inert " we give no definition, but content ourselves with one or two statements merely intended to give the reader a qualitative idea of its meaning. Thus, if the velocity of a body is to be altered, time is always required. If the body is given time, it " willingly " obeys even small forces. If we attempt to alter the velocity of a body within a very short time, the forces which arise are often surprisingly large. These statements may be illustrated by a simple experiment.

Fig. 5 shows a ball suspended by a string o. A handle is attached to the ball by means of a similar string u. We pull the handle straight down. If we pull slowly, the invariable result is that the upper string alone breaks; for the ball " willingly " obeys a *slow* pull, i.e. an attempt to alter the velocity *gradually*. As a result, the upper string has to stand not only the weight of the ball itself but the muscular force of the hand. Hence it gives way before the lower one does, for the latter is subject to the muscular force of the hand only and not to the weight of the ball as well.

If, on the other hand, we give the handle a quick pull, the invariable result is that the lower string alone breaks (even if it is replaced by a much thicker string). For the " inert " ball opposes a *rapid* pull, i.e. an attempt to alter the velocity in a *short* time. The muscles of the hand stretch the lower string far beyond

Fig. 5.—To illustrate inertia

its breaking point before the ball follows it to any appreciable extent and hence before there is any appreciable increase in the force on the upper string.

In order to understand this more clearly we may imagine the strings o and u replaced by rubber bands. The readiness of the ball to " obey " a slow pull is then shown by the increasing length of the rubber band o.

4. The Measurement of Masses and of Forces.

Masses are measured by means of balances and sets of weights.—The masses of two bodies are said to be equal, quite independently of any chemical or physical properties of the bodies, if the bodies are interchangeable as regards their effect on a balance. The choice of a unit of mass, like that of any unit, is arbitrary. As regards *physical* literature it has been agreed internationally to use a platinum weight kept in Paris and to define its mass as a mass of one kilogram. In the last instance all sets of weights in pratical use are compared with this *standard kilogram* by means of balances.*

If spring balances are used this comparison must always be carried out at the same place. Otherwise errors of as much as 0·2 per cent may arise; cf. p. 147.

* [The British standard of mass is the imperial standard pound (avoirdupois), kept in the Standards Department of the Board of Trade, London. (1 kg. = 2·2046 lb.)].

Balances we simply accept as one of the productions of applied science, just as we frankly took modern clocks for granted in measuring time. In reality, of course, the construction of practical balances, like that of clocks, involves quite an extensive knowledge of physics. We intentionally refrain from considering these details, however.

As we have already mentioned, the mass of a body is manifested not only by its weight, but by its inertia. Thus in calibrating a set of weights for measuring mass, we have used only *one* of the properties of mass, i.e. *weight*.

A priori it was by no means certain that the numerical values of the mass obtained from the weight would also suffice for a *simple* discussion of the phenomena of inertia. But even the experience of everyday life suggests that the weight and the inertia of a body to a great extent run parallel with each other. Very heavy bodies are also particularly inert. Hence attempts were at first made to see how far sets of weights calibrated by *weight* alone suit the *simple* numerical conception of the phenomena of *inertia*. Experiment gives the following very remarkable result: the numerical values of the mass obtained from the *weight* are such that the phenomena of inertia may be described by means of a very *simple* equation, with an accuracy unassailed by the most stringent tests of measurement by modern methods. It is only in the

Fig. 6. — Spring balance suitable for projection experiments (here torsional elasticity is used: consider the limiting case). (1/5.)

general theory of relativity that full justice is done to this experimental fact. In the classical mechanics it is merely recorded as a curiosity.

Forces are measured by *dynamometers*; these instruments almost invariably take the form of a spring balance (fig. 6). Two fundamentally different methods may be used to calibrate a dynamometer: (1) the static method, used exclusively in engineering and everyday life; (2) the dynamic method, preferred in most physical textbooks.

We shall first consider the static method; a discussion of the dynamic method follows on p. 35.

In the static method of calibrating a dynamometer, we use the force which we call weight, and define our unit of force * as the *weight of a kilogram*, or kg.-force, i.e. the force with which the earth attracts a kilogram weight placed at its surface (cf. fig. 6).

Strictly speaking, what we use is an *average* value of this force, which varies very slightly with the geographical latitude and the height above sea-level of the

---

* [The corresponding British unit of force is the *weight of one pound*.]

place of observation. That is, we are doing exactly what we do in all physical measurements of time, for these are tacitly based on an *average* value of the solar second, which varies a little in the course of a year. It is only in rare cases that trifling corrections have to be made in order to take account of these niceties.

The student of physics must take care to distinguish clearly between weight, a *piece of metal*, and weight, a *force*.

Forces are *vectors*. We often require to resolve them into two or more components; an example is shown in fig. 7. A roller A is to be held fast on a steep slope by means of a horizontal force F. The arrow W indicates the weight of the roller. We resolve F and W into components parallel and perpendicular to the slope. The latter components, represented by the arrows I, II, merely deform the sloping surface, if only to an inappreciable extent. The former components, $W \cos a$ and $F \sin a$, tend to pull the roller downwards and upwards respectively. For equilibrium $F = W/\tan a$. If the slope is very steep a and $\tan a$ tend to zero together, and F becomes very large. For $a = 5°$, for example, a horizontal force of 11·5 kilograms weight is required to prevent a kilogram weight A from slipping down.

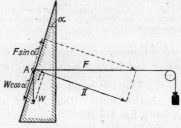

Fig. 7.—Resolution of forces into components

5. The Fundamental Theorems of Dynamics: Newton's Second and Third Laws of Motion.

Now that we can measure masses and forces we are in a position to turn to the fundamental problem of dynamics, that of the *connexion between motion and force*. Its solution may be summed up in two laws or axioms which are due to Isaac Newton: (1) the law of acceleration, commonly called Newton's second law of motion, and (2) the law of action and reaction, or Newton's third law of motion.*

Both laws are based on the following facts of experience. Fig. 8 shows a dynamometer resembling a form used in practice, placed in a horizontal position (so that gravity has no effect on it). It consists of a single hoop of steel spring sliding on a bar. It is merely meant to indicate by its deformation (extension or compression) the *existence* of forces and to make them visible at a distance, so that it has no scale for making actual *measurements*. The right-hand end of the dynamometer is held in the hand, the left-hand end being free. In this position the dynamometer can never register any deflection and the muscles

---

* [Newton's first law of motion (law of inertia) is as follows: Every body continues in its state of rest or of uniform motion in a straight line, except in so far as it is compelled by forces acting on *io* to change its state.]

of the observer's arm never experience the subjective feeling of force. This fundamental fact of experience is familiar to everyone.

It is only when the left-hand end of the dynamometer as well as the right-hand end is constrained, so as to be subject to one of the

Fig. 8. — A steel hoop   Fig. 9.—A dynamometer   Fig. 10.—A dynamometer between a
sliding on a bar, forming a   between two muscles   muscle and an accelerated mass
dynamometer.

two following conditions, that we observe a deflection of the dynamometer and have a sensation of muscular force. The conditions are:

(1) The left-hand end of the dynamometer is kept "fixed", e.g. by another muscle (fig. 9), by any other moving body "pulling" it, or by a fixed wall.

(2) There is an *accelerated mass* at the left-hand end of the dynamometer (fig. 10).

No matter how this acceleration arises, it is invariably associated with a deflection of the dynamometer. We may mention three examples:

(1) The experimenter moves his hand horizontally towards the right.

(2) The experimenter sits on a small truck without moving, and holds the apparatus of fig. 10 in his hand. The *observer* accelerates the truck towards the right by pushing with his foot.

(3) The experimenter swings the apparatus round in a circle. The acceleration of the mass then consists merely in a change in the *direction* of the velocity.

In all three cases the dynamometer exhibits a deflection *during*, and *only* during, the acceleration. It is only *during* the acceleration that the muscle on the right experiences the feeling of force.

These experiences are summed up in the law of acceleration which, qualitatively stated, is as follows: *Acceleration of a body, i.e. change of its velocity in magnitude and direction, never occurs in the absence of forces*.

In *both* the cases illustrated in figs. 9, 10, the fact that the dynamometer shows a deflection leads us to infer the existence of at least *two* forces.

In the first case the left hand is pulled towards the right in the direction of the arrow R, and we regard this arrow as symbolizing a force acting on the left hand; the right hand is also pulled towards the left in the direction of the arrow L and we regard this arrow as symbolizing a force acting on the right hand. Except for the sign, the two forces represented by the arrows R and L are equal. This fact is expressed by the statement that *action and reaction are equal and opposite.*

In the second case the mass on the left is accelerated towards the right in the direction of the arrow R and we regard this arrow as symbolizing a force acting on the mass M. Thus far the beginner will raise no objections, but he will be very dubious about the next statement, that during the acceleration the hand is pulled towards the left in the direction of the arrow L, and that we have to regard this arrow as a symbol of a force acting on the hand. The two forces L and R are again identical except for sign and we again have the theorem: *Action and reaction are equal and opposite.*

This fact, often puzzling to the beginner, is conveniently illustrated by means of two flat trucks, in which the friction is very small, placed on a horizontal floor, so that gravity does not come in. In fig. 11

Fig. 11.—Equality of action and reaction

we see two men joined by a rope, one on each truck. A dynamometer may be inserted in the middle of the rope. With this apparatus we perform three successive experiments:

(1) Both men pull simultaneously.
(2) The man on the right alone pulls (acts as a motor), the man on the left holding the rope passively in his hand or having it round his body.
(3) The man on the left alone pulls (acts as a motor).

In all three cases the trucks meet at the same place. If the symmetry is complete, i.e. if both trucks and both men have the same weight, the trucks meet at a point midway between their initial positions. If we replace the rope by a bar of sufficient length we may repeat the experiment with the accelerations in the reverse direction.

These experiments compel us to assume the existence of the second force L even in the case of fig. 10. Apart from their directions, the two forces R and L acting on the mass and the muscle in fig. 10 are no more to be distinguished from each other than the forces R and L acting on the left and right hand respectively in fig. 9.

Hitherto we have distinguished the forces acting towards the left and right in figs. 9 and 10 merely by the *letters* R and L. Very often, however, *names* are used instead of letters. The phraseology used depends on the naïve and untenable idea that a force must have not only a *point of application* but also a *point of origin*. Attempts are then made to find the " cause " of the force, or as we may in many cases say without risk of being misunderstood, the " motive power ", in this point of origin. For example in fig. 9 we call L the force due to the left hand because it " originates " in the left hand, R the force due to the right hand, because it originates in the right hand; in figs. 10 and 12 we call R the accelerating muscular force, because it originates from the muscle of the arm, and L the reaction which originates in the " inert " mass during the acceleration.

All this phraseology is often very convenient. We must never

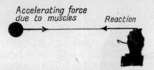

Fig. 12.—Action and reaction when a weight is swung round

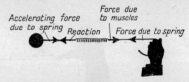

Fig. 13.—Illustrating the naming of forces

forget, however, that from the physical point of view, the point of origin of a force always remains arbitrary. We can see this by a glance at fig. 13, which again shows a stone being whirled round by the hand; this time, however, an arbitrary length of the string is replaced by a spring, to indicate the elastic tension of the string. This sketch is just as correct as that of fig. 12, although other names and other points of origin are assigned to the same forces.

Moreover, the distinction between the accelerating force and the reaction obviously has a meaning only when the point of origin of the accelerating force, i.e. the hand, is connected with the earth or forms part of it. I can, of course, say " the force of my muscles pulls a chair towards me "; " the force exerted by a magnet accelerates a piece of iron "; " the force which we call weight originates at the centre of the earth and accelerates a stone towards the earth "; " the reaction of the whirling stone exerts a pull on my hand " (fig. 12). But the second body may also be movable and may be visibly accelerated. We should then speak only of a *mutual attraction* (or repulsion) of the two bodies. A distinction between the accelerating force and the reaction is not wrong, but is quite arbitrary. The accelerating force and the reaction interchange their rôles if the phenomena are described in reverse order. This is very clearly brought out by the three experiments illustrated by fig. 11.

6. The Quantitative Statement of Newton's Second Law of Motion. The Dynamical Unit of Force.

We first meet with the second law of motion in the form of a qualitative statement. Our next task is to express it quantitatively. Anticipating the results of experiment, we have

$$f = \text{const.} \frac{F}{m}: \quad . \quad . \quad . \quad . \quad . \quad (2)$$

"f, the acceleration of a body of mass m subjected to a force F, is directly proportional to the force F and inversely proportional to the mass m".

We shall test this experimentally in the two limiting cases of acceleration, namely, for acceleration wholly in the direction of motion (in the present section and the following one) and for acceleration wholly at right angles to the direction of motion in §§ 8 to 11 (p. 41).

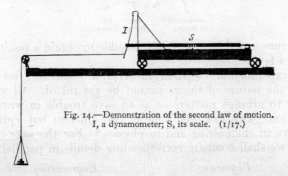

Fig. 14.—Demonstration of the second law of motion.
I, a dynamometer; S, its scale. (1/17.)

In the case of acceleration in the direction of motion, we work with a *constant* acceleration, which may be either horizontal (as in this section) or vertical (as in § 7). For horizontal accelerations in the direction of motion the apparatus sketched in fig. 14 will be found useful for lecture experiments.

The body to be accelerated consists of a long truck on a smooth horizontal base. Its mass (including all accessories) is a kilogram, but may be increased by the addition of weights. The accelerating force F is produced by means of a string and metal block, and its magnitude during the acceleration is recorded by a dynamometer I. The dynamometer is *previously* calibrated in kilograms weight (kg.-force), i.e. in terms of the statical or practical unit of force, with the truck *at rest*. The experimental readings, on the contrary, are taken while the truck is moving. The dynamometer then indicates a constant deflection on the scale S. This deflection is a little less than that obtained while the truck is still held fast (it is released by means of a wire). The force accelerating the truck is therefore *less* than the weight

of the suspended block of metal but is *constant*. Thus the weight of
the block enables us to produce a force which is *constant* during the
acceleration. If instead of the metal block we used our muscles to
pull the string, we should only attain this result after wasting a great
deal of time in preliminary practice.

To measure the acceleration we use equation (4) on p. 21, $s = \frac{1}{2}ft^2$.
Together with equation (2) above, it gives

$$\frac{2sm}{Ft^2} = \text{const.} \quad \ldots \ldots \ldots \quad (2a)$$

We ascertain the time taken by the truck in moving from start
to finish of the path s by means of a stop-watch recording hundredths
of a second. If the truck is suitably constructed the observations
can be carried out very conveniently by projection on a screen. Such
experiments yield the following result:

$$f = 9 \cdot 8 \; F \; / \; m. \quad \ldots \ldots \quad (2b)$$

measured   kg.   kg.
in m./sec.²   (force) (mass)

In the lecture experiment it is only possible to obtain a result accurate
to within a few per cent.

This experimental equation involves a *numerical factor*, which
owing to the nature of things cannot be got rid of. All we can do
is merely to arrange matters so as to save trouble in writing down
the equation. This has been done in two different but equally justifiable ways in engineering and in physics.* For the sake of greater
clearness we shall compare corresponding details in parallel columns.

Physics	*Engineering*

The numerical value of the constant depends only on the choice
of the units employed. We can easily make the constant equal to 1,
so that

$$f = F/m. \quad \ldots \ldots \ldots \quad (2c)$$

To do this we must introduce another unit for the

force	mass

which is 9·8 times

smaller.	larger.

* [The following discussion of the two systems of units employed on the Continent
has been retained as an illustration of the way in which systems of units may be
constructed. In Great Britain the gm.-mass-cm.-sec. system (unit of force, the *dyne*)
described below is used in *physics*. British *engineers* use the lb.-force-ft.-sec. system
(in which the value of g is 32·2 ft./sec.²). The lb.-mass-ft.-sec. system (unit of force,
the *poundal*; 1 poundal = 1/32·2 lb.-force = 0·0141 kg.-force) is very seldom used.
The poundal corresponds to the "large dyne" in the Continental kg.-mass-m.-sec.
system.]

For

| unit of force | | unit of mass |

we accordingly no longer use

| 1 kg.-force | | 1 kg.-mass |

but

| 0·102 kg.-force. | | 9·8 kg.-mass. |

With this unit 9·8 times

| smaller | | larger |

the numerical value of the

| force in the numerator | | mass in the denominator |

of equation (2) is 9·8 times

| greater. | | smaller. |

The quotient F/m no longer requires to be subsequently multiplied by the factor 9·8 in order to make the right-hand side of equation (2) numerically equal to the observed numerical value of f. In both cases we have $f = F/m$, i.e. the constant is equal to unity. Thus one of the *four* fundamental quantities (namely length, time, force, mass) becomes unnecessary. Henceforth a system of units with only *three* fundamental quantities will suffice. The choice of these three out of the four original quantities is arbitrary. We thus obtain the

| physical | | engineering |

system of units with its three fundamental quantities,

Length l measured in metres
Time t measured in seconds

| Mass m measured in kg.-mass | | Force F measured in kg.-weight |

(but usually referred to merely as kilograms).

| Force | | Mass |

is a subsidiary quantity. Its dimensions are

| Mass × Acceleration | | Force ÷ Acceleration |
| or $F = [ml\,t^{-2}]$. | | or $m = [Fl^{-1}t^2]$. |

The unit in terms of which it is measured is

| 1 kg.-mass . m./sec.², | | 1 kg.-force . sec.²/m., |

which we call the

| large dyne. | | practical unit of mass. |

It is the

| force which gives a mass of 1 kg. | mass to which a force of 1 kg. gives |

an acceleration of 1 m./sec.^{2} Thus

| 1 large dyne = 0·102 kg.-force. | 1 practical unit of mass = 9·8 kg.-mass. |

Accordingly, the weight of a man is

| about 700 large dynes | about 70 kg.-force |

and the mass of a man is

| about 70 kg.-mass. | about 7 practical units of mass. |

Thus in both the practical and the physical system of units the factor 9·8 in equation (2) has been got rid of and likewise the unnecessary fourth fundamental quantity of mechanics. Our comparison clearly brings out the fact that either system is equally justifiable. Here in the interests of clearness we have deviated as regards a superficial detail from the commonest usage on the physical side. We took the metre and kilogram as units of length and mass respectively, whereas in physical literature the smaller units centimetre and gram are used in the vast majority of cases. The common physical system is based on the experimentally obtained equation

$$f = 981 \; F \;/\; m,$$

measured in cm./sec.^{2} gm. (force)   gm. (mass)

the new unit of force, the *dyne* ($= 1·02$ mg.-force) being defined in a way analogous to that above. One dyne (1 gm.-mass . cm./sec.^{2}) is the force which gives a mass of 1 gm. an acceleration of 1 cm./sec.^{2}, i.e. raises its velocity by 1 cm./sec. per second. The large dyne is accordingly equal to 10^5 dynes.

In carrying out experiments scarcely anyone pays any attention to the fact that one of the four fundamental quantities is superfluous.

| The physical unit of force 1 kg.-mass . m./sec.^{2} | The practical unit of mass 1 kg.-force . sec.^{2}/m. |

is merely a paper quantity. No measuring instrument is graduated in terms of it. In the vast majority of measurements

| the practical unit of force (kg.-force, &c.) is used in physics, G, the *numerical* value of a weight, being multiplied by 9·8, &c. | the physical unit of mass (kg.-mass, &c.) is used in practice, G, the *numerical* value of a mass, being divided by 9·8, &c. |

The sole difference is one of notation:

the physicist tries to hide his methods of measurement, at least in his calculations; he does not write	the engineer lets his methods of measurement appear even in his calculations; he writes
$$F = 9 \cdot 8\, G$$	$$m = G/9 \cdot 8$$
in his equations.	in his equations.

The physical or engineering notation is preferred according to the circumstances of the case.

Units and systems of units are of very little importance in themselves; it is merely a question of convenience and suitability. No one system of units can claim a monopoly. The important thing is to see that the units employed are clearly stated in each particular case. In the present book we shall carry out numerical calculations in accordance with the following table.

We have to insert	for masses	for lengths	for forces
in the physical gm.-mass-cm.-sec. system	the numerical value in gm.-mass	the numerical value in centimetres	$0 \cdot 98$ times the numerical value in mg.-force (for 1 dyne = $1 \cdot 02$ mg.-force)
in the physical kg.-mass-m. system *	the numerical value in kg.-mass	the numerical value in metres	$9 \cdot 8$ times the numerical value in kg.-force (for 1 large dyne = $0 \cdot 102$ kg.-force)
in the practical kg.-force-m.-sec. system	the numerical value in kg.-mass divided by $9 \cdot 8$ (for the nameless practical unit of mass = $9 \cdot 8$ kg.-mass)	the numerical value in metres	the numerical value in kg.-force
in the British practical lb.-force-ft.-sec. system	the numerical value in lb.-mass divided by $32 \cdot 2$ (for the nameless practical unit of mass = $32 \cdot 2$ lb.-mass)	the numerical value in feet	the numerical value in lb.-force.

* In this system the unit of work is 1 large dyne-metre or 1 kg.-mass m.²/sec.² = 1 watt-second (see p. 75). It is very conveniently related to the international volt-ampere system of units. Thus, for example, an electric field of intensity 1 volt/metre exerts a force of 1 large dyne on a charge of 1 ampere-second or coulomb.

To each *dynamical* statement of a force in dynes or large dynes we shall add a *statical* one in kg.-force, gm.-force, &c., for these statical values are more readily compared with the experience of everyday life.

7. Further Experiments on the Second Law with Acceleration in the Line of Motion.

All the experiments in the last two sections had one thing in common: the motion and the acceleration were always in a *horizontal* direction. The effect of gravity on the masses to be accelerated was avoided by placing the smooth supports or sliding rails in the apparatus as nearly horizontal as possible, thus making the conditions of experiment particularly simple.

In nature, however, the great majority of accelerations take place in a *vertical* direction, especially in the case of the motion of our bodies or limbs. These accelerations are always subject to the effect of gravity.

Fig. 15.—
Vertical acceleration of a body due to the difference of two forces.

For this reason a large number of lecture experiments have been devised in order to demonstrate the second law, both qualitatively and quantitatively, *when gravity is involved.*

All these experiments have a common feature: the accelerating force F is the *difference* between two unequal forces F_1 and F_2 in opposite directions. Forces which are not weights are measured by means of a dynamometer, e.g. I in fig. 15.

Accelerations are *only* observed as long as F_1 and F_2 are unequal. The sign of the acceleration varies with the sign of the difference $F_1 - F_2$. We proceed to give some examples.

Fig. 16.—
f, the downward acceleration of a man bending his knees, arises from the downward force $(F_2 - F_1)$.

(1) A man stands on an ordinary platform weighing machine (fig. 16). F_2 is his *weight*, e.g. 70 kg.-force. F_1 is the *elastic force* opposing the weight, which is recorded by the machine. We make three observations one after the other:

(*a*) The man stands still. The weighing machine then records 70 kg.-force; $F_1 - F_2 = 0$.

(*b*) The man moves with acceleration into the knees-bend position. During the downward acceleration $F_1 < F_2$.

(*c*) The man moves with acceleration back into the normal position. During this movement $F_1 > F_2$.

The changes in F_1 take place very rapidly, so that it is difficult for the eye to detect the direction of the first swing of the pointer. This difficulty is avoided by means of the following artifice. The pointer of the balance can be made to move about its axis of rotation. We make the man stand still on the platform and then place the pointer vertically downwards, and attach to its point a ball shaped like a ball of wool. This ball is easily displaced along a bar and when the pointer moves is slung to one side in the direction of the first deflection.

A variant of this problem is presented by the common riddle: Suppose we have a sensitive balance with a closed bottle on each scale-pan and there is a fly in one of the bottles. Does the balance record the weight of the fly?

The answer is as follows. If the fly flies at a constant height the deflection of the balance corresponds to the weight of the fly (case (*a*)). We simply have to regard the fly as a rather big air molecule. During an accelerated downward motion (if the fly lets itself fall) the deflection of the balance is too small (case (*b*)). During an accelerated upward motion the deflection of the balance is too large (case (*c*)).

(2) The experimenter holds the hoop - spring dynamometer, which we have already met with, vertically in his hand (fig. 17). At the upper end of the dynamometer there is a body M of weight F_2. If the velocity of the hand is constant the deflection (the compression) of the hoop is the same as when the hand is at rest. If the hand is accelerated upwards or downwards the spring is compressed more or less than this, i.e. F_1 is greater or smaller than F_2.

Fig. 17.—To illustrate how the "lift sensation" arises.

Such an arrangement often plays an awkward part in our everyday life. Suppose the hand represents the platform of a lift. Assuming a rather boldly simplified anatomy, we regard the spring as representing the intestines and the mass M as representing the stomach. A downward acceleration causes the spring to be extended as compared with its normal position in a state of rest. This extension is the physical cause of the unpleasant sinking feeling associated with going down in a lift, which on periodic repetition, e.g. on board a steamer, gives rise to sea-sickness.

(3) The mass $2M$ is divided into two equal parts, which are suspended over a pulley by a string (fig. 18). As both weights are the same, $F_1 - F_2 = 0$, so that there is no acceleration. The masses remain at rest, or if struck move with constant velocity. A small additional mass of m gm. is then added, e.g. to the right-hand mass. This makes F_2 greater than F_1 and an acceleration

$$f = \frac{F_2 - F_1}{2M + m} \quad \ldots \quad \ldots \quad (2d)$$

is produced. The apparatus is tolerably well suited for quantitative experiments. The acceleration is measured with the help of the equation $s = \frac{1}{2}ft^2$ by ascertaining the time t taken to describe the path s, and the value thus obtained is compared with that calculated from equation (2d). The following procedure, however, is more instructive.

(4) *Theory* (fig. 19).—A string hangs from a dynamometer (a balance) and carries a body of mass M. An apparatus enclosed in the body (not visible) causes the body to fall to the ground with

a small acceleration which may easily be measured by equation (4), p. 21. During the downward acceleration the dynamometer registers a change of deflection corresponding to $F_2 - F_1$.

Practical Form of Apparatus. — The body is in the form of a flywheel on a thin shaft, suspended by two threads wound round the shaft. The fly-wheel is at first held high up and is let go at the beginning of the experiment by means of a wire release such as is used in photography. The threads unwind and the body falls to the ground with acceleration. The acceleration f is first determined from the equation $s = \frac{1}{2}ft^2$ by ascertaining the time t taken by the body in falling through the distance s. The force $F_2 - F_1$ is then measured. This may be done by an ordinary pair of scales (fig. 20, Plate III). The distance between the two " duck's bills " is easily recorded at a

Fig. 18. — Apparatus for verifying the second law of motion (Atwood's machine).

Fig. 19. — A body with a constant downward acceleration suspended from a dynamometer.

distance when projected on to a screen. They rapidly take up their final position, and the observed distance between them is remarkably constant. The apparatus is subsequently calibrated by means of small weights.

The observed values of $F_2 - F_1$ are in good agreement with the values of mf calculated from the second law.

Numerical example: $m = 0.539$ kg.-mass, $f = 0.048$ m./sec.^{2}, $F_2 - F_1 = 26 . 10^{-2}$ large dynes $= 2.6$ gm.-force (note the table on p. 37); f is calculated from the distance fallen through ($s = 0.83$ m.) and the time ($t = 5.9$ sec.).

After the threads have unwound themselves the " inertia " of the fly-wheel keeps it running. The threads are wound up again and the body rises. The observations should be repeated for this upward motion. In this case also the result recorded by the dynamometer is *smaller* during the acceleration than it is at rest. The vector representing the acceleration of the body is directed *downwards* in both cases, as the body moves upward with diminishing velocity, i.e. with negative acceleration or " retardation ". This experiment often proves surprising even to more advanced students of physics.

Some of the experiments described in this section (1, 2, and 4) may also be explained qualitatively by means of the law of action and reaction. We must, however, proceed in quite a formal way; e.g. in fig. 21 two bodies A (a person) and B (a metal block) are connected by some mechanical means, by rods, strings, limbs, &c. A stands on a balance. If B is accelerated *upwards* A is accelerated *downwards*. A accordingly compresses the balance in the direction of "heavier". If B is accelerated *downwards* A is accelerated *upwards*. A accordingly presses to a less extent on the balance and the latter moves in the direction of "lighter". Thus far everything is in order from the formal point of view. One need not even hesitate to insert arrows representing forces in the directions of the accelerations and acting through the centres of gravity of B or A. But one must beware of proceeding to *name* the forces in the figure according to their points of origin. For from the physical point of view all that we are given is a *body* on which a force is acting. Those who try to supply names and points of origin for the forces will assuredly find themselves in difficulties when B is accelerated downwards.

Fig. 21. — Action and reaction.

8. The Second Law for Normal Accelerations.   Motion in a Circle: Radial Force (Observer at Rest).

First, a piece of good advice for the student: he should never enter into any arguments about circular or rotatory motion without coming to a definite agreement with his partner (possibly the writer of his textbook) about the system of reference to be used. On p. 15 we agreed to refer our observations to the earth or the floor of the lecture-room.

All our verifications of the second law hitherto have been concerned with the limiting case of acceleration wholly *in the direction of motion*. We now proceed to verify the theorem for the other limiting case of wholly *normal* acceleration.

Let a body of mass m describe a circular path of radius r with constant angular velocity ω. According to the kinematic point of view of § 5 of last chapter (p. 23) this motion is *accelerated*. The normal acceleration directed towards the centre of the circle is

$$f_n = \omega^2 r.$$

According to the second law, this acceleration involves the existence of a force

$$F = mf_n = m\omega^2 r, \quad \ldots \ldots \quad (3)$$

where $\omega = 2\pi n$, the number of revolutions in 2π seconds directed towards the centre. We shall call this force the *radial force*.

In order to test equation (3) experimentally, we replace the angular velocity ω by n, the number of revolutions per second:

$$F = 4m\pi^2 n^2 r \quad . \quad . \quad . \quad . \quad . \quad (3b)$$

(for units, see p. 37).

We shall use springs to produce the radial force, i.e. it is to be an elastic force. We shall consider three cases:

(1) Only a limited force is available.

(2) The force available increases proportionally to the radius of the path, i.e. $F = \mu r$ (*linear* law of force).

(3) The force available increases more rapidly than direct proportion to r, e.g. $F = \mu r^2$ (*non-linear* law of force).

(1) *Elastic Force of Limited Magnitude*

Fig. 22 shows a leaf-spring, its lower end being free to revolve and its upper end being behind the stop a. If the spring is bent as

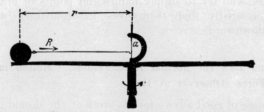

Fig. 22.—A sphere held on a whirling table by the spring to the left of *a*

in a crossbow it can only give an elastic force of limited magnitude. If the maximum value F_{max} is exceeded the spring snaps. This maximum value F_{max} may be determined by means of a string and spring balance.

This spring is made to exert a radial force on a sphere on the edge of a small whirling table. The number of revolutions required to reach the maximum radial force available is calculated from equation (3b):

$$n_{max} = \frac{1}{2\pi}\sqrt{\frac{F_{max}}{mr}}. \quad . \quad . \quad . \quad . \quad . \quad (3c)$$

Numerical Example.—$F_{max} = 0.18$ kg.-force, or about 1.77 large dynes; $m = 0.27$ kg.-mass; $r = 0.22$ m; $n_{max} = 0.87$ revolution per second; shortest time of revolution (T_{min}) $= 1.14$ seconds.

At all smaller rates of revolution the sphere participates in the circular motion of the whirling table, while if the limiting value is exceeded it flies off, leaving the disc in a *tangential* direction. Once free of the normal acceleration it continues to move in a straight line with constant velocity. Unfortunately the phenomenon is generally complicated by the weight of the sphere, which changes the trajectory, originally a straight line, into a parabola. This disturbance, however, is not prominent if the velocity of motion is high. A grindstone in action provides a good example of this type of motion, exhibiting very clearly the way in which the particles fly off *tangentially*. The sparks certainly do *not* fly away *centrifugally* from the centre of rotation (fig. 23, Plate III).

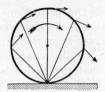

Fig. 24.—Directions of the splashes from a motor-car wheel, as observed by a pedestrian.

The behaviour of a motor wheel scattering mud seems to differ from that of the grindstone. It is possible to cross a smooth road immediately behind a mud-splashing car without being splashed. The explanation is simple. To an observer in the moving car the tyre has the same appearance as the grindstone, i.e. appears to scatter mud tangentially all round. To the pedestrian, on the contrary, it appears as in fig. 24. From his point of view the centre of rotation is the lowest point of the wheel, and all the mud flies off at right angles to the various chords in the directions indicated by the arrows.

(2) *Linear Law of Force*

The elastic force is proportional to the radius of the path:

$$F = \mu r. \qquad \qquad (4)$$

If we introduce this condition into the general equation (3b) we obtain the frequency

$$n = \frac{1}{2\pi} \sqrt{\frac{\mu}{m}}, \qquad \qquad (3d)$$

where μ is the force required to extend the spring by unit length.

This means that the mass only describes a circular path at a *single* rate of revolution. Here the radius of the circular path is quite arbitrary; if the " critical rate of revolution " is maintained the mass will move in *any* circle in which it may be set going.

An apparatus for producing a linear law of force is shown in fig. 25. The mass is divided symmetrically and attached to two bars with as little friction as possible. These bars are for getting rid of gravity effects. The spring

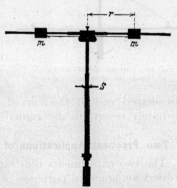

Fig. 25.—Circular motion with a linear law of force. The long spiral spring below S does not show up well in the figure. (1/9.)

is so arranged as to show the extent to which it is stretched, even when it is in rotation.

The spring must be stretched to the amount $F = \mu r_0$ even in the state of rest, where r_0 is the distance between the centre of gravity of the sphere and the axis of rotation in the state of rest.

Our predictions are confirmed by experiment. If the rate of revolution is the right one, we can increase or decrease r, the distance of the mass from the axis of rotation, to any extent we please by touching S, the disc-shaped end of the spring, lightly with the finger. No matter what the radius is, the masses continue to move in a circular path. At this critical rate of revolution the masses are in " neutral equilibrium ", like a ball on a horizontal table-top.

(3) Non-linear Law of Force

For example, let the elastic force increase proportionally to r^2 ($F = \mu r^2$). If we introduce this condition into the general equation for the radial force (3b) we obtain the frequency

$$n = \frac{1}{2\pi} \sqrt{\frac{\mu}{m} r}, \quad \ldots \ldots \ldots \quad (3e)$$

μ being the force required to extend the spring by unit length.

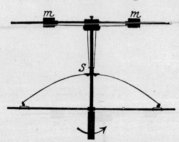

The frequency n depends on the radius r; to each rate of revolution there corresponds only one possible radius of path. When moving in this path the mass is in "stable equilibrium", like a ball at the bottom of a rounded bowl.

A non-linear law of this kind may be experimentally realized, e.g. by means of a curved spring as in fig. 26. If while the apparatus is in rotation we lightly touch S, the

Fig. 26.—Circular motion with a non-linear law of force. (1/9.)

disc-shaped end of the curved spring, the radius of the path immediately reverts to its original value.

9. Two Practical Applications of Motion in a Circle.

The two experiments illustrated in figs. 25 and 26 are frequently utilized for practical purposes; e.g. both forms of apparatus are used as *governors* for machines of all kinds.

In the case of the linear law of force (fig. 25) the revolving masses react even to *small* changes of frequency with extremely *large* deviations. If the frequency deviates in either direction from the

" critical " one given by equation (3d) the circular path is no longer possible at all. The masses approach as near to the axis of rotation or move as far away from it as is compatible with the construction of the particular apparatus.

In the case of a non-linear law (as in fig. 26), on the other hand, a *small* change in the rate of revolution involves merely a *small* change in the distance of the masses from the centre. If the rate of revolution is raised or lowered stable motion in a circle is again attained, the radius r being somewhat increased or diminished.

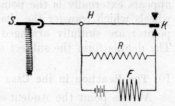

Fig. 27.—Governor of a shunt-wound electric motor.

In both cases the change in the distance of the revolving masses from the axis may be used to move the regulators of machinery of any kind. For example, we may imagine S, the upper terminal disc of the spiral spring, fixed between the forked ends of the regulating lever H in fig. 27.

Governors depending on a *linear* law of force (represented diagrammatically in fig. 25) are distinguished by their remarkable sensitiveness.

Practical example.—The rate of revolution of a shunt-wound electric motor *falls* if the current in its field coil F is *increased* by short-circuiting a resistance R. If the critical rate of revolution is *exceeded* the right-hand end of the regulating lever in fig. 27 is tilted downwards. This motion is made to short-circuit the resistance R by means of the contact K. After a few revolutions the rate of revolution falls *below* the critical value, the contact at K is broken, the current in the field coil falls, and the rate of revolution rises until the process begins all over again.

By means of a governor of this kind the average rate of revolution of an electric motor may be held constant to within a few hundred-thousandths of its value. On the average a direct-current electric motor so regulated runs as regularly as the very best watches, by which the 100,000 seconds or so which make up a day are measured correct to within a second.

These very useful governors depending on a linear law of force, however, possess one disadvantage: they will maintain only one constant rate of revolution, namely, the " critical " rate given by equation (3d). In order to alter the critical rate of revolution it is necessary to change the masses or the spring in accordance with equation (3d).

Governors with a non-linear law of force are less sensitive but enable the rate of revolution which is to be maintained to be varied more conveniently. For example, all that is required in order to adjust the regulator in fig. 27 for a higher rate of revolution is to move the contact K downwards. So much for governors.

Circular motion with a non-linear law of force is also put to practical use in the construction of the well-known *speed counter* or *tachometer*. We imagine the right-hand end of the lever H in fig. 27 moving past a scale. Every rate of revolution corresponds to a definite height of the upper end of the spring. In the practical forms of accurate speed counters (cf. fig. 2 and fig. 7, Plate II) all that appears externally is the pointer and scale. The levers and toothed wheels which transfer the change in the length of the spring to the pointer are skilfully arranged so as to occupy only a small space. The details form the subject of numerous patents.

10. The Reaction in the Case of Circular Motion.

At this point the student should re-read the first paragraph of § 8 (p. 41).

According to the law of action and reaction, the radial force must always be accompanied by a *reaction* of equal magnitude, but in the

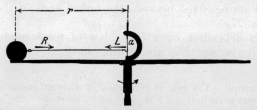

Fig. 28.—Reaction in circular motion

opposite direction. About the point of application of this reaction there can never be any doubt. Examples:

(1) Fig. 28 is the same as fig. 22 (p. 42) with a trifling addition; the mass is accelerated radially towards the axis of rotation by a force represented by the arrow R, and the spring is acted on by the reaction represented by L.

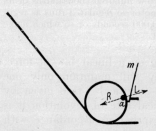

Fig. 29.—Reaction in circular motion

(2) Suppose that the ball in fig. 29 is let loose at the top left-hand corner and traverses a circular path as in the well-known loop-the-loop toy. At the point *a* of its path the acceleration of the ball towards the centre of the circle must be nearly horizontal. The necessary radial force is symbolized by the arrow R. The rail is acted on by the reaction, which is equal in magnitude and opposite in direction and is represented by the arrow L. This force of reaction deforms the rail, and the deformation may be made visible at a distance by means of a light lever giving a large magnification.

Such examples may be multiplied indefinitely. All have one feature in common: one body (the hand in whirling a weight, a shaft, a rail) is " rigidly " fixed to the earth, and is acted on by the reaction as long as the rotating body is being accelerated towards the centre of the circle by the radial force.

(3) Instead of a fixed body and a rotating body we now take two bodies, both of which may be accelerated. It is convenient to use two balls of unequal mass connected dumb-bell wise by a steel bar (fig. 30).

We set the dumb-bell spinning, supported on our upraised hand. The two masses describe circular paths of radii r_1 and r_2 about the hand, which must, of course, be placed under S, the common centre

Fig. 30.—Arbitrariness of the distinction between the force and the reaction

of gravity * of the two masses. For the elastic forces in the bar under tension, which give rise to the normal accelerations, must be the same for both spheres. Hence, by equation (3) (p. 41),

$$m_1 \omega^2 r_1 = m_2 \omega^2 r_2$$

or
$$r_1/r_2 = m_2/m_1, \quad \ldots \quad \ldots \quad (3f)$$

so that the distances of the masses from S are inversely proportional to the masses, as must be the case for the centre of gravity.

In this case of two bodies moving freely we can again speak only of a *mutual* attraction of the two bodies if we are to avoid arbitrariness. The circumstances are entirely analogous to those illustrated in fig. 11 (p. 31) with the two trucks, except that here the acceleration wholly in the direction of motion is replaced by a wholly normal acceleration. In the present case of the " mutual attraction " of two bodies the distinction between radial force and reaction is just as arbitrary as the distinction between accelerating force and reaction in the case of the trucks in fig. 11. The arrow L may be taken as symbolizing a radial force and the arrow R as symbolizing the corresponding reaction, or just the reverse; it merely depends on the order in which the phenomena are described.

It is quite unnecessary to give a special name to the reaction for the case of circular motion. Above all, it should not be referred to as " centrifugal force ". An observer from the ordinary standpoint, i.e. one who regards the earth or lecture-room floor as fixed, has no right to use the term centrifugal force at all. To avoid perpetual confusion it is necessary that the term centrifugal force should be definitely reserved for observers participating in the circular motion. In short, the term centrifugal force implies the use of an accelerated system of reference (Chapter VIII, p. 134).

* [See pp. 85, 99.]

11. The Radial Force in the Motion of an Endless Chain.

Our previous experiments on normal accelerations due to radial forces all involved moving bodies or masses of very simple form (" small " balls or blocks). Their diameter could be neglected in comparison with r, the radius of the path, without giving rise to appreciable error. In short, they were " particles "; we could regard the masses as concentrated in their centres of gravity.

The question of the reaction in the case of normal acceleration was discussed in last section, and we are now in a position to illustrate the utility of the idea of radial force in the case of the revolution of other less simply-shaped bodies. As an example we take a slack endless chain.

To begin with, the chain is put on to the fly-wheel, fitting closely, for a preliminary experiment (fig. 31). If they were not held together the individual links of the chain would fly off tangentially, like sparks from a grindstone, when the fly-wheel was set in motion.

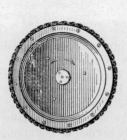
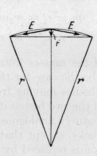

Fig. 31.—Chain on fly-wheel, to exhibit dynamical stability. (1/8)

Fig. 32.—Radial force in an endless chain under tension.

Fig. 33.—Oval form of a bicycle chain before it is thrown off the toothed wheel.

Here, however, they all co-operate in producing a state of tension in the chain. Each link of the chain is accelerated in the direction of the centre of the chain by the normal component of the elastic force arising in the chain (F in fig. 32). Without breaking, the chain may be subject to an elastic force sufficient to force the links (of mass m) into a circular path (of radius r). While the fly-wheel is rotating rapidly the chain is thrown off it by a push sideways. It does not fall slackly together but runs along the table like a rigid hoop, and even jumps over obstacles in its path. In this form the experiment gives a good qualitative example of " dynamical stability ".

An extension of the experiment, however, is even more instructive. If we introduce the linear velocity $v = \omega r$, equation (3), giving the radial force, becomes

$$F = mv^2/r. \quad \ldots \ldots \ldots (3a)$$

Thus for one and the same linear velocity v the radial force must be proportional to $1/r$. This assumption may be very prettily confirmed by means of the endless chain, in which, of course, all the links have the *same* linear velocity.

A small arc of the chain is shown in fig. 32; the arrows E indicate the elastic force, the short arrow F its component in the direction of the centre of curvature (the centre of the circle). This is the smaller the longer the chain is. By the geometrical figure, it diminishes as $1/r$ does. The endless chain should accordingly be stable when moving *not only as a circular ring but in any other form*, e.g. in the oval shown in fig. 33. This is confirmed by experiment. For the chain it is convenient to use a bicycle chain. If the rate of revolution is sufficiently high it is thrown off the toothed wheel.

This experiment sometimes takes place unsought in factories when a driving belt slips off.

CHAPTER IV

Simple Harmonic Motion and Central Orbits

1. Preliminary Remarks.

In our kinematical discussions in Chapter II we confined ourselves to the *simplest* motions, those which depend on an acceleration wholly in the line of motion or wholly at right angles to it. These cases of motion in a straight line and motion in a circle sufficed to introduce us in Chapter III to the fundamental laws of motion, in particular, to Newton's second law, and enabled us to verify them by a series of experiments. We shall take it that the fundamental laws of dynamics have thereby been investigated and verified to a sufficient degree. Henceforth they shall form the basis for the discussion of the dynamics of more complicated motions. In the present chapter we shall chiefly deal with simple harmonic motion and central orbits. In both cases we shall assume that a sufficient degree of accuracy is maintained if the bodies are taken as concentrated into heavy " particles ".

2. Linear Vibrations; Simple Harmonic Motion.

In this book any body capable of vibrating about its equilibrium position, e.g. a sphere between two light spiral springs (fig. 1), will be referred to as a " pendulum ". The distance of the body at any instant from its equilibrium position is called the displacement.

Fig. 1.—Simplest form of spring " pendulum "

The well-known gravity pendulum, consisting say of a ball on a string, is a very specialized type of pendulum. In fact, the essential features of vibrating motion are not exhibited by the gravity pendulum; it is not a suitable introduction to the study of vibrations and is not dealt with till § 2 of next chapter (p. 65).

A vibrating pendulum may describe a complicated path in space (§ 6, p. 60). In the first instance, however, we shall confine ourselves to the special case of vibrations in a straight line. Such vibrations are said to be "linearly polarized". But even if the path is a straight line the "form" of the vibration (fig. 2) may vary very greatly. In addition, the frequency depends in general on the *amplitude* of the vibration, i.e. on the maximum displacement of the body from the position of equilibrium. A limiting case, however, exists, which is distinguished by its great simplicity and its particular importance. This is the case of *sinusoidal* or *simple harmonic motion*, in which the frequency is independent of the amplitude. At present we shall confine ourselves to this type of motion.

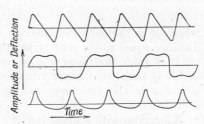

Fig. 2.—Examples of vibration curves which are not sinusoidal

The course of a simple harmonic motion as time goes on is graphically represented by the simplest wave-form, that of the sine curve. We have already seen that simple harmonic motion and circular motion are closely related; in short, we may say, a circular motion when looked at sideways appears as a simple harmonic motion (p. 11). We can accordingly start directly from our discussion of circular motion as we left it in last chapter. We found that there is a special case of circular motion in which n, the frequency, and r, the radius of the circle, are independent of one another. The necessary condition for this special case is the linear law of force

$$F = \mu r, \qquad \dots \dots \dots \dots \quad (1)$$

which gives the frequency

$$n = \frac{1}{2\pi} \sqrt{\frac{\mu}{m}},$$

which is independent of r (μ is the force required to extend a spring by unit length).

Looked at sideways, r, the radius of the circle in which the body is moving, appears as x, the distance of the body from its position of equilibrium O (fig. 3.) Similarly, the arrow Z denoting the centripetal force along the *radius* appears as the arrow F denoting a force *in* the line of motion, directed towards the position of equilibrium. Hence for the linear vibration we have the linear law of force

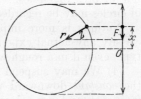

Fig. 3.—To illustrate the deduction of the period of vibration in the case of a linear law of force.

$$F = \mu x. \qquad \dots \dots \dots \dots \quad (1)$$

Here again, moreover, the frequency of the pendulum is independent of the amplitude of the vibration and is given by

$$n = \frac{1}{2\pi} \sqrt{\frac{\mu}{m}}, \qquad \dots \dots \quad (2)$$

where μ is the force required to displace the pendulum through a distance equal to the unit of length employed.

It is from the linear *law of force* (2) (and not from the straight or linear path) that linear vibrations take their name. The constant $\mu = F/x$ is the force required to produce a displacement through unit distance.

Vibrations with a linear law of force can be realized experimentally with considerable accuracy. The simplest way is to make use of *elastic* forces; for moderate extensions (x) they are, broadly speaking, proportional to the magnitude of the extension; all that is required in fig. 1, for example, is to make the springs sufficiently long.

In very slack springs, for which μ is small, the weight of the sphere sometimes gives rise to error by making the spring hang down. This external difficulty can be met in various ways, of which we mention four:

(1) We may support the sphere by means of a very smooth glass plate scarcely causing any friction.

Shaft

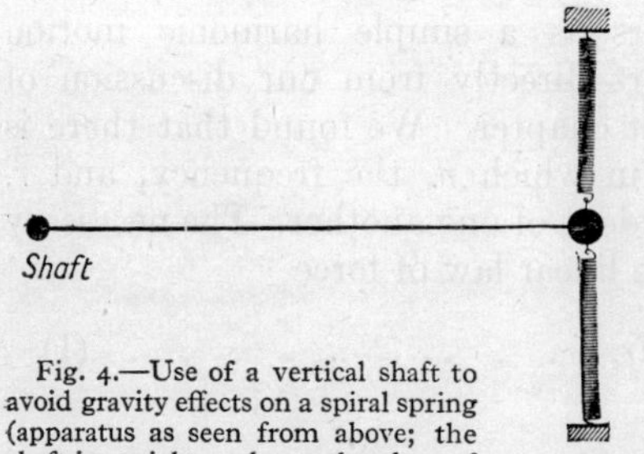

Fig. 4.—Use of a vertical shaft to avoid gravity effects on a spiral spring (apparatus as seen from above; the shaft is at right angles to the plane of the paper).

(2) We may suspend the sphere by a thin thread *several metres* long.

(3) We may, as in fig. 4, fix the sphere to the end of a light bar which is fixed spokewise to a vertical shaft. (Strictly speaking, in cases (2) and (3) the vibrations no longer take place in a straight line but in arcs of circles. Within reasonable limits, however, the arcs of circles may be regarded as straight without much loss of accuracy.)

Nothing is more important for the beginner than the power to classify any actual experimental arrangement rapidly under a known type, even if in a rough way only.

(4) We may suspend the sphere vertically by a spiral spring, for the *constant* downward pull of gravity makes no difference to the frequency and general course of the vibrations; it merely shifts the position of equilibrium.

Illustration.—Imagine the right-hand spring in fig. 1 to be prolonged indefinitely. Then its tension is practically unaltered when the sphere is displaced through a distance x. It is only the left-hand spring that exerts a restoring force tending to bring the sphere back into the position of equilibrium. The constant force towards the right due to the right-hand spring remains without effect on the ball.

The very important equation (2) may be verified experimentally by any of the above arrangements, but the simple scheme of fig. 1 is usually adopted. Even qualitative experiments reveal the decisive influence of the mass on the period of vibration of the pendulum, the latter increasing with the mass. The effect of an increase in mass can be compensated by an increase in the tension of the spring, and vice versa. By means of a stop-watch and a spring-balance for finding the force μ the equation (2) may be verified for any desired direction. In view of the fundamental importance of linear vibrations this is one of the most important exercises in practical physics.

The linear law $F = \mu x$ is, however, a special case. Nevertheless, it is of the greatest importance; for in the case of any body capable of vibration the law of force, no matter how complicated, may be replaced by the linear law of force; all that is necessary is that the motion should be confined to *sufficiently small* amplitudes.

Mathematically this means that any law of force $F = f(x)$ may be expanded in a power series

$$f(x) = \mu_0 + \mu_1 x + \mu_2 x^2 + \ldots \ldots \ldots \quad (3)$$

The constant μ_0 must be zero since the force must vanish for $x = 0$. For sufficiently small values of x the series may be broken off at the first term, so that we obtain $F = \mu_1 x$.

Later we shall consider the vibratory motion expressed by equation (2) in great detail and shall continually be making use of it. Here we shall confine ourselves in the first place to two extensions of it.

The first is of rather a superficial nature. The apparatus of fig. 1 may be replaced by that of fig. 5, i.e. a coiled-up spring is used instead of a spiral spring. The vibrating mass then moves in the arc of a circle. The displacement x is to be measured along this *arc*; it is an angular quantity. The direction of the restoring force coincides with that of the tangent

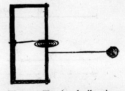

Fig. 5.—Torsional vibrations with shaft and coiled spring

to the path at any instant; the force may be measured by means of a spring-balance placed along the tangent. Formally these vibrations along the arc of a circle are dealt with exactly as those along a straight line are. In the limiting case of vibrations of small amplitude the arcs of circles practically degenerate into straight lines. The well-known gravity pendulum is a particular case of this type of arrangement, the force exerted by the spring being replaced by a component of the weight (see fig. 1, p. 66).

The second extension is of a fundamental nature. Once excited, none of the pendulums mentioned continues to vibrate undamped, i.e. with constant amplitude (as in the upper curve in fig. 6). Owing to unavoidable frictional resistances all vibrations of pendulums are *damped* and die away (lower curve in fig. 6). In practice there exists no such thing as an undamped pendulum. The sine wave of constant amplitude is an idealized limiting case. It belongs to the same category as the rectilinear path described with constant velocity by a body acted on by no forces.

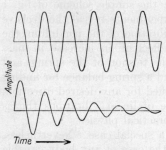

Fig. 6.—Damped and undamped sinusoidal vibrations

Linear vibrations are the first type of motion we have met with in which the acceleration is *not* constant. By Newton's second law $f = F/m$ and by the linear law of force $F = \mu x$; hence $f = \mu x/m =$ const. x. The acceleration at any instant is proportional to x, the distance of the body from its position of equilibrium, so that its variation as time goes on, like that of x, is represented by the sine wave.

3. Central Orbits. Definition.

In the case of linear vibrations the path is a straight line, although the acceleration is no longer *constant*. The additional velocity dv arising in time dt is always in the direction of the already existing velocity v, either increasing it (fig. 7a) or diminishing it (fig. 7b); that

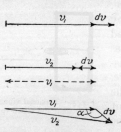

Fig. 7. — Illustrating the definition of resultant acceleration.

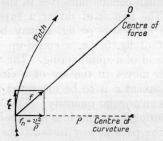

Fig. 8.—Resolution of an acceleration into two components

is, the acceleration is always *in the direction of motion*. In the most general cases of motion, however, the arrow dv makes an arbitrary angle α with the arrow v (fig. 7c). Then both tangential and normal accelerations exist simultaneously as components of the resultant acceleration f (fig. 8). The tangential acceleration f_t changes the *magnitude* of the velocity in the direction of the path. The normal acceleration f_n causes the *curvature* of the path. By equation (6), p. 23, its magnitude is given by v^2/ρ, where ρ is the length of the " radius of curvature " joining the point to the " centre of curvature " at that instant. The latter is the centre of the circle which may be made to coincide as closely as we like with the part of the path under consideration. From the almost hopeless complexity of such motions (consider, for example, the motions of our limbs) we select in the first instance a single group, that of motions in central orbits.

These are motions of a body (or particle) in any plane path, in which an acceleration varying in magnitude and direction is always directed towards the same point (centre). The line joining the body and the centre is called the radius vector. According to this definition motion in a circle and linearly polarized vibrations are limiting cases of motion in a central orbit, the tangential acceleration being absent in the first case and the normal acceleration in the second.

4. The Theorem of the Conservation of Areas.

There is a simple theorem, known as the theorem of the conservation of areas, which applies quite generally to motion in a central orbit: *The radius vector describes equal areas in equal times.* This theorem is a purely kinematical one, being a geometrical consequence of the assumption of an acceleration which, though variable, is always directed towards the same point, provided the acceleration does not degenerate into a wholly tangential one. We see this from fig. 9, which is an extension of fig. 12, p. 23. *Three* portions of the orbit are replaced by the *three* straight lines *xa*, *ac*, *ce*. The acceleration towards the centre increases from left to right. The faint arrows *ab*, *cd* represent the motions which would occur in equal intervals of time (*dt*) if the velocity were constant, the motion of the

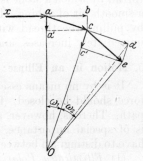

Fig. 9.—To illustrate the conservation of areas

body along the immediately preceding element of the path being continued along the tangent. The heavy arrows *aa'*, *cc'* denote the accelerations towards the centre O.

As by hypothesis $ac = cd$,

$$\triangle Oac = \triangle Ocd,$$

and as their altitudes $cd = c'e$ are equal,

$$\triangle Ocd = \triangle Oce;$$

$$\therefore \triangle Oce = \triangle Oac.$$

In demonstrations of the conservation of areas it is usual to be
content with experiments which are merely qualitative, e.g. we may
repeat the experiment shown in fig. 23, p. 14. The weight represents
the revolving mass and its distance from the axis of the body or stool
the length of the radius vector. The variable acceleration towards
the centre is produced by the muscular force of the experimenter, or,
to put it more simply, the length of the radius vector is altered by
bending or stretching the arms. The angular velocity promptly reacts
to every change in the length of the radius vector. But here, of
course, the theorem of the conservation of areas can obviously be
satisfied only qualitatively. The approximation to a " particle " is
a very poor one, the arms are not bars of practically no mass, and
so on. On occasion, however, it is permissible to use quite crude
methods. Qualitatively these usually give us what we want, and
in the present case it is certainly not worth while to cut out the errors
indicated. Such attempts only succeed in the case of a moderately
good approximation. Later we shall meet with a general theorem
which includes the theorem of the conservation of areas as a limiting
case and in addition is immediately applicable to extended masses
such as the rotating stool, observer, and weight together. The following demonstration, on the other hand, is free from objections. The
thread by which a stone is whirled round is led to the left hand through
a smooth bit of tubing. The length of the radius vector r is decreased
by pulling the thread with the right hand. The angular velocity ω
immediately increases, and in proportion to $1/r^2$.

5. Motion in an Ellipse: Elliptically Polarized Vibrations.

It is by no means essential that an orbit described under a central
force should be closed; for example, a body may move in a spiral
path. There is, however, a type of motion under a central force which
is of special importance, namely, motion in an elliptical path. We
have to distinguish between two cases which we shall discuss separately.

(1) *Elliptically Polarized Vibrations.*—The centre towards which
the revolving body is accelerated is the *centre* of the ellipse, i.e. the
point of intersection of the major and minor axes.

(2) *The Kepler Ellipses.*—Here the centre towards which the
revolving body is accelerated is one of the two *foci*.

In this section we shall deal with elliptically polarized vibrations.
Kinematically these arise from the superposition of two linearly
polarized sinusoidal vibrations in directions at right angles to one
another and of the same frequency. This may be verified not only
graphically and by calculation, but in a variety of experimental ways.

A method which is particularly easy to follow depends on the
close connexion which, as we already know, exists between simple
harmonic motion and circular motion (p. 12). The two simple harmonic

PLATE III

Chap. III, Fig. 20.—A flywheel with constant downward
acceleration (Maxwell's disc) is suspended from a pair of
scales. The balance has an oil damping apparatus concealed
behind the plate marked 10 kg.

Chap. III, Fig. 23.—Sparks from a grindstone

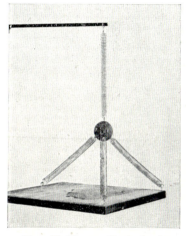

Chap. IV, Fig. 25.—Elastic vibrations
in space (the most general case of Lissajous
figures).

Chap. IV, Fig. 10.—Apparatus for demonstrating
elliptic vibrations and Lissajous figures

Chap. IV, Fig. 15.—Bevel wheels
for altering the difference of phase
between the rotating discs in fig. 10.

motions are produced by two rods revolving at right angles to one another at the same rate (fig. 10, Plate III). The two motions are projected practically *sideways* on to a screen by means of an arc lamp at a sufficiently great distance. The shafts A and B are geared to one another and may be driven simultaneously by any form of motor.

The shadows of the two rods at rest form a black cross (fig. 11). During the motion we carefully observe their point of intersection. While the shadows are swinging backwards and forwards we see— here comes something surprising—a *white* path on a grey background. This means that during each half-revolution every point of the screen is in shadow twice, except for the path described by the point of intersection of the bars, which is in shadow only once.

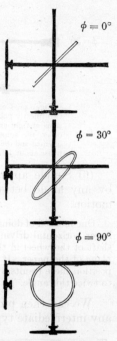

The distance of the point of intersection from the equilibrium position at any instant is the amplitude of the resultant vibration. We begin our experiments with two limiting cases:

(1) In fig. 11 we see both rods in their mean positions. Both vibrations start from there simultaneously. The *difference of phase* of the two simple harmonic vibrations is zero. The point where the dark shadows of the rods cross executes a linearly polarized vibration in an oblique direction (fig. 11).

(2) We displace the circular support of the horizontal rod through 90° (to do this we have only to loosen the screw Z). The vertical rod now begins to move at the exact moment when the horizontal one is reversing its direction at the position of maximum deflection. " The difference of phase is 90° " and a white circle is described (fig. 13). The amplitude remains constant and the line representing it revolves in a clockwise direction. (If the difference of phase is 270° it goes round counter-clockwise.)

Figs. 11–13.—Combination of two linearly polarized vibrations at right angles to each other, the amplitudes being equal but the difference of phase varying. (1/8.)

We now pass to the general case:

(3) We set the rods so that they make an angle of 30° with one another. With this difference of phase we get the white elliptical path shown in fig. 12.

(4) Setting the rods at any other angle likewise gives rise to an oblique ellipse.

(5) In all these ellipses the principal axes make an angle of 45° with the vertical. The position of the axes is only altered if we make

the amplitudes of the two component vibrations unequal, which is
done in practice merely by altering the distance of one rod from its
axis of rotation. For this purpose the ends of the rods are fixed to the
discs by means of a slot and a screw in such a way that they can be
slid towards or away from the centres of the discs (cf. fig. 14).

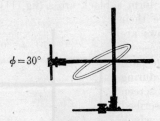

$\phi = 30°$

Fig. 14.—Combination of
two linearly polarized vibrations at right angles to each
other, the amplitudes being
unequal and the difference of
phase about 30°.

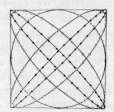

Fig. 16.—Envelope of the
elliptical vibrations when the
two component vibrations at
right angles to one another
have equal amplitudes.

(6) Simple appliances enable us to alter the difference of phase
by any angle between 0° and 360° while the rods are actually in
motion.

One method of doing this is to insert the bevel gear of fig. 15 (Plate III) into the
lower horizontal driving shaft in fig. 10 between the two middle supports. The
shaft of the wheel in the middle can be moved in a plane at right angles to the
plane of the paper. If this movable shaft is set to make an angle with its mean
position there results a difference of phase between the two horizontal axes equal
to twice this angle.

We can then produce the cases shown in figs. 11 to 13, as well as
any intermediate type, in as rapid succession as we like. All the paths
which occur are enveloped by a square
(fig. 16). If the two amplitudes are
different the square is replaced by a
rectangle (fig. 17).

Fig. 17.—Envelope of the elliptical
vibrations when the two component
vibrations at right angles to one another
have unequal amplitudes.

We may summarize as follows. An
elliptically polarized vibration of any
form can be represented kinematically
by two linearly polarized vibrations in
a straight line at right angles to one
another and of the same frequency, the
difference of phase being variable. If the difference of phase is 0° or
180° the ellipse degenerates into a straight line. If the difference of
phase is 90° or 270° a circularly polarized vibration, i.e. a circular
path, may arise; for this it is necessary that the two amplitudes
should be equal.

There is also another way of representing an elliptically polarized vibration kinematically. This too may be carried out conveniently with the apparatus of fig. 10. The difference of phase between the two component vibrations is set once and for all at 90°, but the amplitudes are varied. In this kinematical representation of the ellipse the major axes are horizontal or vertical, and we obtain figures of the type shown in figs. 18, 19.

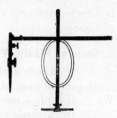

These two kinematical methods of describing elliptically polarized vibrations are of great importance in all branches of physics. Here in mechanics they show us immediately how elliptically polarized vibrations of a body (or particle) can be realized *dynamically*: each of the component vibrations arises from a *linear law of force* (equation (1), p. 51). All that is required is the simple apparatus sketched in fig. 20. If struck in a vertical direction the ball executes vertical vibrations only; if struck

Fig. 18.—Elliptical vibrations when the amplitudes of the two component vibrations at right angles are unequal and there is a phase difference of 90°.

horizontally, it executes horizontal vibrations only. In both cases the frequency is the same (as we may test by a stop-watch). By means of this apparatus we may make an actual body, in this case a ball, traverse the ellipse represented kinematically in figs. 11 to 13. The decisive factor is the direction of the original impulse in the plane of the diagram.

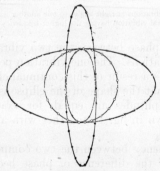

Fig. 19.—To show how the shape of the ellipse varies with the amplitudes of the two component vibrations, the phase difference being 90°.

Fig. 20.—Apparatus for producing elliptical vibrations

Instead of the three springs in fig. 20 we may use four springs, one pair horizontal, one pair vertical. This arrangement, however, is misleading to the beginner. The frequency of the horizontal or vertical vibration is by no means wholly due to the horizontal or vertical pair of springs. As a result of the crossbow-like way in which they are stretched the springs also contribute a component to the force associated with the vibration taking place transversely to their lengths.

The essential feature of the apparatus shown in fig. 20 is that both component vibrations depend on a linear law of force. Otherwise the vibrations are not simple harmonic. The elastic springs used are by far the most important means for realizing the linear law of force in practice. Hence the elliptical path in which the acceleration is directed towards the centre of the ellipse may be called the " ellipse of elastic vibration ".

6. General Elastic Vibrations of a Particle.

The methods and results of last section enable us to deal with the most general case of elastic vibrations of a particle without difficulty. We shall confine ourselves to a brief summary.

(1) In carrying out the experiment illustrated in fig. 20 it is never possible to make the frequencies of the component vibrations exactly

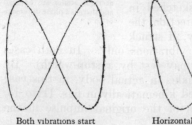 

Both vibrations start together   Horizontal vibration 30° behind the vertical   Horizontal vibration 45° behind the vertical

Fig. 21.—Lissajous figures; the component vibrations at right angles to one another have their frequencies in the ratio 2 : 1, the vertical vibration having the higher frequency

the same. Hence the difference of phase between the two vibrations is subject to regularly recurring variations. The one vibration periodically " gets ahead of " the other. As a result of this continual change in the difference of phase we find that the shape of the ellipse is continually changing. When the amplitudes are equal, for example, we get the succession of forms shown in fig. 16 together with all the intermediate forms.

(2) When the difference of frequency between the two component vibrations is greater the change of the difference of phase becomes noticeable during a single revolution. The ellipse is distorted into the characteristic form of a plane Lissajous figure. Some examples of these Lissajous figures are shown in figs. 21, 22. Their form depends on two things, (a) the ratio of the frequencies of the two component vibrations, (b) the difference of phase between the two vibrations on starting from their position of equilibrium.

Both these quantities can be varied quite simply using the apparatus of fig. 10. For (a) we have to change the driving wheels and use wheels with the number of teeth in a simple ratio, e.g. 20 : 40,

20 : 30, &c.   For (b) we displace the disc with the lower rate of revolution through an angle of 30°, 90°, &c., relative to its zero position.   The Lissajous figures shown in figs. 21 and 22 were obtained in this way.

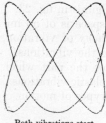 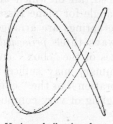 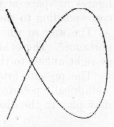

Both vibrations start   Horizontal vibration about   Horizontal vibration about
together   20° ahead of vertical   30° ahead of vertical

Fig. 22.—Lissajous figures; the component vibrations at right angles to one another have their frequencies in the ratio 3 : 2, the vertical vibration having the higher frequency

The device mentioned on p. 58 again enables us to alter the difference of phase while the rods are actually in motion.   We can thus display the figures from left to right in figs. 21 and 22 in rapid succession.   The envelope of the successive curves is very clearly shown. In the series of figures given the two component vibrations at right angles have *equal* amplitudes, so that the envelope of the successive curves is a *square*. In the more general case, where the amplitudes are unequal, it is a *rectangle* (fig. 23).

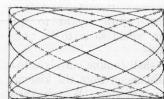

Fig. 23.—Rectangular envelope of Lissajous figures with varying difference of phase.

Fig. 24.—Production of Lissajous vibrations by means of elastic forces.

(3) *Dynamically*, as elastic vibrations of a particle, the succession of Lissajous figures may be obtained by e.g. the apparatus of fig. 24.   The frequencies of the horizontal and vertical component vibrations are in the ratio say of 2 : 3.   They can easily be observed separately by striking the ball horizontally or vertically.   The succession of Lissajous figures is that shown in fig. 22.

(4) Hitherto the two linearly polarized vibrations at right angles to one another have lain in one plane.   The resultant vibrations, whether they take the form of ellipses or of Lissajous figures, lie in the same plane.   That is, we confined ourselves to " plane vibrations ". In the general case we have to consider " vibrations in space ".   In addition to the two original component vibrations we have a third

linearly polarized vibration in a direction at right angles to both of them. Dynamically this can always be realized by the use of *four* springs (see fig. 25, Plate III). In general the frequencies in the three principal directions are unequal. A periodic succession of Lissajous figures in space is produced, all of which are enclosed in a " box " corresponding to the rectangle for plane Lissajous figures in fig. 23.

The way in which the springs are arranged in space is of no importance. There are always three *principal directions* of vibration, at right angles to the faces of the " box " enclosing the vibrations.

The present brief summary may suffice to show the part which " elliptically polarized vibration " or the " ellipse of elastic vibrations " may play in the disentangling of very complicated types of motion.

7. The Kepler Ellipse.

In the Kepler ellipse the centre towards which the acceleration is directed is one of the two foci of the ellipse. In the history of physics the Kepler ellipse has twice attained a position of quite fundamental importance. To the teacher of experimental physics it is a regular bugbear. Kinematically it can be realized in a lecture experiment only imperfectly, and dynamically not at all. In view of the failure of numerous attempts, a state of resignation with regard to this problem would seem to be quite justifiable.

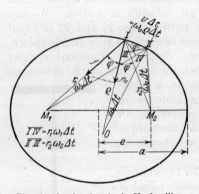

Fig. 26.—Accelerations in the Kepler ellipse

We shall confine ourselves to a brief discussion from the kinematical point of view. We take three geometrical properties of the ellipse for granted:

(1) In fig. 26 the normal bisects the angle between the focal radii r_1 and r_2;

(2) $\omega_1 \Delta t + \omega_2 \Delta t = 2\omega_0 \Delta t$; (I)

(3) $r_1 + r_2 = 2a$. (II)

The portion of the path traversed in time Δt may be regarded approximately as a circular path with radius ρ. Such a circular path involves the existence of a normal acceleration

$$f_n = \frac{v^2}{\rho}$$

(where v is the velocity in the path) directed towards the centre of

curvature O. This is the component $f \cos \phi$ of the resultant acceleration f directed towards the centre of attraction M_1, so that

$$\frac{v^2}{\rho} = f \cos \phi.$$

The surface described by the radius vector r_1 in the interval Δt is
$\frac{1}{2}r_1 . r_1 \omega_1 \Delta t$; by the theorem of the conservation of areas this is equal
to a constant, which we shall call k. Further, we see from the figure
that

$$r_1 \omega_1 \Delta t = r_2 \omega_2 \Delta t = v \cos \phi \, \Delta t,$$

and we hence obtain

$$r_1 v \cos \phi \, \Delta t = 2k. \quad \ldots \ldots \quad (4)$$

Hence the resultant acceleration directed towards M_1 is given by

$$f = \frac{4k^2}{r_1{}^2 \rho \cos^3 \phi \, (\Delta t)^2}. \quad \ldots \ldots \quad (5)$$

We proceed to eliminate the radius of curvature ρ and the angle ϕ
from this equation. By equation (I),

$$2\omega_0 = \omega_1 + \omega_2,$$

or

$$\frac{2v}{\rho} = v \cos \phi \left(\frac{1}{r_1} + \frac{1}{r_2} \right),$$

so that

$$\rho \cos \phi = \frac{2r_1 r_2}{r_1 + r_2}. \quad \ldots \ldots \ldots \quad \text{(III)}$$

Further, if we apply the cosine formula to the triangle $M_1 I M_2$, we
have

$$4e^2 = r_1{}^2 + r_2{}^2 - 2r_1 r_2 \cos 2\phi,$$

or, as

$$\cos 2\phi = 2 \cos^2 \phi - 1 \quad \text{and} \quad r_1 + r_2 = 2a,$$

$$\cos^2 \phi = \frac{a^2 - e^2}{r_1 r_2}. \quad \ldots \ldots \ldots \quad \text{(IV)}$$

Substituting from (III) and (IV) in (5), we have

$$f = \frac{1}{r_1{}^2} \frac{4k^2 a}{(a^2 - e^2)(\Delta t)^2}.$$

All the quantities occurring in the factor multiplying $1/r_1{}^2$ are constant,
so that

$$f = \frac{\text{const.}}{r_1{}^2}.$$

" In an elliptic path described under an acceleration towards a centre
the resultant acceleration towards one of the foci of the ellipse is
inversely proportional to the square of the focal radius."

CHAPTER V

Weight and Attractions

1. Special Features of Weight as an Accelerating Force.

Among the forces available in the laboratory, weight occupies a peculiar position in three respects.

(1) So long as electrical and magnetic apparatus is not available, weight is the only experimental means of producing a force which remains constant throughout the process of acceleration.

This lack of constant forces often imposes very troublesome limitations on experiment.

(2) The accelerations caused by weight alone are not only independent of all physical constants of the body undergoing acceleration, but are also independent of its mass.

(3) Relative to the system of reference which we have hitherto exclusively used, namely, the earth, it is only in the case of accelerations caused by weight that *no* reaction exists; Newton's law, actio = reactio, is *not* fulfilled.*

The second statement is very important in practice. It is most familiarly exemplified by the fact that all bodies fall at the same rate. We have already demonstrated this in several lecture experiments, in which, however, we were merely concerned with methods for measuring the kinematical quantity " acceleration ". We shall now explain the connexion between this experimental fact and Newton's second law $f = F/m$.

By Newton's second law the acceleration of a body due to a given force is proportional to $1/m$, where m is the numerical value of the mass found by the balance. On the other hand, the force, called the weight, with which a body is attracted and accelerated towards the earth, is proportional to m. Hence when the acceleration is *solely* due to the weight and does not depend on other forces acting on the body at the same time, the mass m cancels out.

The cogency of the evidence furnished by our previous experiments,

* Another way of putting this is as follows: the earth is not an inertial system of reference for the force which we call weight. In an inertial system of reference both the second and the third law of motion must hold.

however, must not be overestimated. Differences between the accelerations of different bodies amounting to several parts in a thousand or even more would be entirely compatible with the degree of accuracy observed on those occasions.

It has been found possible, however, to refine the methods of experiment considerably. To this end periodically recurring motions have been investigated, since in their case even quite small deviations add up in the course of time to quantities which are capable of definite measurement.

In these periodically recurring motions, however, the accelerating force can no longer be supplied by the weight alone, and elastic forces have to be used in addition. This is the case in the well-known gravity pendulum. Accordingly, gravity pendulums of the same geometrical dimensions but of very different masses are compared with one another, the observations being extended over many thousands of vibrations. If the experiment is properly conducted the pendulums all move at the same rate, no variation in the period of oscillation being recorded even to the sixth decimal place. Hence the varying accelerations of the pendulums at various points on their paths agree to six places of decimals. Thus the second statement at the beginning of this section is one of the best-established experimental facts. This experimental law forms the best evidence in favour of Newton's second law $f = F/m$. The remarkable feature of this theorem has already been strongly emphasized: the acceleration of a body is in opposition to the *inertia* of its mass. Yet in the second law we may use values for the mass which are deduced not from phenomena of inertia but with the help of *gravity*, that is, by means of a balance *at rest*. Might not the effect of a mass *at rest* on a balance-pan in the last resort be an inertia effect also? Might not earth + balance + mass be jointly subject to an acceleration whose nature has still to be explained? The general theory of relativity answers this surprising question in the affirmative, but to a large extent abandons all the physical concepts which have been built up hitherto.

2. Linear Vibrations of the Gravity Pendulum.

Linear vibrations take their name (p. 52) from the *linear law of force*

$$F = \mu x.$$

The vibrating body (or particle) is pulled back into the position of equilibrium by the force F, which is proportional to x, the distance of the body from its position of equilibrium.

Linear vibrations have three characteristic properties:

(1) The time-graph of the vibrations is a sine wave (fig. 20, p. 12).

(2) The frequency is independent of the amplitude of the vibration.

(3) The frequency is given by the equation

$$n = \frac{1}{2\pi} \sqrt{\frac{\mu}{m}},$$

where μ is the force required to deflect the pendulum through unit length.

Fig. 1.—The gravity pendulum

The linear law of force is also realized in the case of the gravity pendulum, provided the angle of deviation from the vertical is small and the path of the body therefore practically a straight line. This follows geometrically from fig. 1, in which the weight W acting on the pendulum bob is resolved into two components. One of these, $F_1 = W \cos a$, tends to stretch the thread. The other, $F_2 = W \sin a$, accelerates the inert bob in the direction of its motion. If the angle of deviation a is small, we may put $\sin a = x/l$; for angles under 4·5° the error so made does not attain one part per thousand. We have $F_2 = Wx/l$. Thus the force F_2 *is* proportional to the deflection x, the factor of proportionality being W/l. In dynamical units of force W, the weight of the mass m, is equal to mg, where g is the acceleration due to gravity, equal to 9·81 m./sec.² or 981 cm./sec.². Substituting $W = mg$ and $\mu = W/l$ in the equation for the frequency, we have

$$n = \frac{1}{2\pi} \sqrt{\frac{g}{l}}, \quad \ldots \ldots \quad (1a)$$

so that the period of oscillation (τ) is given by

$$\tau = \frac{1}{n} = 2\pi \sqrt{\frac{l}{g}}. \quad \ldots \ldots \quad (1b)$$

Numerical Examples.—(1) $l = 1$ m.; $\tau = 2$ seconds, i.e. one half-swing per second (the so-called seconds pendulum). (2) $l = 10$ m.; $\tau = 6·3$ seconds.

In contradistinction to all other cases of vibrating bodies, the frequency of a gravity pendulum is independent of its mass, as we already saw in last section.

Equation (1b) is of great practical importance, as it enables us to calculate the value of g, the acceleration of gravity, very accurately merely by making observations of the frequency of a pendulum, provided that we make as close an approximation as possible to the case of a " particle " suspended by a " light " (i.e. mass-less) string.

By § 5 of last chapter (p. 56), vibrations under a linear law of force need by no means be confined to vibrations in a straight line (the case of linear polarization). A more general case is that of elliptic polarization, the path being a plane ellipse. Hence the gravity pendulum with a linear law of force may also be made to execute elliptical vibrations. For example, a gravity pendulum may be struck in rapid succession in two directions at right angles to one another and to the thread. The ellipse may degenerate into a circle: we then speak of the conical pendulum.

3. The Variation of Weight with Distance from the Centre of the Earth. The Motion of the Moon.

Our discussions in this chapter have hitherto merely brought the experimental fact that all bodies fall at the same rate into relationship with Newton's second law. As regards weight itself, that mysterious force, our experiments on acceleration have shown us nothing beyond what any spring balance will tell us, namely, that *weight is a force which is practically constant over considerable regions.*

The next advance, quite a striking one, in our knowledge of weight depends not on laboratory experiments but on astronomical observations. Here the decisive step was the application of Newton's second law to the motion of the moon.

The moon goes round the earth in an orbit which is nearly a circle, with a radius sixty times (note this number) that of the earth. On p. 23 we gave a kinematical interpretation of the astronomical observations. The moon moves through about 1 km. per second at right angles to the radius, moving a little " away " from the earth. At the same time it is subject to a normal acceleration (f_n) of 2·7 mm./sec.^{2} which brings it radially " nearer " the earth by an amount (s) of 1·35 mm. per second $(s = \frac{1}{2} ft^2)$.

According to the second law $(F = mf$, equation $(2c)$, p. 34) this acceleration requires the existence of a radial force directed towards the centre of the orbit, that is, towards the earth. *This force must act without any mechanical means of connexion.* A force satisfying this condition is weight. Perhaps the force required is simply provided by the *weight* of the moon? This idea is at first sight contradicted by the magnitude of the observed acceleration. At the surface of the earth the acceleration due to the weight of a body is 9800 mm./sec.^{2}, that is, brings it 4900 mm. nearer the centre of the earth in the first second.

The way out of this difficulty was discovered by Isaac Newton: 9800/2·7 or 4900/1·35 is equal to 3600 or 60^2. From this Newton concluded that in contradiction to all terrestial observations, the weight of a body is *not* constant. It decreases to 1/3600 of its value if the distance of the body from the centre of the earth is multiplied by 60. Instead of equation (1), p. 26, we have in general to write

$$\text{Weight } W = \frac{m}{R^2} \cdot \phi \text{ (Earth)}. \quad \cdots \quad (2)$$

And now the law of action and reaction makes the last step almost inevitable. It interprets the as yet unknown effect of the earth in bringing about the existence of weight, represented in equation (2) by the symbol ϕ (Earth). If the earth attracts the moon with the force which we call the *weight* of the moon, the reverse must also hold: the moon must attract the earth. For an observer on the moon (here we have a new system of reference) the earth has weight. An imaginary observer on the sun may apply the law of action and reaction (another change of the system of reference). For these observers the two forces or weights must be equal and opposite. Thus in general we have instead of the weight the mutual attraction of two bodies with the force

$$F = \text{const.} \frac{mM}{R^2}, \quad \ldots \ldots \quad (3)$$

where m, M are the masses of the bodies and R the distance between their centres of gravity. In the case of homogeneous masses of spherical form this law holds for all values of R. In the case of bodies of arbitrary shape, R must be large compared with the dimensions of the bodies.

This is Newton's famous *law of gravitation*, the law of force which applies to attracting masses in general. The constant occurring in the equation is called the *gravitational constant* (γ). Its numerical value depends merely on the units employed. A method for determining it will be discussed in the next section.

4. Verification of the General Law of Attraction by Laboratory Experiment.

This requires the use of a very sensitive instrument for measuring force, preferably in the form of a torsion balance. We have already made use of a torsion balance in proving the existence of " internal " friction (fig. 3, p. 26). In that case we used a coiled-up spring and a horizontal axis of rotation supported by cup bearings.

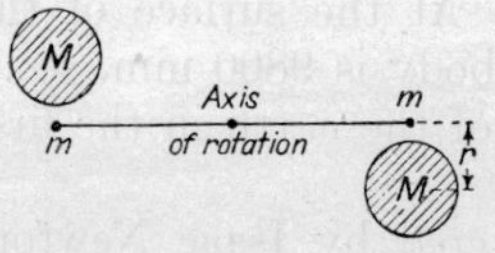

Fig. 2. — Diagrammatic sketch of a torsion balance for testing the law of attraction.

In the most sensitive forms of torsion balance the axis of rotation is vertical and bearings are entirely omitted in order to avoid friction there. The " balance beam " is suspended by a fine metal wire or band, so that we make use of the forces arising from the twisting of the wire.

In the diagrammatic sketch of fig. 2 we see two small masses (e.g. let each mass marked m be 10 gm.) fixed symmetrically to the ends of a balance beam. Two large masses (e.g. let each mass marked M be 10^4 gm.) are made to slide along rails towards the small masses.

In practical forms of the apparatus it is convenient to use a balance beam only a few centimetres long, from which the two spherical masses are suspended at different heights. A commonly used type of apparatus is shown in fig. 3. It has the following advantages:

(1) The period of oscillation of the balance is still passably short (only about 8 minutes).

(2) The attraction between one pair of masses is disturbed only to a trifling extent by the large mass of the other pair.

The balance beam, the two small spheres, and the suspending wires are enclosed in a double-walled case consisting chiefly of metal tubes. The tubes have no screening effect whatever on the attraction between the masses.

Whenever they are brought into the neighbourhood of the large masses the small masses are set in motion (*accelerated* motion). This motion can be followed out in detail by means of a mirror and beam of light giving a linear magnification of about 200. To begin with we may neglect the reaction of the twisted suspension-wire and regard the distance (*r*) between the centres of the two spheres as constant. Meanwhile the acceleration of the small masses is also constant, and we have the equation

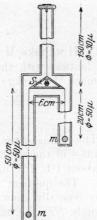

Fig. 3.—Practical form of a torsion balance for determining the gravitational constant (weight of balance beam 0.5 gm.)

$$s = \tfrac{1}{2}ft^2.$$

With stop-watch in hand we observe the distance (*s*) described say in one minute, and calculate the acceleration *f*. For the given experimental data *f* is found to be about 10^{-7} m./sec.^{2}.

This method of observation has two advantages. In the first place, it is not necessary to know the sensitiveness of the balance, so that this quantity has not to be calculated from the dimensions and elastic properties of the suspension-wire. In the second place, it is not necessary to wait for the final deflection of the balance, which, owing to the long period of vibration of this very sensitive instrument, would waste a great deal of time.

Using values of the acceleration obtained in this way, a lecture experiment gives about $6 . 10^{-11}$ large dyne-m.^{2} (kg.-mass)$^{-2}$ or $6 . 10^{-8}$ dyne-cm.^{2} (gm.-mass)$^{-2}$ for the constant* in the law of attraction (equa-

* There is a touch of fanaticism about the way in which many physicists fight against numerical factors in physical equations. Only those numerical factors which can exhibit some relationship to the number π, say $\tfrac{1}{4}\pi$ or $16\pi^2$, find favour with them. Remarkable to relate, however, the above numerical value of the gravitational constant has been allowed to pass and has not been replaced by a dimensionless number like 1 or 4π. Such a substitution is in itself easy to carry out. All that is required is to define a fresh unit of mass. Thus, for example, the constant becomes

tion (3)). Accurate experiments give $\gamma = 6.66 \cdot 10^{-11}$ large dyne-m.^{2} (kg.-mass)$^{-2}$ $= 6.66 \cdot 10^{-8}$ dyne-cm.^{2} (gm.-mass)$^{-2}$ or $6.52 \cdot 10^{-10}$ (kg.-force)$^{-1}$ m.^{4} sec.$^{-4}$

Using Newton's second law we obtain the general law of gravitation

$$f = \gamma \frac{M}{r^2}. \qquad \qquad (3a)$$

If we take M and r as the mass and radius of the earth, f must be identical with the " gravitational acceleration " obtained by experiment. In the kg.-mass-m.-sec. system this is 9·81 m./sec.^{2}, and we have

$$9.81 = 6.66 \cdot 10^{-11} \frac{M}{r^2}. \qquad \qquad (3b)$$

The radius of the earth is about 6400 km. $= 6.4 \cdot 10^6$ cm. Hence the mass of the earth is equal to $6 \cdot 10^{24}$ kg. The volume of the earth is about $1.1 \cdot 10^{21}$ m.^{3}

The quotient of mass by volume, i.e. the mass per unit volume, is generally known as the *density*. The density of the earth is accordingly 5500 kg./m.$^3 = 5.5$ gm./cm.^{3} This, of course, is an average value. The average density of the rocks in the earth's crust is only 2·5 gm./cm.^{3} Hence we must assume that in the interior of the earth there are masses of very great density. Many indications point to the conclusion that the core of the earth contains a great deal of iron.

5. The Law of Gravitation and Elements of Astronomical Dynamics.

The discovery of the law of gravitation is rightly counted as one of the greatest achievements of the human intellect. Newton's equation (3) not only reproduces the motion of the moon but holds throughout celestial mechanics, applying to the motion of planets, comets, and double stars.

equal to unity, if we take $1.5 \cdot 10^7$ gm. as unit mass. The choice of this unit of mass at the same time alters the " system of units ", leading to a new absolute system of units with only two fundamental quantities, namely length l (cm.) and time t (sec.). Mass has the dimensions $l^3 t^{-2}$ and its unit is 1 cm.^{3}/sec.^{2} ($= 1.5 \cdot 10^7$ gm. in the ordinary absolute system of units). So the physicist has not to buy 1·5 kg. of brass, but 10^{-4} cm.^{3} sec.$^{-2}$ of brass. Here, of course, we have a very awkward divorce of theoretical matters from ordinary everyday life. In analogous cases, however, this is not considered a drawback. For example, in electricity we have Coulomb's law for the attraction between two electric charges e at a distance r apart,

$$F = \text{const.} \frac{e \cdot e}{r^2},$$

which is completely analogous to the law of attraction between masses. To make this constant equal to unity the absolute electrostatic system of units has been established. In this system electric charges are not measured in terms of coulombs but in terms of the peculiar unit 1 cm. $\sqrt{\text{dyne}} = 1$ gm.$^{\frac{1}{2}}$ cm.$^{\frac{3}{2}}$ sec.$^{-1}$ ($= \frac{1}{3} \cdot 10^{-9}$ coulomb). This fight against numerical factors and the use of " absolute electrical units " are not infrequently regarded as symptoms of unusual erudition. Other opinions on this point, however, are equally well justified.

The observations which had previously been made on the motion of the planets were summarized by Johannes Kepler (1571–1630) in three laws. " Kepler's laws " are as follows:

(1) Each planet revolves round the sun in an elliptical orbit, the sun occupying one focus of the ellipse.

(2) The radius vector from sun to planet describes equal areas in equal times.

(3) The squares of the times of revolution are proportional to the cubes of the semi-major axes of the orbits.

The ellipses in which the major planets move differ only very slightly from circles, the discrepancy being greatest in the case of Mars. If the orbit of Mars is represented on paper by an ellipse with a major axis of 20 cm., it nowhere deviates from the enveloping circle by as much as 1 mm. These figures enable one to arrive at a better appreciation of Kepler's achievement.

By means of his law of gravitation, Newton was able to give a unified explanation of these three statements by his great predecessor.

(1) An elliptic orbit necessitates the existence of an acceleration towards a centre. In the ellipses observed by Kepler one focus is distinguished from the other. It follows by the kinematical considerations of § 7, p. 63, that the acceleration must be proportional to $1/r^2$. But by equation (3), p. 68, this is true for the force of attraction between masses in general.

(2) Kepler's second law is the theorem of the conservation of areas, which is true for all cases of central orbits.

(3) Kepler's third law likewise follows from equation (3), p. 68. This is easily seen in a special case. We let the Kepler ellipse degenerate into a circle. By p. 42 we have

$$F = 4\pi^2 n^2 mr = \frac{4m\,\pi^2 r}{\tau^2}$$

(τ being measured in seconds) for the circular path. For F we substitute the value given by the law of gravitation (equation (3)) and then obtain

$$\text{const.}\ \frac{m}{r^2} = \frac{4m\,\pi^2 r}{\tau^2},$$

or $\tau^2 = \text{const.}\ r^3.$ . . . . . (4)

In contradistinction to planets, comets are often found to move in extremely long ellipses in which the major axis may be a hundred times as long as the minor axis. But Kepler's third law may also be deduced from Newton's law of gravitation in the general case of an ellipse of any eccentricity, although the argument is necessarily somewhat more complicated.

In order to emphasize the most important facts of astronomical dynamics we shall conclude with a simple example.

Suppose a shot is fired in a horizontal direction near the surface of the earth, and that the atmosphere (and hence the resistance due to the air) is absent. What must the velocity of the bullet be in order that it shall continue to move round the earth at a constant distance from the earth's surface, like a miniature moon?

By equation (6), p. 23, motion with velocity v in a circular path involves the existence of a normal acceleration $f_n = v^2/r$. Now the weight of the bullet gives rise to an acceleration of 9·81 m./sec.² directed towards the centre of the earth. The distance of the surface of the earth from the centre of the earth is about 6·4 . 10⁶ metres. Hence we have

$$9 \cdot 8 = \frac{v^2}{6 \cdot 4 \cdot 10^6},$$

or $$v = 8000 \text{ m./sec.} = 8 \text{ km./sec.}$$

Thus if the muzzle velocity is 8 km./sec. in a horizontal direction, we have the case of fig. 4a, where the bullet revolves round the earth close to its surface like a small moon.

If the initial velocity exceeds or falls short of this value we obtain

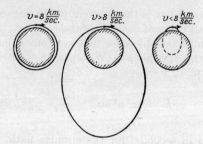

Fig. 4.—Elliptic orbits described about the centre of the earth, the initial velocities varying

elliptic orbits, as shown in figs. 4b, 4c. If the velocity exceeds 8 km./sec. the bullet revolves round the earth as a planet or comet in an ellipse, the centre of the earth occupying the focus *nearer* the bullet. If the velocity exceeds 11·2 km./sec. the ellipse becomes a hyperbola and the bullet leaves the earth never to return.*

For velocities less than 8 km./sec. we also obtain an ellipse, but in this case only the undotted part of the curve is actually realized. This time the centre of the earth occupies the focus of the ellipse *farther* from the bullet (hence the attraction exerted on it by the

* For the sun the corresponding value is 618 km./sec.

earth is the same as if the whole mass of the earth were concentrated at the centre of the earth).

The smaller the initial velocity the longer the ellipse. Finally we obtain the limiting case of fig. 5. The centre of the earth, towards which the acceleration is directed, appears to be practically at an infinite distance and the radii vectors pointing to it are practically parallel. The remnant of the elliptical orbit above the surface of the earth may be regarded as a *parabola*, the well-known parabola obtained when a body is projected horizontally. These considerations are not altogether futile, although

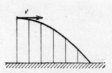

Fig. 5.—Parabolic trajectory resulting from horizontal projection of a body.

practical verification is impossible on account of the resistance of the air. Even at ordinary velocities of say 100 m./sec. the retarding effect of air resistance is very considerable, and the parabola can only be regarded as a very rough approximation to the actual trajectory, the so-called ballistic curve.

CHAPTER VI

Three Useful Concepts: Work, Energy, Momentum

1. Preliminary Remarks.

The gravity pendulum was discussed in § 2 of last chapter (p. 65). Among other things, a child's swing may be regarded as a gravity pendulum. Everyone knows how a child works up to swings of large amplitude: he periodically alters the length of the pendulum, i.e. the distance of his centre of gravity from the point of suspension of the

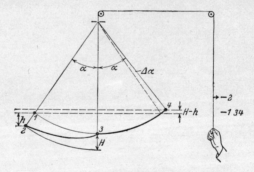

Fig. 1.—Increasing the amplitude of vibration of a swing

swing. When the swing is at its farthest-back point he bends his knees; when moving forwards he stretches his legs straight out again. This may be imitated in a lecture experiment by the apparatus shown in fig. 1, which requires no further explanation.

The quantitative discussion even of this and many simple mechanical examples like it, by means of the second law, involves a fair amount of calculation. To avoid this calculation three useful concepts have been developed: *work, energy,* and *momentum.* These enable us to give a rapid quantitative discussion not only of the above-mentioned simple case, but also of many very complicated examples.

2. Work: Definition and Examples.

The definition of *work* is illustrated by fig. 2; m is a body, F a force, no matter of what origin, acting on it. A dynamometer I may be provided to measure the latter. The distance through which the body moves is x. Then we define the *work done* as the product of the length of the path and the component of the force along the path. In particular, the force in the direction of the path x may be constant; then the work W is given by Fx. This is by no means

Fig. 2.—To illustrate the definition of work

true in general. Suppose the dynamometer I indicates the forces $F_1, F_2, \ldots, F_n$ along the successive elements of path Δx. Then the work is defined as the sum $F_1\Delta x_1 + F_2\Delta x_2 + \ldots + F_n\Delta x_n$ or $\Sigma F\Delta x$; or, in the "limit",

$$W = \int F\,dx. \qquad \ldots \ldots \ldots \quad (1)$$

In this definition of force there may also be an opposing force F_0 (the dotted arrow) applied to the body in addition to the force F. This makes no difference provided $F_0 < F$. The displacement of the body (x), however, is indispensable. If the body does not move, no work is done.

As *unit of work* we may use the product of any unit of force we like and any unit of distance we like. The following are examples:

1 large dyne-metre $=$ 1 watt-second $=$ 1 joule $=$ 1 kg.-mass . m.^{2}/sec.^{2} (physical kg.-mass-m.-sec. system);

1 kg.-force-metre $=$ 9·8 watt-seconds (practical kg.-force-m.-sec. system);

1 foot-pound * $=$ 1·356 watt-second (practical lb.-force-ft.-sec. system);

1 dyne-centimetre or erg $=$ 1 gm.-mass . cm.^{2}/sec.^{2} $=$ 10^{-7} watt-second (physical gm.-mass-cm.-sec. system);

1 kilowatt-hour $=$ 1 Board of Trade Unit (B.T.U.) † $=$ 3·6 . 10^6 watt-seconds $=$ 3·67 . 10^5 kg. - force - metres $=$ 2·66 . 10^6 foot-pounds.

In practical examples on work, two *limiting cases* must be distinguished; these may again be conveniently illustrated by fig. 2.

* [British and American unit of work for *mechanical* purposes.]

† [*Electrical* unit of work. Often simply called *unit*; the name *kelvin* has also been proposed.]

(1) In addition to the force "doing work" there exists a second force F_0 opposing it and nearly equal to it. "Work is done against" this force. The body is practically not accelerated at all. Its velocity at the end of its path x is practically the same as at the beginning and we may without loss of accuracy put $F = F_0$.

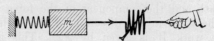

Fig. 3.—Work done against elastic forces

(2) The force F_0 is entirely absent. The body is accelerated along the path x. The force F merely increases the kinetic energy of the body (p. 80).

Of course there are any number of intermediate cases in which F_0 is a fraction of F.

We shall begin with examples of the first limiting case.

(a) *Work done against an Elastic Force.*—In fig. 3 we see on the left a spiral spring. Initially it is slack. Otherwise fig. 3 is the same as fig. 2. The force F recorded by the dynamometer is not constant. If the spring is long enough the force increases linearly with the displacement x and we have the linear law of force $F = \mu x$ already familiar to us. An extension of the spring by x, by fig. 4, requires an amount of work given by

$$W = \Sigma \Delta W = \Sigma F \Delta x = \tfrac{1}{2}\mu x^2, \quad \ldots \ldots \quad (3)$$

or
$$W = \tfrac{1}{2}F_x . x, \quad \ldots \ldots \ldots \quad (3a)$$

where F_x is the force required to extend the spring through a distance x.

Numerical Examples.—Crossbow for sporting purposes:

$x = 0.4$ m.; $F = 20$ kg.-force; $W = \tfrac{1}{2} . 200 . 0.4 = 40$ watt-seconds.

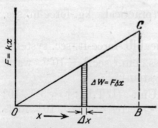

Fig. 4.—To illustrate the calculation of the work done against an elastic force

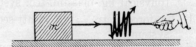

Fig. 5.—Work done against friction

(b) *Work done against Friction.*—In fig. 5 the body m is moved with constant velocity along a horizontal plane. The dynamometer records a constant force F. It is practically equal to the force F_0 which we call the force of friction.* The work done is simply the product Fx. For example,

---

* By the law of action and reaction, the body over which m is dragged is acted on by a force of the same magnitude but in the opposite direction, which is also called the force of friction (cf. fig. 2, p. 25).

a body may be dragged round in a circular path of radius r. Then the work done against friction is $2\pi r F$. (It is by no means equal to *zero* even when the path is a closed curve.)

(c) *Work done in Raising a Weight*, as in fig. 6.
—We suppose that the pulley is frictionless. A dynamometer inserted in the cord would register a constant force F, numerically equal to the force F_0, which we call the weight of the body (w). The work done in raising the weight is

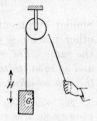

Fig. 6.—Work done in raising a weight

$$W = wh, \quad . \quad . \quad . \quad . \quad (4)$$

that is, the product of the weight w and the perpendicular height h through which it is raised. We shall measure w in kg.-force or lb.-force and h in metres or feet; the work done is then measured in kilogram-metres or foot-pounds.

Numerical Example.—Suppose a man weighing 150 lb. leaps over a rope 5 ft. from the ground. He does not have to do work amounting to 750 ft.-lb.,

Fig. 7.—Expert athletes jumping over a rope

but far less. For he does not jump over the rope with his legs extended straight down. The centre of gravity of the human body when the legs are stretched out straight is about 40 in. from the ground. In vaulting the legs are thrown to one side (fig. 7), so that the centre of gravity of the body flies over the rope about 67 in. above the ground. Thus the work done is only 337·5 ft.-lb. = about 460 watt-seconds.

(d) *Work done in Dragging a Weight up an Inclined Plane* (fig. 8).—The raising of weights is seldom done directly, some form of "machine" being generally used. The simplest machine for raising a weight is the well-known inclined plane. The work required does not have to overcome the whole weight w of the body, but only its

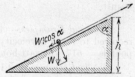

Fig. 8.—Work done in dragging a weight up an inclined plane

component $w \cos \alpha$ *parallel* to the surface of the inclined plane. On the other hand, the distance x through which the body has to be moved is greater than the perpendicular height h through which it is raised: $x = h/\cos \alpha$. Hence the work done in dragging the body right up the slope is

$$W = w \cos \alpha . h/\cos \alpha = wh \text{ kg.-metres.}$$

Similar arguments apply in the case of slopes curved in any way or other machines such as the block and tackle. The result is always the same. *The work done in raising a weight is independent of the path and is equal to the product of the weight and the perpendicular distance through which it is raised.*

The component of the weight at right angles to the surface of the inclined plane, $w \sin \alpha$, is neglected in finding the work done. For if the inclined plane is sufficiently rigid, the amount by which the plane is deformed is negligible, so that the product of force and distance remains practically zero in its case.

So much for our examples of the first limiting case, where work is done without masses being accelerated, the body being acted on by two opposing forces F and F_0 nearly equal in magnitude. In the second limiting case the force F_0 is entirely absent. The one force F does work in accelerating the body and this work amounts to

$$W = \tfrac{1}{2}mv^2, \quad \ldots \ldots \ldots \quad (5)$$

provided the body was previously at rest. This may be proved by quite an elementary calculation in the case of a body *initially at rest and moving with constant acceleration.*

We have

$$W = Fx = mfx = mf . \tfrac{1}{2}ft^2 = \tfrac{1}{2}m(ft)^2 = \tfrac{1}{2}mv^2.$$

In general

$$W = \int F dx = \int mf dx = \int m \frac{dv}{dt} . v\, dt = \int mv\, dv = \tfrac{1}{2}mv^2.$$

The work done in accelerating a body initially at rest is the product of one-half the mass and the square of the velocity attained by the body. The length of the path described under acceleration and the variation of the acceleration with time make no difference whatever to the result.

Numerical Examples.—(1) An express train has a mass of 490 tons. The work required to reach a velocity of 50 ft./sec. (34·1 miles per hour) is $1372 . 10^6$ foot-pounds = 517 kilowatt-hours.

(2) A fast steamer of $3·10^4$ (metric) tons displacement (i.e. of mass $3·10^7$ kg.), in order to attain a velocity of 25 knots = about 13 m./sec., requires to do work amounting to $2·5 . 10^9$ watt-seconds = 700 kilowatt-hours.

(3) A 38 cm. shell weighing 750 kg., muzzle velocity 800 m./sec.: work required = 66 kilowatt-hours.

(4) The pistol bullet of p. 16: $m = 3·26$ gm. $= 3·26 . 10^{-3}$ kg.; $v = 225$ m./sec.: work required = 82 watt-seconds.

Contrary to our previous assumption, the body may already have a velocity v_0. The work done in accelerating it is then

$$W = \tfrac{1}{2}m(v^2 - v_0{}^2). \qquad \qquad (5a)$$

3. Muscular Work.

If the body does not move, no work (in the *physical* sense of the word) *is done.* Here a peculiar physiological fact readily leads beginners into misunderstanding. " Work causes fatigue ", but this statement is not reversible. By no means *every* fatiguing use of our muscles is work. Suppose we hold up a body at a constant height or merely reach up and touch a hook fixed in the ceiling. There is no displacement of the body and hence no work is done, yet the muscles of the arm become tired.

This fatigue is no doubt partly related to the fact that the state of rest of our muscles is only an apparent one. In actual fact some small amount of work in the physical sense *is* done. A stretched muscle is subject to small contractile motions ("tetanus") in rapid succession (about 50 per second). These twitching motions of tense muscle may be heard as a deep " humming " when the jaws are pressed closely together.

We may also consider the case of an electromagnet which is merely acting as a " hook " for holding up a weight. Even this leads to the using-up of its indispensable source of energy (e.g. an accumulator).

The work done by our muscles as prime movers is extraordinarily trifling in amount as measured by engineering standards. Take a very exaggerated case: let a man drag his body of mass 70 kg. up a mountain 7000 metres high in one day. He does work amounting to $70.7000 =$ about 5.10^5 kg.-force-metres or about $1\tfrac{1}{2}$ kilowatt-hours. The commercial value of this " day's work " is a little over a farthing!

4. Power.

The *work done in unit time*, i.e. as a rule the work done per second, is called *power*. The commonest units of power * are

$$1 \text{ watt} = 10^7 \text{ erg/sec.} = 0 \cdot 102 \text{ kg.-metre/sec.} = 0 \cdot 738 \text{ ft.-lb./sec.} \quad (6)$$

and

$$1 \text{ kilowatt} = 10^{10} \text{ erg/sec.} = 102 \text{ kg.-metre/sec.} = 738 \text{ ft.-lb./sec.} \quad (6a)$$

For the space of a few seconds a man can easily do work at the rate of a kilowatt. For example, one can run up a stair 6 metres high in 3 sec. The power required is $70.6/3 = 140$ kg.-metre/sec. $= 1 \cdot 37$ kilowatts. [1·8 H.P.]

* [The British and American unit of power in mechanical engineering is the *horse-power* (H.P.).
1 H.P. = 550 ft.-lb./sec. = 746 watts.
The *Continental* horse-power is 75 kg.-force-metres/sec. = 735 watts.]

In the commonest human motion, *walking* on level ground, the work done is trifling, but it increases rapidly with increasing velocity. The work done in walking chiefly consists of two parts, (1) that due to a periodic raising of the centre of gravity of the body (walk along a wall holding a piece of chalk against one's side, and notice the wavy line produced), (2) that done in moving the legs.

It has been proved experimentally that a man weighing 70 kg. when walking at the ordinary pace of 5 km. per hour (= 1·4 m./sec.) does work at the rate of about 60 watts. In rapid walking at 7 km. per hour work is actually done at the rate of 200 watts.

In cycling, the centre of gravity of the body is raised less and the work done in moving the legs is also less. Only about 30 watts are required for a speed of 9 km. per hour, and 120 watts for a speed of 18 km. per hour. Such numbers enable one better to realize the increased power made available by engineering progress.

5. Capacity for doing Work, or Energy.

In § 2, p. 76, we considered various cases of mechanical work. According to the *conditions of experiment* we distinguished between two idealized limiting cases:

(1) Work done by one force " against " another (e.g. against the weight of a body, the force of a spring, or the force of friction).

(2) Work done in accelerating a body.

According to the *result* of the work done we have now to make another classification, again under idealized limiting cases:

(a) Work which merely results in the body and its path being *heated*. If we rub one hand over the other we distinctly feel the warmth produced. In other cases a more refined method of measuring the temperature is required. From the crude, strictly mechanical point of view the work done against friction in fig. 5 has no result.

(b) Work which gives rise to *energy* (capacity for doing work).

If a mass is raised through a height h above the ground against its weight w we say that it has " potential energy " wh.

If a spring is elastically deformed (extended or compressed) by the amount x under a linear law of force, it has " potential energy " $\frac{1}{2}\mu x^2$. This may always be assumed provided we confine ourselves to sufficiently small deformations (p. 53).

If a mass is accelerated from rest up to the velocity v, we say it has " kinetic energy " $\frac{1}{2}mv^2$.

In all three cases a thick steel block placed on the ground and a steel ball suffice to verify the existence and magnitude of the capacity for doing work.

In fig. 9 work wh has been done against w, the weight of the ball, by a muscular force. When let go, the ball shows that it is able to do work; it is able to accelerate itself (by means of its weight). The work done in accelerating the ball gives the ball a new capacity for doing work in the form of the *kinetic* energy $\frac{1}{2}mv^2$. For this energy is capable of *deforming* the ball and the surface of the steel block. In doing work against elastic forces it can supply the *potential* energy of an elastic deformation.

In order to prove that elastic deformation does take place during impact, we make a steel ball bounce on a glass plate dusted with soot. We find that the circular marks where the ball hits the plate rapidly decrease in diameter with decreasing height of the rebound and gradually merge into the track left by the ball as it merely rolls over the plate.

This potential energy of the elastically deformed plate may in its turn give the ball an upward acceleration. The ball then re-acquires the *kinetic* energy $\frac{1}{2}mv^2$. For this kinetic energy is again able to do the original amount of work *wh*: the ball (to a close approximation) regains the height from which it started, and hence its initial *potential* energy.

The experiment will repeat itself a large number of times without our intervention as the ball continues to bounce. Gradually, however, a failure to reach the initial height becomes apparent. The experiment is not completely "reversible". Here too a part of the original

Fig. 9.—To illustrate the principle of energy: a steel ball is made to bounce on a steel plate.

work done is gradually transformed "irreversibly" into a *heating* of the ball, plate, and air. That is why we spoke above of "idealized limiting cases". In practice we have to take account of "frictional losses" in all laboratory experiments on mechanics; so also here, where energy is continually changing from one form into another, we have to speak of the inevitable "losses of energy owing to friction".

6. The Principle of Energy.

What is the use of the concept of energy? The answer is that in all mechanical processes energy is an invariant. Energy is subject to a *conservation* law. Limited to the mechanical case, this law states that *in any system of moving bodies at any instant the sum of the potential energy and the kinetic energy is constant*, provided that no energy is introduced from outside (no work is done on the system) and no energy is given away to outside (no work is done by the system). Otherwise, the sum of the energies is altered by the amount of energy taken in or given out.

This theorem will save us a lot of calculation and we shall use it frequently in future. Here, therefore, we shall confine ourselves to three quite simple examples.

(1) *Galileo's Check Pendulum* (fig. 10).—The bob of a gravity pendulum is let go at *a*, and the string catches on a nail at *b*. *Observation*: no matter what the position of the nail the bob always rises to the level of *a*. *Explanation*: at the points where the motion is reversed, the velocity of the pendulum is zero. Hence all its energy is in the form of *potential* energy. If the bob overshot or failed to reach the dotted horizontal line, this potential energy would be changed. But the work required to do this is not forthcoming, for the nail is rigid and is held fast.

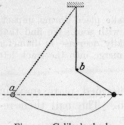

Fig. 10.—Galileo's check pendulum

(2) *Increasing the Amplitude of Swing of a Gravity Pendulum.*—This was mentioned at the beginning of the present chapter and illustrated in fig. 1 (p. 74). The reader should now take note of the horizontal dotted lines. While the path 1 . . . 2 is being described the pendulum supplies the muscle with potential energy *wh*. While the path 2 . . . 3 is being described, however, the muscle gives back the *greater* amount of energy *wH*. In the position 3, as compared with the position 1, the pendulum has acquired energy amounting to $w(H - h)$. At the turning-point 4 this energy is again exclusively represented by potential energy. There the gain of energy $w(H - h)$ kilogram-metres must manifest itself by the weight rising $(H - h)$ metres past the horizontal plane through 1.

Here we have assumed that the amplitudes are small, otherwise the work done against the centrifugal force must be added to $w(H - h)$.

(3) Given any pendulum subject to a linear law of force (§ 2, p. 51), and suppose it passes its equilibrium position (mean position) with velocity v_0: what is its maximum displacement x_0?

In the mean position the whole energy of the pendulum is in the form of kinetic energy $\frac{1}{2}mv_0^2$; at the turning-points the whole energy is potential, and amounts to $\frac{1}{2}\mu x_0^2$ (where μ is the constant of the linear law of force (p. 51)). By the principle of energy the two amounts of energy must be the same, so that

$$\tfrac{1}{2}mv_0^2 = \tfrac{1}{2}\mu x_0^2$$

or

$$v_0 = x_0 \sqrt{\frac{\mu}{m}},$$

or, if we use equation (2), p. 52,

$$v_0 = \omega x_0 \quad \text{or} \quad x_0 = v_0/\omega, \quad \ldots \ldots \quad (7)$$

where $\omega = 2\pi n = 2\pi/\tau$, τ being the period of oscillation; v_0 and x_0

may be measured in terms of any unit of length but it must be the same for both.

This relationship is frequently used (see e.g. p. 92).

7. Definition of Momentum.

In nature many motions take place very suddenly; the forces at work are rapidly changing in magnitude and direction. The entire period during which acceleration takes place may be compressed into a small fraction of a second (for a numerical example, see § 12, p. 96). In the discussion of this type of motion another concept besides energy is of very great service, namely, *momentum. Momentum is defined as the product of mass and velocity*, i.e. *mv*.

Newton originally called the product *mv* "quantity of motion", but this has now been superseded by the term " momentum ".

The arbitrary curve in fig. 11 indicates the variation of any force with time. The whole area enclosed between this curve and the *x*-axis is the " time-integral " of the force,

$$\Sigma F_n \Delta t_n = F_1 \Delta t_1 + F_2 \Delta t_2 + \ldots + F_n \Delta t_n,$$

or, in the " limit ",

$$\int F\,dt.$$

This time-integral of the force is called an *impulsive force*, or *impulse*. It is measured in kg.-force-seconds, dyne-seconds, &c.*

The effect of an impulse acting on a body (heavy particle) may be calculated from the second law of motion. Let the body have the velocity v_1 before the impulse. During every interval of time Δt_n the acceleration is $f_n = F_n/m$. During the interval Δt_n this acceleration gives rise to an increase of velocity

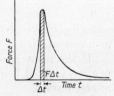

Fig. 11.—Time-integral of the force, or impulse

$$\Delta v_n = f_n \Delta t_n = \frac{1}{m} F_n \Delta t_n;$$

that is　　　　　$m\,\Delta v_n = F_n \Delta t_n.$

If we sum over all the intervals of time Δt_n, $\Sigma \Delta v_n$ is the total change of velocity of the body, $(v_2 - v_1)$, caused by the acceleration. Hence

$$m(v_2 - v_1) = \int F\,dt, \quad \ldots \ldots \quad (8)$$

* Similarly in electricity we have the time-integral of current $\int i\,dt$, measured in ampere-seconds (coulombs), and the time-integral of pressure $\int E\,dt$, measured in volt-seconds (*Physical Principles of Electricity and Magnetism*, pp. 34, 102).

(the units to be chosen according to the table on p. 37). The impulse $\int F\,dt$ has the effect of changing the momentum from its initial value mv_1 to its final value mv_2. In many cases there is no initial velocity v_1. We shall then call the total velocity produced by the impulse $\int D\,dt\ v_0$ instead of v_2 and write

$$\int F\,dt = mv_0. \qquad \ldots \ldots \ldots \quad (8a)$$

Here we make an abrupt digression in order to clear up a matter which has a fundamental bearing on the structure of mechanics.

Our train of thought has led us from Newton's second law of motion to the definition of momentum $(8a)$. Of course the reverse process is equally justifiable (and indeed was at first used by Newton). We put the definition of momentum at the beginning and say: " The change of momentum with time is proportional to the force acting on the body ", or, in mathematical notation,

$$\frac{d}{dt}(mv) = \text{const. } F. \qquad \ldots \ldots \ldots \quad (9)$$

If the mass m is constant we may write

$$m\frac{dv}{dt} = \text{const. } F, \qquad \ldots \ldots \ldots \quad (9a)$$

so that we obtain the second law

$$mf = \text{const. } F$$

as a consequence.

If the mass is *constant* both methods are equally justifiable, the one we adopted being the one better suited to the teaching of *experimental* physics.

As a result of the developments in physics during the last ten years, however, it has been found that the assumption that mass is *constant* is only an *approximation*, albeit one holding within wide limits. The realm of its validity is that of " classical " mechanics. In the next approximation (principle of relativity) we have to replace m by

$$\frac{m_0}{\sqrt{1 - v^2/c^2}},$$

where c is the velocity of light $(3 \cdot 10^{10}$ cm./sec.$)$. With this correction equation (8) holds, but not Newton's second law of motion. In the region of extremely high velocities the limits of validity of this remarkably simple law are overstepped. So much for our digression.

8. The Principle of Momentum and some Simple Examples.

The usefulness of the concept of *energy* depends on the possibility of formulating a "conservation law" for energy (see § 6, p. 81). This law of conservation is not a new fact to be added to the second law, but a *consequence* of that law.

The same applies to the concept of momentum, for which a "conservation law" may also be formulated. This principle of the conservation of momentum also represents no new *fact* but on the contrary is just a *quantitative statement of the law of action and reaction.*

As an example we shall consider a simple "system" consisting of two balls (particles) M and m. Their common centre of gravity is defined by the equation

$$Ma = mb. \qquad \ldots \ldots \quad (10)$$

The balls are joined by a stretched spring. They are initially at rest (fig. 12), so that the momentum of each by itself is zero. Hence Σmv, the resultant momentum of the whole "system", is zero. The spring is then let go by means of some releasing arrangement. Each ball receives an impulse $\int F\,dt$. According to the law of action and reaction, the impulse must be the same for both balls, except for sign. Hence by equation (8a) both balls acquire equal momenta, and we must have

$$Mv_1 = mv_2. \qquad \ldots \ldots \quad (11)$$

These momenta, like the associated velocities, are vectors. The arrows representing Mv_1 and mv_2 are of the same length but in opposite directions. The resultant momentum obtained by combining them is zero. The zero total momentum existing before the impulse has been "conserved".

Considerations such as these can readily be generalized. We thus obtain a general form of the principle of momentum, which is as follows. *The sum of all the momenta for any system of bodies moving in any way remains constant, provided there are no "external" forces acting on the system.*

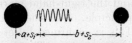

Fig. 12.—To illustrate the conservation of momentum

The principle of momentum may be expressed in yet another way. We again start from the simple example above. After receiving the impulse the two bodies move off with *constant* velocities v_1, v_2 (as usual, we neglect the inevitable losses due to friction). Hence the velocities in equation (11) may be replaced by the distance s described in any interval of time. We obtain

$$Ms_1 = ms_2. \qquad \ldots \ldots \quad (11a)$$

Combining this equation with the equation defining the centre of gravity (10), we obtain

$$\frac{a + s_1}{b + s_2} = \frac{m}{M} \; : \quad \ldots \ldots \quad (12)$$

in words, the distances of the balls from the centre of gravity are once more inversely proportional to the masses of the balls. That is, the common centre of gravity of the two balls also remains unaltered by the impulse; or, more generally, *the centre of gravity of a system cannot be altered except by external forces.*

Up to this point of our discussion the law of action and reaction has remained very much in the background as compared with the second law. Here and there, it is true, in discussing a case of acceleration we have made clear at what point the reaction acts, but this is all. In the principle of momentum the *law of action and reaction* now appears in the quantitative form appropriate to its outstanding importance in all branches of physics. In this form the law of action and reaction will now play a central part in our account of mechanical phenomena.

The principle of momentum, like the principle of energy, requires to have its full meaning brought out by a few simple experiments.

(1) Two similar flat trucks of the same mass ($M = m$) are initially at rest, joined by a strong stretched spring. On either side of the trucks to left and right there is a mark, say a bell-jar, at a distance of 3 metres. The spring is then released. According to the principle of energy both trucks should acquire the same velocity. In actual fact they do describe equal distances in equal times, as is shown by their hitting the marks at the same instant.

(2) The experiment is repeated in a modified form. The mass of the left-hand truck is doubled by loading it with a block of metal, i.e. $M = 2m$. The left-hand mark is set at a distance of 1·5 metres and the right-hand mark at a distance of 3 metres. According to the principle of momentum we should have $2mv_1 = mv_2$ or $v_1 = \frac{1}{2}v_2$. This is confirmed by observation. Both trucks hit the mark at the same instant. Thanks to its doubled velocity the right-hand truck is enabled to cover double the distance of the left-hand one in the same time.

(3) Given a flat truck about 2 metres long, at rest. A man stands on its right-hand end (fig. 13). Truck and man form one system. The man begins to run towards the left. He thereby acquires a momentum directed towards the left. At the same time the truck begins to run towards the right. According to the principle of momentum the truck acquires a momentum of equal magnitude, but in the opposite direction. The man continues to run and steps off the truck at its left-hand end, taking his momentum with him.

Meanwhile the truck continues to move towards the right with
constant velocity, for, apart from sign, it has acquired a momentum
equal to that of the man.

Fig. 13. — To illustrate the conservation of
momentum. A man on a truck accelerates himself
and thereby gives the truck momentum in the
opposite direction.

(4) To test this quantitative statement we make another man run
towards the empty moving truck (fig. 14), the mass and velocity of
the second man being arranged to be the same as those of the first.
The second man steps on to the truck and remains standing on it.
The truck immediately stands still. The momentum brought by the

Fig. 14.—The momentum of the truck in fig. 13 is equal to the
momentum of the man

man and transferred to the truck is equal and opposite to that of the
truck running empty.

(5) The truck is at rest. A man comes running from the right
with constant velocity. He steps on to the truck at its right-hand end
and steps off at the left-hand end (fig. 15). The truck remains at rest.
The man takes his whole supply of momentum with him, and it is
not altered to any appreciable extent while he is on the truck. Hence
the momentum of the truck cannot vary from its initial value of zero
either.

The truck has rubber tyres and is practically incapable of motion except in the direction of its long axis. It thus enables us to demonstrate the vector character of momentum. If the man runs on to the truck at an oblique angle α and stops on the truck, the

Fig. 15.—To illustrate the conservation of momentum. The runner has not noticeably altered his momentum in passing the truck

momentum in the longitudinal direction of the truck is only the component $P \cos \alpha$. If $\alpha = 60°$ the truck reacts with only half the velocity ($\cos \alpha = 0.5$); if $\alpha = 90°$ the velocity of the truck remains zero ($\cos 90° = 0$).

9. Further Applications of the Principle of Momentum. Elastic and Inelastic Impact.

When two bodies collide the forces which arise last only for a very short time. Without the principle of momentum it would be difficult to deal with such phenomena, but the use of this principle makes a discussion of collisions very simple.

When two bodies collide we distinguish the two limiting cases of *elastic* and *inelastic* impact. To define these limiting cases we utilize the idea of *energy*. We make one body a solid wall at rest, and let the second body impinge on it at right angles. The impact is said to be *elastic* if none of the kinetic energy of the impinging body is transformed into heat and *inelastic* if all its kinetic energy is transformed into heat. Or, in other words: if we take one of the bodies as our system of reference, elastic impact changes the sign, but not the magnitude, of the velocity of the impinging body. Or, thirdly: *elastic* impact leaves the relative velocity of the two bodies *unaltered* except for *sign*, whereas *inelastic* impact reduces it to *zero*.

Both limiting cases can to a large extent be realized in practice. For elastic impact we use a steel wall and a steel ball. For inelastic impact we cover the wall with lead or some similar plastic substance.

The bouncing steel ball in fig. 9, p. 81, regains 95 per cent of its initial height after each impact. That is, at each impact it loses only 5 per cent of the energy it had prior to that impact, or 2·5 per cent of its velocity.

In the general case the opposed body is not a fixed wall, but is itself movable, and, owing to the momentum transferred to it, also acquires a velocity v_x. We shall calculate the values of the velocity and of the transferred momentum for a special case, that of two balls of unequal mass colliding directly with one another. This example is of considerable importance in the kinetic theory of gases and in atomic physics. The calculation depends on an application of the principle of momentum. For the sake of simplicity we shall assume that before the collision the larger ball is at rest.

Before the collision the sum of the momenta is given by

$$P_0 = 0 + \underset{\leftarrow}{mv}.$$

After collision we have to consider the two limiting cases singly:

(1) *Elastic Impact.*—After the impact the large ball moves with velocity v_x towards the *left*, and the small ball with velocity $(v - v_x)$ towards the *right*. For the small ball was reflected from a " wall " retreating with velocity v_x towards the left. Hence the sum of the momenta after impact is

$$P_1 = \underset{\leftarrow}{Mv_x} + \underset{\rightarrow}{m(v - v_x)}.$$

By the principle of momentum $P_0 = P_1$, so that

$$mv = Mv_x - m(v - v_x),$$

or

$$v_x = \frac{2mv}{M + m}. \qquad \ldots \ldots \ldots (12a)$$

(2) *Inelastic Impact.*—After impact the two balls move together with velocity v_y towards the left, the small one not being reflected towards the right. After impact the sum of the momenta is

$$P_1 = \underset{\leftarrow}{v_y(M + m)}.$$

By the principle of momentum P_0 must again be equal to P_1, so that

$$mv = v_y(M + m),$$

or

$$v_y = \frac{mv}{M + m}. \qquad \ldots \ldots \ldots (12b)$$

Combining equations (12a) and (12b), we have the simple result $v_x = 2v_y$; or, in words: if the body collided with was previously at rest, it acquires in the elastic case *double the velocity*, and hence *double the momentum*, that it does in the inelastic case.

In order to test this theoretical statement, we have to compare the velocities with which the ball struck moves off after elastic and inelastic collisions. This may be done by means of the apparatus shown in fig. 16. The large ball at rest is the bob of a gravity pendulum, the thread of which is several metres long, so that we may assume that a linear law of force holds and may apply equation (7), p. 82.

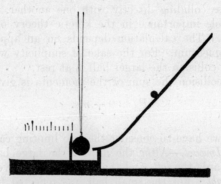

Fig. 16.—A steel ball is made to run against a ballistic pendulum; on the left a light pointer which may readily be slid along (length of thread about 4½ m.). (1/16.)

This means that *the ballistic deflection (throw) is proportional to the velocity with which the ball leaves its position of equilibrium.*

The small ball comes running down a steep groove and impinges directly on the large one centrally. For the inelastic collision we fix a small piece of lead foil to the point where the impact takes place. The bob of the pendulum when deflected makes a light cardboard pointer slide along a rail. The pointer remains at the farthest point reached by the pendulum bob and so makes it easy to ascertain the maximum deflection. The experiment is found to corroborate the statement above, an elastic collision giving a deflection practically double that of an inelastic collision, so that in the former case the struck ball receives double the momentum.

10. Exchange of Momentum in the Elastic Impact of Balls of Equal Mass.

If $M = m$, i.e. if the two balls are of equal mass, it follows from equation (12a) that the velocity of the *struck* ball after impact is v, and that of the *striking* ball after impact *zero*. That is, *the ball*

*previously at rest takes up the velocity of the ball striking it, while the
latter remains at rest after the impact.* This experiment may be repeated
many times in succession with the apparatus of fig. 17.
If there are *three* equal balls, two at rest and one colliding
with them, the middle one plays first the part of struck
ball and then that of striking ball, the alternations taking
place in rapid succession within a very small interval of
time. It too remains at rest and only the third ball flies
up. The phenomena in the case of a large number of balls
may be discussed in a similar way.

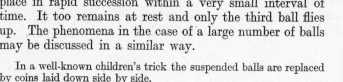

In a well-known children's trick the suspended balls are replaced
by coins laid down side by side.

Fig. 17
Elastic col-
lision of two
steel balls of
equal mass.

11. The Ballistic Pendulum: Measurement of the Velocity of a Bullet.

In the example of last section we throughout used the principle
of the conservation of momentum: when there are no external forces
the sum of the momenta of the system remains constant. In what
follows, however, we shall go back to the equation defining momentum.
For a body previously at rest it is

$$\int F\,dt = mv_0.$$

This equation plays an important part in connexion with physical
measurements. But we anticipate a little. Some of the most important
measuring instruments used in physics consist, mechanically speaking,
of a pendulum obeying a linear law of force. Externally they exhibit
a pointer moving over a scale. After a few swings backwards and
forwards the pointer takes up a steady position and exhibits a " permanent deflection ". The permanent deflection x is proportional to
Q, the quantity to be measured. The instrument has equidistant
graduations, has a " linear scale ". The factor of proportionality
$\mu = Q/x$ is called the " sensitiveness " of the instrument.* That is,
the sensitiveness of a measuring instrument is the value of the quantity
Q which is to be measured (force, pressure, current, &c.), required
to give a deflection of *one* scale-division. The number of scale-
divisions read off, multiplied by the sensitiveness, gives the value of
the quantity to be measured.

The *linear* scale is the distinguishing feature of measuring instruments depending on a pendulum with a *linear* law of force.

In this section we shall confine ourselves to an example from
mechanics and discuss it in a rather more detailed way which will

---

* This physical term is peculiar in that *small* numerical values for it imply *great*
sensitiveness.

make it easier for us to pass on to consider *electrical* instruments in the following section.

In fig. 18 we see an ordinary pendulum transformed into a dynamometer (balance), with a scale divided into decimetres. In fact, this primitive balance has the great advantage of a linear scale. We find that its sensitiveness is say 20 gm.-force per scale division or 2 large dynes per metre.

The measurement of forces, however, is not the sole possible application of this balance. It may be applied in a different way to measure

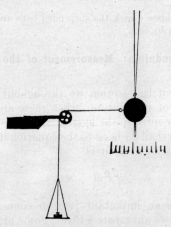

Fig. 18.—The gravity pendulum acting as a balance with linear scale; length of cord about $3\frac{1}{2}$ m.

time integrals of force (impulses), e.g. in dyne-seconds, large dyne-seconds, or lb.-force-seconds, i.e. it may be used as a ballistic pendulum.

Let an impulse $\int F\,dt$ act on the pendulum at rest. It gives momentum mv_0 to the balance; that is,

$$\int F\,dt = mv_0. \qquad \ldots \ldots \quad (8a)$$

$\int dt$, the duration of the impulse, must be small compared with τ, the period of the balance. Then the pendulum is practically still in its equilibrium position when the impulse ceases. We may therefore regard v_0 as the velocity with which the pendulum used as balance leaves the *equilibrium position*. We may apply equation (7), p. 82, and substitute it in equation (8a). We obtain

$$\int F\,dt = m\omega x_0 = Bx_0; \qquad \ldots \ldots \quad (13)$$

or, in words: *the ballistic deflection* (*throw*) *of the measuring instru-*

ment is proportional to the time-integral of the force, or impulse. The factor of proportionality $B = m\omega$ may be called the ballistic sensitiveness.*

Application of the Ballistic Pendulum to the Measurement of the Velocity of a Bullet.—In fig. 19 we see a simple gravity pendulum used as a ballistic pendulum; it consists of a cigar-box filled with lead and sand, of mass M equal to 2 kilograms, suspended by two cords. The pendulum has a period τ equal to 4·19 seconds, so that $\omega = 2\pi/\tau = 1\cdot5$. Hence the ballistic sensitiveness $B = 2 \cdot 1\cdot5 = 3$ large dyne-seconds per metre.

We shall now measure the impulse $\int F\,dt$ produced when this apparatus is used to stop a pistol bullet.

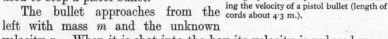
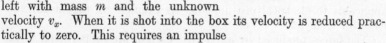

Fig. 19.—Ballistic pendulum for measuring the velocity of a pistol bullet (length of cords about 4·3 m.).

The bullet approaches from the left with mass m and the unknown velocity v_x. When it is shot into the box its velocity is reduced practically to zero. This requires an impulse

$$\int F\,dt = Bx_0 \text{ large dyne-seconds,}$$

where x_0 is the observed throw of the pendulum.

Numerical Example. — $x_0 = 0\cdot25$ metre; $\int F\,dt = 3.0\cdot25 = 0\cdot75$ large dyne-seconds, or 77 gm.-force-seconds.

From the impulse $\int F\,dt$ which we have just measured by the ballistic pendulum we can now calculate the velocity v_x from equation (8a) above.

For the impulse $\int F\,dt$ must be equal to the momentum of the bullet, so that

$$Bx_0 = mv_x$$

and we only require to know m, the mass of the bullet.

Continuation of the Numerical Example.— $m = 3\cdot3$ gm.-mass; 0·75 large dyne-seconds $= 0\cdot0033\,v_x$, so that $v_x = 227$ metres per second (which is in good agreement with the value we found previously (p. 16)).

But just look at the simplification brought about by the use of the idea of momentum! In the former case we needed a chronograph marking off intervals of time, an electric motor, a regulating resistance, and counting apparatus, not to mention some means of finally in-

---

* The ballistic sensitiveness B may be calculated from the ordinary sensitiveness μ with the help of equation (2), p. 52. We find that $B = \mu/\omega = \mu\tau/2\pi$. In this calculation, however, it is assumed that the damping of the vibrations is small.

tercepting the bullet: whereas now that the principle of momentum is available all that we require to carry out the same measurement is a cigar-box filled with sand, a watch, and some string.

Completion of the Numerical Example.—In the box the bullet is brought to a standstill after about 2 cm. in about $0.02/200 = 10^{-4}$ sec.; 77 gm.-force-seconds give 770 kg.-force as the average value of the force required to stop the bullet. This force, exerted by a bullet of cross-section about 0.3 sq. cm., exerts on the wood a pressure of about 2500 kg./cm.^{2} or atmospheres!

Beginners sometimes attempt to apply the principle of energy to the measurement of the velocity of a bullet by the ballistic pendulum. They equate $\frac{1}{2}mv_x^2$, the kinetic energy of the bullet, to $\frac{1}{2}M(\omega x_0)^2$, the kinetic energy of the pendulum. This is quite wrong, as the impact is not elastic (p. 89), all but about 0.16 per cent of the kinetic energy of the bullet being transformed into heat during the collision.

12. The Ballistic Galvanometer: Its Application to Measure the Duration of a Collision.

The whole procedure followed out in last section is immediately applicable to electrical instruments. As example we take a galvanometer or ammeter.

The forces produced by an electric current are proportional to the current (i, measured in amperes),* so that galvanometers can be made with a linear scale, e.g. the well-known moving-coil galvanometer. Here instead of $F = \mu x$ we have

$$i = \mu_i x, \quad \ldots \ldots \ldots \quad (14)$$

where μ_i is called the sensitiveness of the galvanometer (e.g. 10^{-8} ampere per scale-division). The structure of such galvanometers is no concern of ours here, and may be represented, so far as principles

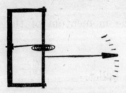

are concerned, by the diagram of fig. 20. By means of some arrangement of coils (not shown in the figure) the current is made to move a mass in the form of a pointer against the force of a coiled spring. From the mechanical point of view we are merely dealing with a pendulum with a linear law of force.

Fig. 20. — Diagram of an electrical instrument with linear scale.

In fig. 19 we used the mechanical balance as a *ballistic pendulum*; its ballistic deflections (x_0) gave us *time-integrals of force* (*impulses*) $\int F\, dt$, say in dyne-seconds. In an entirely analogous way the galvanometer mentioned above may be used to measure *time-integrals of current* $\int i\, dt$, in ampere-seconds (coulombs). Again an assumption is implied, namely, $\int dt$, the time of flow of the current, must be small as compared with τ, the period of the galvanometer. In equation (13) we have to replace the force F by the current, which is propor-

* See *Physical Principles of Electricity and Magnetism*, p. 145.

tional to it, and indicate this substitution by adding a suffix to the factor of proportionality (ballistic sensitiveness) B. We thus have

$$\int i\,dt = B_i \cdot x_0. \quad \ldots \quad \ldots \quad (15)$$

B_i is the number of ampere-seconds corresponding to a throw of one scale-division (cf. footnote, p. 83). If the current is constant the area enclosed by the current-time graph is a rectangle (see *Physical Principles of Electricity and Magnetism*, fig. 22 *b*, p. 34). It is easily proved that the time-integral of the current is proportional to the throw. We have

$$it = Bx_0. \quad \ldots \quad \ldots \quad \ldots \quad (15a)$$

Measurements of this kind are carried out in the book just referred to. Here, on the other hand, we shall regard equation (15) as established by the simple way in which we derived it.

We shall now describe an *application of the ballistic galvanometer*, namely, to the *measurement of the duration of the impact of a steel ball on a steel wall.*

In fig. 21 a thick steel plate acts as a wall; a steel ball hangs by a wire at a distance of a few milli- metres from it. Wall and ball play the part of a " switch " in an elec- tric circuit, which contains a source of current at 100 volts pressure (a high - tension wireless battery) and a mirror galvanometer with a period of about 30 sec. We let the steel ball swing against the wall from a distance of about 30 cm.

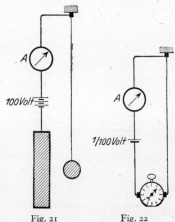

Fig. 21   Fig. 22

Illustrating the measurement of the duration of impact by the electrical method

and rebound from it. We then catch it again. While the wall and the ball are in contact a current flows in the circuit. We are not interested in its value in amperes. The current gives rise to a throw x_0;

$$it_x = B_i x_0. \quad \ldots \quad \ldots \quad \ldots \quad (15b)$$

We then replace the ball and plate by a " stop-watch switch " and replace the source of current by one giving a pressure of only $\frac{1}{100}$ volt (fig. 22). The current only flows while the watch is going. The strength of the current is 10,000 times as small as when the pressure was 100 volts.

When this small current flows for 1·30 sec. it produces the same throw x_0 as above. Hence

$$10^{-4}i \,.\, 1\text{·}30 \text{ sec.} = B_i \,.\, x_0. \quad . \quad . \quad . \quad . \quad (15c)$$

Comparing equations (15b) and (15c), we find that t_x, the duration of the elastic collision of ball and plate, is $1\text{·}30 \,.\, 10^{-4}$ sec. Thus in our example the effects of the elastic forces, the deformations and the accelerations in varying directions, are all comprised within this minute fraction of time! Without the ballistic galvanometer, i.e. in the last resort without the idea of momentum, we should have had considerable difficulty in measuring this interval of time. In fact, it could scarcely be done except by a photograph on a rapidly moving plate.

CHAPTER VII

The Rotation of Rigid Bodies

1. Preliminary Remarks.

Any arbitrary motion of a body is in general a combination of two motions, a *translation* and a *rotation*. Hitherto we have concerned ourselves exclusively with translatory motion. *Formally* we have dealt with bodies as if they were particles with no extension in space. *Experimentally* we have avoided rotatory motions by means of two devices. In the case of motion in a straight line we made the line of action of the accelerating force pass through the *centre of gravity* of the body. In the case of motion in a curved path we made all the linear dimensions of the body small com-pared with the radius of curvature of its path. To be sure, a stone or other body whirled round makes one complete *rotation* about its centre of gravity during each complete revolution. But the kinetic energy of this rotation (§ 4, p. 102) is small compared with the kinetic energy of the translatory motion. Hence we may neglect the rotatory motion in comparison with the translatory motion.

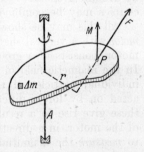

In the present chapter we now consider the other limiting case: the body does not advance as a whole, but its motion consists

Fig. 1.—To illustrate the definition of turning moment

exclusively of rotations. In the first instance the positions of the axes about which these rotations take place shall be determined by fixed supports.

2. Definition of Turning Moment (Torque, Couple).

Fig. 1 shows a rigid body of any form free to turn about an axis A supported by bearings. When the body rotates every point in it, or better, every element of mass Δm, moves in a plane at right angles to the axis, called the *plane of rotation*. The body is to be capable of remaining at rest in any position. For this it is necessary that the effect of its weight must be got rid of. What we have to do is to make the axis of rotation exactly vertical; then the plane of rotation of every point is horizontal.

Not every force gives rise to a rotatory motion; it must have an effective *turning moment* (*torque, couple*) about the given axis. That is, *qualitatively speaking*, the line of action of the force must neither pass through a point of the axis of rotation nor be parallel to it.

Quantitatively, the turning moment M is defined by the equation

$$M = Fr, \qquad \ldots \ldots \ldots \quad (1)$$

where r is the perpendicular (shortest) distance between the line of the force F and the axis of rotation. The turning moment M is a vector, represented by an arrow at right angles to both F and r. The length of the arrow denotes the magnitude of the turning moment, measured e.g. in large dyne-metres, foot-pounds, &c. If we look in the direction of the arrow the direction of rotation is *clockwise*.

According to this definition only those couples are effective with respect to a given axis of rotation which have a component *parallel* to the axis of rotation.

In most cases a body capable of rotation is acted on simultaneously by a number of forces exciting widely differing couples on it. All the couples may be combined into one resultant couple.

All this may be shown very conveniently by means of an *electric motor* with a vertical shaft. The rotor, or movable part of the electric motor, consists of a grooved iron core and wires carrying a current. In the magnetic field of the stator (the fixed case of the motor) the individual particles of the iron core and the individual wires are acted on by forces differing in magnitude and direction. Together these give rise to a resultant turning moment or torque, the torque of the motor, in a direction parallel to the axis of rotation. In order to *measure* the magnitude of this turning moment we use a known turning moment of equal magnitude but tending to produce rotation in the opposite direction, which we produce by means of an arm and dynamometer (spring balance) attached to the shaft of the motor. We observe the lengths of arm and directions of the force for various points of application. We thus demonstrate the vectorial nature of the turning moment and its resolution into components.

These facts do not need to be derived afresh from experience. They may all be deduced from our previous knowledge, say from the principle of energy.

To take a very simple example, imagine the rotary motion of fig. 2 used to raise a weight. Let the path of the weight be so arranged that the angle of rotation α remains small, i.e. the acting force F acts at right angles to the arm r before the beginning and after the end of the rotation. We keep h, the height through which the weight is raised, the same, and vary the length r_2. Then, by the principle of energy, the work done in all cases must be the same, so that

$$\alpha r_1 F_1 = \alpha r_2 F_2 = \alpha r_n F_n$$

or $\qquad\qquad\qquad r_1 F_1 = r_2 F_2.$ . . . . (2)

If the force acts obliquely on the arm of the lever we proceed in the same way, resolving the force into components.

Fig. 2.—To illustrate the principle of moments

In fig. 1 the axis of rotation is vertical. In this limiting case the weight of the body, i.e. that of its individual elements of mass Δm, can give rise to no turning moment *parallel* to the axis and hence to no effective couple. It is otherwise in the second limiting case, where the axis of rotation is horizontal. Here the weight of each individual element of mass Δm gives rise to a turning moment proportional to $r\,\Delta m$ (fig. 3). In general the body rotates no matter what its initial position. Only in a special case does it remain at rest in any position. In this case the axis passes through the centre of gravity of the body. That is, for an axis passing through the centre

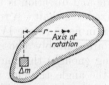

Fig. 3.—Couple due to the weight of a body

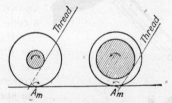

Fig. 4.—Couples acting on reels of thread

of gravity the resultant couple and hence the sum $\Sigma r\,\Delta m$ must vanish. This equation contains a definition of the centre of gravity, which we shall make use of later. Apart from this we shall, as before, regard the centre of gravity of a body and methods for determining it as known, seeing that, along with levers, balances, and simple machines, it is treated in great detail in school courses.

In the case of an axis determined by *fixed bearings* there can hardly ever be any doubt about the direction, magnitude, and sense of a couple. In other cases the beginner occasionally runs up against difficulties. For example, there is the case of " obedient " and " disobedient " reels of thread. A reel is dropped on the floor and rolls under the sofa, and attempts are made to recover it by pulling the thread. Some reels follow obediently, while others roll farther into their hiding-places. The explanation is given in fig. 4. It is not the axis of symmetry of the reel that is to be regarded as the axis of rotation, but the line along which it touches the floor at any moment, which is denoted by A_m in fig. 4 ("instantaneous axis"). If the thread is held "low" enough even the most obstreperous reel may be compelled to obey. Here, as happens so frequently in life, a little knowledge of physics is more effective than the most violent outbursts of temper.

3. Production of Known Couples. Torsional Rigidity. Angular Velocity as a Vector.

Forces of known magnitude and direction may be produced in a very simple way by means of spiral springs. If the dimensions are suitably chosen (if the spring is long enough) the forces are proportional to the changes in length of the spring, and we have the linear law of force

$$F = \mu x$$

(equation (1), p. 98), where μ is the force required to extend the spring by unit length (centimetre or metre).

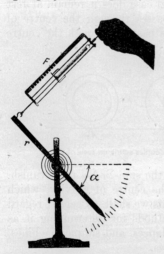

Quite analogously, a particularly straightforward method of producing turning moments or couples (M) of known magnitude and direction is to use a coiled spring fixed to a shaft. A rotator like this is shown in fig. 5 (Plate IV). If the dimensions are suitably chosen (if the spring is long enough) the couple is proportional to the angle through which the spring is twisted, and we again have a linear law of force

$$M = \mu' a,$$

where the constant μ' is the couple required to twist the spring through unit angle and is called the *torsional rigidity*. Here the angles are measured in radians, i.e. we write 2π for 360°, $\pi/2$ for 90°, &c., so that unit angle in radian measure is $360/2\pi = 57.3°$.

That is, the torsional rigidity of a spring is the couple required to twist it through an angle of 57.3°. In future we shall often need both spiral springs with a

Fig. 6.—Calibration of the rotator of fig. 5 in a horizontal position: e.g. $r = 0.1$ m., $a = 180°$ ($= \pi = 3.14$ radians), $F = 0.175$ kg.-force $= 1.71$ large dynes; $Fr = 0.0175$ kg.-m. $= 0.171$ large dyne-m.; $\mu' = 0.0175/3.14 = 0.0056$ kg.-m./radian $= 0.055$ large dyne-m./radian. ($1/100$.)

known value of μ and coiled springs with shaft with a known value of μ'. We accordingly calibrate the rotator shown in fig. 5 by a method which will easily be understood from fig. 6 and the annexed numerical example.

To produce known *forces* we may also use cylindrical *rods* or *wires* instead of *spiral springs*. Similarly, in order to produce known *couples*, the coiled spring attached to the shaft may be replaced by a rod- or wire-like shaft which can be twisted. In both cases, however, the use of springs makes for greater clearness, as with given forces or couples of the same size they give deflections which are larger and more easily seen at a distance.

Beginners are prone to underestimate the ease with which even thick steel rods may be twisted. Fig. 7 shows a steel rod 1 cm. across and only 10 cm. long, fixed in a vice. This body appears rigid enough and yet one can twist it with one's finger-tips to an extent which may be made visible. To verify this we have merely to use a ray of light about 10 metres long reflected from the mirrors a and b.

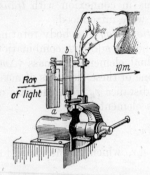

Fig. 7. — A short thick steel rod twisted between the thumb and finger.

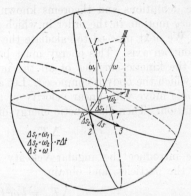

Fig. 8.—Angular velocity as a vector

Forces alter the linear velocity of a body, couples the angular velocity. The linear velocity (v) is not completely defined unless both its magnitude and its direction are known; it is a vector. The same is true of the angular velocity (ω or Ω). Its magnitude is the angle described in unit time (measured in radians), its direction that of the axis of rotation. The arrow representing the angular velocity vector must be drawn in the direction of the axis of rotation. This is illustrated in fig. 8. A point P simultaneously revolves about the axis I with angular velocity ω_1 and about the axis II with angular velocity ω_2. During a sufficiently small interval of time Δt, the point describes the practically rectilinear path $\Delta s = P \ldots 3$, which we may construct as the resultant of the two component paths $\Delta s_1 = \omega_1 r \Delta t$ and $\Delta s_2 = \omega_2 r \Delta t$.

We may, however, obtain the path $P \ldots 3$ in another way. Along the axes I and II we draw arrows representing the magnitudes of the angular velocities ω_1, ω_2 respectively. We combine these vectors graphically to form the resultant angular velocity ω. This determines a new axis *III*, about which we make the body rotate with angular velocity ω; it then traverses the path $\Delta s = \omega r \Delta t$ in time Δt. Here the vector addition of two angular velocities follows immediately from the obvious similarity of the triangles used in the constructions.

4. Moments of Inertia: Torsional Vibrations.

With the help of the idea of the turning moment M we can readily complete the passage from translatory motion to rotatory motion. We make use of the table opposite. The two upper horizontal lines contain the same kinematical concepts, velocity and acceleration. Following these up we have inserted into the left-hand vertical column the definitions and theorems known to us in connexion with *translatory* motion, in the order in which they were introduced.

We next proceed to calculate the *kinetic energy* of a body rotating about an axis. This energy must be found by additive combination of the kinetic energies of all the individual elements of mass (Δm) of which the body is composed. Let any one of these elements of mass (Δm_n) move with linear velocity v_n at a distance r_n from the axis of rotation. Then the kinetic energy of this element is

$$\Delta T_n = \tfrac{1}{2} v_n{}^2 \Delta m_n.$$

We introduce the angular velocity $\omega = v/r$, which is the same for all the particles, and obtain

$$\Delta T_n = \tfrac{1}{2} \omega^2 (r_n{}^2 \Delta m_n).$$

Summing for all the individual particles, we obtain the kinetic energy of the whole rotating body, which is

$$T = \tfrac{1}{2} \Sigma (r_n{}^2 \Delta m_n) \omega^2. \quad \ldots \ldots \quad (3)$$

We write

$$I = \Sigma (r_n{}^2 \Delta m_n) \quad \ldots \ldots \ldots \quad (4)$$

and call this quantity the *moment of inertia*. Using this abbreviated notation, the kinetic energy of a body rotating with angular velocity ω is

$$T = \tfrac{1}{2} I \omega^2. \quad \ldots \ldots \ldots \ldots \quad (5)$$

We thus reach the seventh line of the table on p. 103. The corresponding equation on the left for translatory motion is

$$T = \tfrac{1}{2} m v^2,$$

or, in words: in the case of rotation the angular velocity ω and the moment of inertia I take the place of the linear velocity v and the mass m. We insert I in the third line of our table on the right.

Motions of Translation.	Motions of Rotation.
1. Velocity: $$v = \frac{dx}{dt}. \quad \text{(p. 16)}$$	Angular velocity: $$\omega = \frac{d\alpha}{dt}. \quad \text{(p. 22)}$$

$$r$$
$$d\alpha = \omega\, dt \qquad dx = v\, dt \qquad v = \omega r$$

Motions of Translation.	Motions of Rotation.
2. Acceleration: $$f = \frac{dv}{dt}. \quad \text{(p. 18)}$$	Angular acceleration: $$\dot{\omega} = \frac{d\omega}{dt}.$$
3. Mass: $$m. \quad \text{(p. 26)}$$	Moment of inertia: $$I = \Sigma r^2 \Delta m. \quad \text{(p. 102)}$$
4. Acceleration f $$= \frac{\text{Force } F}{\text{Mass } m} \quad \text{(p. 34)}$$	Angular Acceleration $\dot{\omega}$ $$= \frac{\text{Couple } M}{\text{Moment of Inertia } I}$$

$$M = Fr$$

Motions of Translation.	Motions of Rotation.
5. $$\frac{\text{Force } F}{\text{Length } x} = \mu. \quad \text{(p. 51)}$$	$$\frac{\text{Couple } M}{\text{Angle } \alpha} = \mu'. \quad \text{(p. 100)}$$

Linear law of force.

Motions of Translation.	Motions of Rotation.
6. Period of vibration: $$\tau = 2\pi\sqrt{\frac{m}{\mu}}. \quad \text{(p. 51)}$$	Period of vibration: $$\tau = 2\pi\sqrt{\frac{I}{\mu'}}. \quad \text{(p. 105)}$$
7. Kinetic energy: $$T = \tfrac{1}{2}mv^2. \quad \text{(p. 80)}$$	Kinetic energy: $$T = \tfrac{1}{2}I\omega^2. \quad \text{(p. 102)}$$
8. Momentum: $$P = mv. \quad \text{(p. 83)}$$	Angular momentum: $$G = I\omega. \quad \text{(p. 110)}$$

P and G are not of the same dimension.

Motions of Translation.	Motions of Rotation.
9. Force: $$F = \frac{dP}{dt}. \quad \text{(p. 84)}$$	Couple: $$M = \frac{dG}{dt}.$$

The next thing is to make ourselves at home with the idea of the moment of inertia by considering a few examples.

In the case of bodies of simple geometrical construction there is no difficulty in calculating the moment of inertia. The summation required can usually be carried out in a few lines. Examples:

(1) *Uniform circular disc about an axis passing through the centre at right angles to its plane.* If M is the mass of the disc, h its thickness, R its radius, ρ its density or mass per unit volume ($= M/\pi R^2 h$),

$$I = \tfrac{1}{2}MR^2 = \tfrac{1}{2}\pi h\rho\, R^4. \quad \ldots \ldots \quad (6)$$

Details of Calculation: for the elements of mass we take concentric rings of radius r and width dr. Such a ring has mass

$$\Delta m = 2\pi rh\rho\, dr, \quad \ldots \ldots \ldots \quad (7)$$

and its moment of inertia is

$$2\pi rh\rho\, dr \cdot r^2.$$

Summing for all the rings from radius zero to radius R, we get

$$I = 2\pi h\rho \int_0^R r^3\, dr = \tfrac{1}{2}\pi h\rho\, R^4.$$

(2) *Uniform sphere about an axis passing through the centre.* If the mass of the sphere is M $(=\tfrac{4}{3}\pi\rho R^3)$,

$$I = \frac{2}{5}MR^2 = \frac{8}{15}\,\pi\rho\, R^5. \quad \ldots \ldots \quad (8)$$

(3) *Uniform elongated bar of any section about an axis passing through the centre of gravity at right angles to the longitudinal axis.* Here

$$I = \tfrac{1}{12}Ml^2. \ldots \ldots \ldots \quad (9)$$

(4) *Steiner's Theorem.*—Given I_g, the moment of inertia of any body of mass M about an axis passing through its centre of gravity: what is the moment of inertia I_0 about an axis parallel to the first at a distance a from it? Answer:

$$I_0 = I_g + Ma^2. \quad \ldots \ldots \quad (10)$$

Details of Calculation (see fig. 9):

$$I_g = \Sigma r_1^2\,\Delta m;$$
$$I_0 = \Sigma r_2^2\,\Delta m;$$
$$r_2^2 = r_1^2 + a^2 - 2r_1 a \cos \alpha;$$
$$\therefore \quad I_0 = I_g + Ma^2 - 2a\Sigma r_1 \cos \alpha\,(\Delta m).$$

But $\Sigma r_1 \cos \alpha\,(\Delta m) = \Sigma r\Delta m = 0$, by the equation on p. 99 which defines the centre of gravity.

Much more important than the calculation of moments of inertia, however, is their measurement, for if the body is complicated in form the summation raises unprofitable difficulties.

Moments of inertia are usually measured by means of *torsional vibrations*. We have merely to replace the mass m by the moment of inertia I and the constant μ for a spiral spring by the constant μ' for a coiled spring, in the sixth line of our table. We know the value of the torsional rigidity μ' for the rotator shown in fig. 5 (Plate IV). To the upper end of the shaft we fix the body under observation, in such a way that its axis is in the same line as the shaft. We then

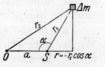

Fig. 9.—To illustrate Steiner's theorem

twist the body through an angle of 90° from its equilibrium position and observe the period of vibration by means of a stop-watch. Then

$$I = \mu' \, \frac{\tau^2}{4\pi^2}. \quad \ldots \quad \ldots \quad (11)$$

The constant μ' was found (p. 100) to be $5 \cdot 5 \cdot 10^{-2}$ large dyne-metre per radian (57·3°). Hence

$$I = 1 \cdot 4 \cdot 10^{-3} \, \tau^2 \text{ kg.-mass-m.}^2 = 1 \cdot 4 \cdot 10^{-4} \, \tau^2 \text{ kg.-force-sec.}^2\text{-m.}$$

Examples:

(1) *Verification of a Moment of Inertia obtained by Calculation.*—For a circular wooden disc of 0·8 kg.-mass and radius 0·2 m. equation (6) gives a moment of inertia of $1 \cdot 6 \cdot 10^{-2}$ kg.-m.^{2} about an axis through the centre at right angles to the plane of the disc. Experiment gives $\tau = 3 \cdot 37$ sec., so that $I = 1 \cdot 58 \cdot 10^{-2}$ kg.-m.^{2}, which is in good agreement with the calculated value.

Fig. 10.—Disc and sphere with the same moment of inertia

Fig. 11.—Hollow and solid cylinders of equal mass (metal and wood) but of unequal moment of inertia

(2) *Verification of Steiner's Theorem.*—We displace the axis away from the centre parallel to itself at a distance of $a = 10$ cm. By Steiner's theorem (equation (10)) this should increase the moment of inertia by the amount $Ma^2 = 8 \cdot 10^{-3}$ kg.-mass-m.^{2}. Experiment gives $\tau = 4 \cdot 15$ sec., $I_0 = 2 \cdot 41 \cdot 10^{-2}$ kg.-m.^{2}, so that $I_0 - I_g = 8 \cdot 1 \cdot 10^{-3}$ kg.-m.^{2}.

(3) *Disc and Sphere with the same Moment of Inertia.*—Fig. 10 shows on the same scale a disc and a sphere of the same material, their masses being in the ratio of 1 : 3·2. By equations (6) and (8) these should have the same moment of inertia. In actual fact they are found to have the same period of vibration when placed on the rotator of fig. 5 (Plate IV).

(4) *Moments of Inertia of Hollow and Solid Cylinders of Equal Mass.*
—Fig. 11 shows a hollow metal cylinder and a solid wooden cylinder
of the same mass, diameter, and length. Using the rotator we find
that the hollow cylinder has a considerably greater moment of inertia.

This is the explanation of a fact which often proves surprising.
We lay both cylinders side by side on an inclined plane, say a sloping
board, the axes of both cylinders being in one straight line. We then
let the cylinders go at the same moment. The solid wooden cylinder

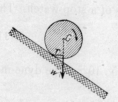

Fig. 12.—Couple $M = Wr$ for a
cylinder on an inclined plane

Fig. 13.—Large rotator for measuring
the moments of inertia of a person in
various positions. F is a strong coiled
spring, for which the torsional rigidity is
about 2·5 large dyne-metres/radian. The
moment of inertia of the man lying down
is about 17 kg.-mass-m.^{2} or 1·7 kg.-force-
sec.^{2}-m.

reaches the bottom long before the other. *Explanation*: in rolling
down the slope both cylinders are accelerated by the same couple wr
(fig. 12), for the masses and radii are the same for both cylinders.
Hence the hollow cylinder with its larger moment of inertia acquires
a smaller angular acceleration $\dot{\omega}$ and angular velocity ω (fourth line
of the table).

(5) *Moments of Inertia of the Human Body.*—We determine the
moment of inertia of the human body for several different positions
of the body and several different axes of rotation by means of the
large rotator shown in fig. 13. Some of the results are summarized
in fig. 14. We shall find these numerical results useful later on.

5. The Importance of the Moment of Inertia in Connexion with the Gravity Pendulum.

Both in ordinary life and in technology, torsional oscillations of
bodies about a vertical position of equilibrium are very frequent.
In the vast majority of cases they arise from the weight of the body.
Any body not suspended or supported exactly at its centre of gravity
may oscillate when once struck. We then have the general case of the
" compound " gravity pendulum of any form. We use this name
in contradistinction to the " simple " pendulum we discussed earlier,
consisting of a mass concentrated at a point and suspended by a
" light " thread.

The *compound pendulum* is important in many physical problems, especially in connexion with physical measurements. We shall accordingly discuss some of its important properties. Although we

$I = 1\cdot2$ 8 $2\cdot3$ kg.-mass-m.^{2}
$l = 0\cdot12$ $0\cdot8$ $0\cdot23$ kg.-force-sec.^{2}-m.

Fig. 14.—Moments of inertia of a man in three different positions. The arrows indicate the direction of the axis of rotation

shall find these useful, a knowledge of them is not absolutely necessary for an understanding of the later chapters.

(1) *Period of Oscillation of a Compound Pendulum: the Equivalent Simple Pendulum.*—Fig. 15 shows a board of any shape suspended as a pendulum; O indicates its axis of rotation, G its centre of gravity, and s the distance between them.

The period of this compound pendulum is given by the formula which holds for any torsional vibration,

$$\tau = 2\pi \sqrt{\frac{I_0}{\mu'}}, \qquad \ldots \ldots \ldots \quad (11)$$

where I_0 is the moment of inertia for the axis passing through O and μ' is again the torsional rigidity, i.e. the couple required to produce a twist through unit angle, so that $\mu' = M/a$. The couple, by fig. 15, is given by

$$M = mgs \sin a. \qquad \ldots \ldots \ldots \ldots \quad (15a)$$

For small values of the angle a we may again put $\sin a = a$, so that

$$\mu' = M/a = mgs.$$

For the "simple" pendulum, i.e. a heavy particle suspended by a light string, we found on p. 66 that

$$\tau = 2\pi \sqrt{\frac{\bar{l}}{g}}.$$

Thus in the compound pendulum the length of the simple pendulum is replaced by the quotient I_0/ms. We call this quantity the *length*

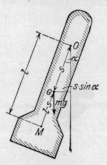

of the equivalent simple pendulum. That is, the length of the equivalent simple pendulum is the length of a simple pendulum with the same period as the compound pendulum. The length of the equivalent simple pendulum (l) is indicated in fig. 15. Its lower extremity M is called the *centre of oscillation*. If the whole mass of the pendulum were concentrated at this point the period of oscillation would remain the same.

Using Steiner's theorem (equation (10), p. 104), we obtain the following expression for the length of the equivalent simple pendulum:

Fig. 15.—The compound pendulum

$$l = \frac{I_g + ms^2}{ms} = \frac{I_g}{ms} + s, \quad \ldots \ldots \quad (12)$$

where I_g is the moment of inertia about an axis passing through the centre of gravity. The length $(l - s)$ $(= I_g/ms)$ is indicated in fig. 15.

In order to *calculate* the length of the equivalent simple pendulum for a pendulum of arbitrary construction it is necessary that *three* quantities, a moment of inertia, a mass, and the distance between the centre of gravity and the axis of rotation, should be found either by calculation or by experiment. This roundabout process may be replaced by a single measurement with the help of the *reversible pendulum*.

(2) *The Reversible Pendulum.*—The period of oscillation of a pendulum remains unaltered when the pendulum is made to swing about an axis passing through the *centre of oscillation*.

Proof.—If the axis of suspension passes through the centre of oscillation (M in fig. 15), we have to substitute $(l - s)$ for s, the distance between axis of suspension and centre of gravity, in equation (12). We then find that the length of the equivalent simple pendulum for this new axis is

$$l' = \frac{I_g}{mI_g/ms} + \frac{I_g}{ms} = s + \frac{I_g}{ms} = l;$$

$$l' = l,$$

so that $\qquad \tau' = \tau.$

In order to determine the length of the equivalent simple pendulum *experimentally* we use a reversible pendulum which in addition to the original fixed axis of suspension has a variable axis of suspension which can be displaced along the pendulum. The period τ for the first axis is found, and the pendulum is then turned upside down and swung from the second axis of suspension and the period τ' ascertained. By trial and error we find a position for the second axis for which $\tau' = \tau$. The second axis then passes through the centre of oscillation corresponding to the first axis, and the distance between the two axes is the required length of the simple equivalent pendulum.

The reversible pendulum is the most important instrument for determining g, the gravitational acceleration of the earth, and its local variations. The values of l and τ for the reversible pendulum are found by experiment and g is then calculated from the formula $\tau = 2\pi \sqrt{l/g}$.

(3) *The Compensated Pendulum used in Accurate Clocks.*—A gravity pendulum, either suspended by leaf-springs or supported on knife-edges, is used in all astronomical clocks. The fundamental problem of their construction is the maintenance of the length of the equivalent simple pendulum,

$$l = \frac{I_g}{ms} + s, \quad \ldots \ldots \ldots \quad (12)$$

at a constant value. Changes in s (the distance between the centre of gravity and the point of suspension) as a result of temperature variations are easy to compensate. The most serious changes in s, however, arise from the perpetual strain on the springs or knife-edges, for the 43,000 oscillations per day inevitably lead to wear and tear. The changes in s produced in this way must be compensated by an opposing change in I_g. This is done by means of an auxiliary mass H fixed above the point of suspension. We thus have the *compensated pendulum* shown in fig. 16. In ordinary clock pendulums s is approximately equal to l, i.e. the centre of gravity of the pendulum is near the lower end of the rod of the pendulum. In the compensated pendulum, on the other hand, the centre of gravity G is brought nearly to the middle of the pendulum rod. The length of its equivalent simple pendulum is thereby made independent of small changes in s.

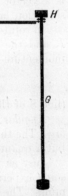

Fig. 16.—Compensated pendulum (1/12.)

This we see from fig. 17, in which l, the length of the equivalent simple pendulum, is plotted against s, the distance between the point of suspension and the centre of gravity. Here I_g/m has been arbitrarily made equal to unity. For $s = \frac{1}{2}l$ the curve has a flat minimum.

$$dl/ds \equiv -k^2/s^2 + 1 \text{ equated to zero gives } l = 2s = 2k. \quad \ldots \quad (13)$$

In the neighbourhood of this minimum l alters only slightly with s

A simple form of compensated pendulum is shown in fig. 16.   A practical example, however, will be useful also.

The so-called *seconds pendulum* makes a half oscillation in one second and is about a metre long. Let its point of suspension be dis-

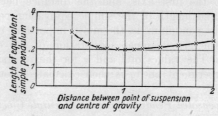

Fig. 17.—To illustrate the principle of the compensated pendulum

placed through 0·1 mm. With an ordinary pendulum the clock would then be 4·3 sec. out in a day, but with a compensated pendulum only 0·0002 sec.

6. Angular Momentum (Moment of Momentum).

In translatory motion momentum is defined as mv. Momentum is a vector and the momentum of a "system" is subject to a conservation law.

In rotatory motion the mass m is replaced by the moment of inertia I, and the linear velocity v by the angular velocity ω. Hence the momentum of a rotatory motion, or angular momentum, is

$$G = I\omega \quad . \quad . \quad . \quad . \quad . \quad . \quad (14)$$

(line 8 of the table on p. 103).

Angular momentum is also a vector and also satisfies a conservation law. The theorem of the conservation of momentum, in fact, is nothing but a quantitative statement of the law of action and reaction.

Just as in the case of translatory motion, we shall bring forward some experimental examples to impress these facts on the reader's mind. The truck which we used in the case of translatory motion is replaced by a revolving stool, capable of rotating with very little friction (owing to the use of ball-bearings) about an axis which is placed accurately vertical. The stool thus reacts only to momenta in a vertical direction, or only to the vertical components of momenta in an oblique direction.

We have still to make up our minds about the direction of rotation of the momentum. In the figures the direction from the feathers to the head of the arrow is to denote rotation in the clockwise direction. The directions of rotation given in the text apply to an observer *looking down from above.*

(1) A man sits on the stool at rest. In his left hand, at about eye-level, he holds a large gyrostat or gyroscope (p. 116) with a vertical axis, also at rest (the rim and spokes of a bicycle wheel weighted with lead). The angular momentum is initially zero. The man grips the spokes from below with his right hand and sets the gyrostat rotating. The gyrostat receives angular momentum $I_1\omega_1$ in a counterclockwise direction. By the theorem of conservation of momentum the man must acquire angular momentum $I_2\omega_2$ of equal magnitude but in the opposite sense. The man, in fact, begins to rotate in a clockwise direction. His angular velocity ω_2 is considerably smaller than that of the gyrostat, for his moment of inertia is much greater.

(2) The man presses the rim of the moving wheel against his chest, thus braking the wheel. The wheel and the man simultaneously cease to rotate. Both angular momenta are again zero.

(3) The man is sitting on the stool at rest and holds the gyrostat, also at rest, with its axis horizontal. He sets the gyrostat in rotation. Here the arrow representing the angular momentum of the gyrostat is horizontal. Both stool and man remain at rest, for they do not react to a horizontal momentum.

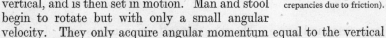

Fig. 18.—To illustrate the conservation of angular momentum (if the acceleration is small there are discrepancies due to friction).

(4) The gyrostat, initially at rest, is held so that its axis makes an angle of 60° with the vertical, and is then set in motion. Man and stool begin to rotate but with only a small angular velocity. They only acquire angular momentum equal to the vertical component of the angular momentum of the gyrostat.

(5) The gyrostat, rotating clockwise, is handed to the man at rest. The man remains at rest although we have given him the gyrostat with its angular momentum. The man now turns the axis of the gyrostat through 180°, bringing its lower end uppermost. He thereby alters its angular momentum from $+G$ to $-G$, in all, therefore, by $2G$. The man himself rotates clockwise with angular momentum $2G$. The man then turns the gyrostat back into its original position and hands it back to his assistant. Stool and man are again at rest. Thus one can play with borrowed momentum for a time and then give it back again.

(6) The man sits on the stool at rest. In his hand he holds a mallet (fig. 19). By swinging the mallet in a horizontal direction the man is to make one complete revolution about his vertical axis. During the swinging the man does rotate, though with smaller angular velocity than his arm and the mallet. Mallet and arm can only be swung through about 180°. When the mallet ceases to rotate so does the man, for

the man and the mallet cannot separately possess angular momentum. To get a second swing the man must bring the mallet back to its starting-point. He can do this by the same path as formerly. But then he loses all the turning he has gained. Thus in order to repeat the swinging process he must bring the mallet back *another* way. He must raise it out of its final position in a *vertical* plane, and then lower into it the initial position, still in a *vertical* plane. His body, which is only free to rotate about a vertical axis, does not react to the angular momentum of this rotatory motion. The process can be repeated afresh, the man is now turned through twice as great an angle

Fig. 19.—Angular momenta in different directions may be produced by means of a long wooden mallet.

Fig. 20.—An electric fan capable of rotation about the vertical, used to demonstrate the conservation of angular momentum.

as at first, and so on. The three separate movements can, of course, be combined into a single one. The arm and mallet are made to describe the surface of a cone whose axis is as nearly as possible vertical.

(7) Fig. 20 shows an ordinary electric fan supported on a vertical axis. The shaft of the driving motor can also be set at various angles α to this vertical axis. Let the angle α be at first 90°. The fan is set agoing by switching on the current. It rotates clockwise, looking through the blades towards the motor. The vertical support of the motor remains at rest. We then decrease α to about 80°. The vertical support begins to rotate counterclockwise with small angular velocity. If the angle α is diminished further, e.g. to 30°, this angular velocity increases.

Explanation: The blades of the fan not only drive the sucked-in air forwards but give it angular momentum. The fan acquires equal angular momentum in the opposite direction. Only the *vertical component* of this angular momentum is observed.

(8) We set the fan going at an angle α of 30° and after some little

time switch off the current. The fan gradually comes to rest. Meanwhile the support loses its angular momentum in a counterclockwise direction. First it comes to rest and then it begins to revolve *clockwise*.

Explanation: After the current is shut off the rotating parts of the motor and the blades of the fan are gradually retarded by friction at the bearings and come to rest relative to the casing of the motor. All their angular momentum is transferred during the braking process to the casing and support. What we observe is the vertical component of this angular momentum.

Finally, we make the fan run pointing slightly downwards ($\alpha = 100°$). Looked at from above, the support rotates clockwise.

(9) We replace the stool by the large rotator of fig. 13, p. 106. A man lies on it at full length, holding on to a handle on either side

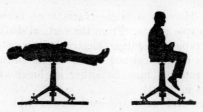

Fig. 21.—To illustrate swinging exercises

(fig. 21). The man is given a push and executes torsional vibrations of small amplitude. *Problem*: the man is to work up to full revolutions of 360° without outside assistance. *Solution*: the man must periodically alter his moment of inertia about the vertical axis. In passing the zero position he bends his legs and raises the upper part of his body. He thereby decreases his moment of inertia and increases his angular momentum. In the extreme position he stretches his legs and reverts to the position with greater moment of inertia. In passing the zero position he repeats the motion. In a short time he will perform revolutions through 360°. This experiment is an excellent illustration of the use of the horizontal bar, except that the axis of the horizontal bar is replaced by the vertical rotating shaft and the couple on the bars due to the weight is replaced by the couple due to the coiled spring of the rotator. We thereby obtain the advantage of a slower motion which is more easily followed by observation. The experiment just described, translated into the language of the gymnast, was the " clear circle ".

The gymnast on the horizontal bar is able to diminish his moment of inertia at the right instant in many ways. In the clear circle alone three ways are common: bending the arms, bending the legs, or spreading the legs widely apart.

7. Free Axes of Rotation.

In all the rotatory motions we have considered hitherto the axis of rotation of the body was determined by an actual axis, a cylinder or knife-edge supported on bearings. We now abandon this limitation.

We thus come to the rotation of bodies about *free* axes. To illustrate the meaning of this term we proceed to consider some experimental examples.

(*a*) A well-known jugglers' trick is shown in fig. 22. A flat dish is made to rotate on the tip of a bamboo cane. Its free axis of rotation is the axis of symmetry.

(*b*) As every child knows, a flat plate, skilfully set in motion, is capable of rotating about a diameter as free axis (fig. 23).

Fig. 23.—Plate rotating freely about a diameter.

Fig. 22.—The axis of symmetry of a dish acting as a free axis.

(*c*) Here is a slight variant of both these experiments. From the vertical shaft of a rapidly rotating electric motor we hang a cylindrical rod by one of its ends. Its free axis of rotation may be either its longitudinal axis, or its transverse axis as in fig. 24.

(*d*) In practice free axes are used as "flexible" axes. Fig. 25 shows an emery-wheel at the end of a wire about 20 cm. long and only a few millimetres across. The wheel is made to rotate about 50 times per second by an electric motor. The wheel rotates stably about its axis of maximum moment of inertia and lies loosely against the piece of work pressed against it.

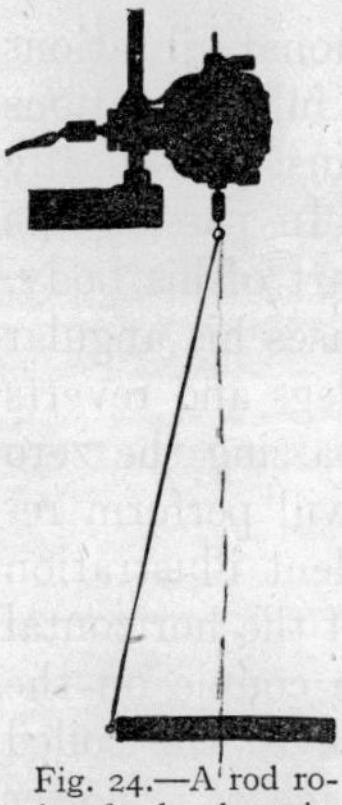

All these examples have two things in common.

(1) The bodies used are *surfaces of revolution*. All of them could theoretically be produced on a lathe. In all of them one axis of symmetry is prominent.

(2) One free axis coincides with the axis of symmetry, while the other is always at right angles to it.

Fig. 24.—A rod rotating freely about its axis of greatest moment of inertia.

Fig. 25.—Flexible shaft of an emery-wheel (shaft somewhat too thin for practical purposes). (1/9.)

In the experiments which now follow the bodies are not surfaces of revolution. As an example we consider a shallow cigar box (fig. 26). Its three pairs of surfaces are each painted a different colour.

(*e*) A ring is fixed to the smallest face of the box, by means of which the box is suspended by a wire from the shaft of the motor just as the cylindrical bar was in fig. 24. Experiment shows that the central lines A and C can act as "free" axes, that is, the body can rotate stably about these axes. The two free axes are at right angles to one another. On the other hand, the third central line B, which is at right angles to A and C and also passes through the centre of gravity, cannot act as a free axis at all. If made to rotate about it the body invariably reverts to one or other of the two stable positions of rotation.

(*f*) The same result is attained in a variant of the experiment. We throw up the box in the air, giving it a rotatory motion by holding it between the fingers in a particular way (fig. 27). A and C can again act as free axes. The same face of the box is always turned towards the observer, as may be seen from its colour. Attempts to make the

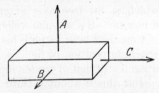

Fig. 26.—A the axis of greatest, B that of medium, C that of least moment of inertia of a box.

$$I_A = 6\cdot5$$
$$I_B = 5\cdot6 \Big\} \ . \ \mathrm{10}^{-3} \ \mathrm{kg.\text{-}mass.\text{-}m.}^2 \ \mathrm{or}$$
$$I_C = 1\cdot4 \Big) \ . \ \mathrm{10}^{-4} \ \mathrm{kg.\text{-}force\text{-}sec.}^2\text{-}\mathrm{m.}$$

Fig. 27.—Method of throwing a box so that it will rotate about its free axis of greatest moment of inertia.

box rotate about the axis B invariably lead to a wobbling motion and the observer then sees varying colours in succession.

These or similar experiments lead us to a single physical definition of free axes: the axes of greatest and of least moment of inertia can act as free axes.

In the selected simple examples (*a*) to (*f*) this is perhaps obvious in each case from purely geometrical considerations. In other cases we may make use of the rotator (fig. 5, Plate IV, and fig. 13, p. 106) and *measure* the moments of inertia for the various directions of axis. To make sure, we carried out these measurements for the shallow cigar-box and gave the results under fig. 26.

8. Free Axes of Human Beings and Animals.

For free axes of rotation to exist it is by no means necessary that the body should be a surface of revolution capable of being constructed on a turning-lathe. This was shown by our experiment with the painted cigar-box. It is still more clearly brought out by the use made of free axes by human beings and animals. Examples:

(a) *A Ballet-dancer Pirouettes on her Toes.*—She rotates about the longitudinal axis of her body, thus making use of her axis of *smallest* moment of inertia as a *free* axis, and rotating about it with large angular velocity ω and angular momentum $I\omega$. To stop herself at any instant she increases her moment of inertia by taking up the position shown in fig. 14b. Her new moment of inertia is about seven times what it was before, so that her angular velocity is reduced to a seventh of its previous value. She lowers her heels to the ground, stops turning, and brings her centre of gravity over her feet.

a   b

Fig. 28.—Changing the moment of inertia in turning a somersault.

(b) *An Acrobat turns a Somersault.*—Bending slightly forward, usually with upraised hands, he gives himself angular momentum about the axis indicated in fig. 28a, which is his free axis of *greatest* moment of inertia. His angular velocity is still small. A moment later he pulls his body into the crouching position of fig. 28b. The axis of rotation is still that of *greatest* moment of inertia. But this is about a third of its former value. Hence, by the principle of the conservation of angular momentum, the angular velocity is three times as great as before. This large angular velocity enables the acrobat to turn one or two complete somersaults, or sometimes even three. Then at a suitable moment the acrobat increases this moment of inertia again by stretching himself, and he lands on the ground, his angular velocity again becoming small. The technique of skilled acrobats is very instructive to a student of physics. The main requirement of an acrobat is courage. Turning somersaults is a question of nerve; the necessary rotations arise automatically from the fact that angular momentum is conserved.

(c) *A cat held upside down by its feet and then let go invariably falls on its feet.*—The animal rotates about its free axis of least moment of inertia, using it as a substitute for the axis on bearings of the revolving stool of fig. 19. Its tail and hinder extremities are swung round instead of the mallet. A human being can readily imitate this trick of the cat after his own fashion: he too can execute rotations about his axis of least moment of inertia while jumping, but in his case it is the *vertical* axis of the body.

9. Definition of the Gyrostat and of its Three Axes.

In the rotations we first considered the axis of rotation is fixed in the rotating body and is supported outside the body by bearings.

In the next group of rotations, those about free axes, the axis of rotation is still fixed in the rotating body, but the bearings are absent.

In the most general case of rotation neither bearings nor a fixed position of the axis of rotation relative to the rotating body exist. The axis of rotation, to be sure, still passes through the centre of gravity of the body, but its direction relative to the body continually varies.

This last mentioned general type of rotation is known as *gyrostatic* or *gyroscopic* motion. Rotations about free axes or about axes on bearings are particular cases of this general type of motion.

Gyrostatic motion in its most general form gives rise to the most difficult problems in the whole of mechanics. Even with the use of powerful mathematical tools the solutions obtained are only approximate. Yet all the essential phenomena are illustrated by the special

Fig. 29.—Two "oblate" tops: the axis of symmetry is the axis of greatest moment of inertia.

case of the "symmetrical" top or gyrostat. This special type is shown in fig. 29. In the examples shown there the axis of symmetry is always the axis of *greatest* moment of inertia. Physically speaking, we have to deal with "oblate" or "flattened" tops, i.e. "tops" in the ordinary sense of the word.

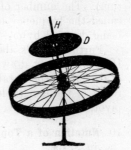

Fig. 30.—Gyrostat for demonstrating the instantaneous axis (back wheel of a bicycle, with the rim and spokes of another wheel fixed below in order to bring the centre of gravity into an easily accessible position). (1/20.)

For the discussion and understanding of all gyroscopic phenomena, it is essential that three different axes should be sharply distinguished.

(1) The *axis of symmetry*, i.e. the axis of greatest moment of inertia, in the case of our tops (fig. 29); it is immediately obvious in both the tops.

(2) The *instantaneous axis*, i.e. the axis about which the rotation is taking place at a given moment; it may be made visible by suitable devices.

(3) The *axis of momentum*, or *invariable line*, which lies between the axis of symmetry and the instantaneous axis in the plane determined by these two. All three axes intersect at the centre of gravity of the gyrostat. The axis of momentum is by far the most important of the three axes. Unfortunately it is inaccessible to direct observation.

We shall now illustrate these statements by diagrams and experiments. We shall use the gyrostat shown in figs. 30 and 33. It is pivoted at its centre of gravity (on a point inside the large ball-bearing). Thus it is "acted on by no forces" and in equilibrium for every position of its axis of symmetry.

To carry out the experiment we set the moving gyrostat on the pivot and give the axis of symmetry a push to one side. In order to make the instantaneous axis visible, a cardboard disc D covered with printed matter (newspaper) is placed above the body of the gyrostat itself but rigidly fixed to it. Owing to the rotation of the gyrostat the print is blurred into a uniform grey. It is only where the instantaneous axis of rotation Ω intersects the plane of the disc that the print is approximately at rest and can readily be made out. The instantaneous axis is thus marked out in an agreeably clear way, and we can watch the axis moving round in the gyrostat. The instantaneous axis and the axis of symmetry move round one another like a pair of dancers, each axis describing a circular cone about the invariable line, which is fixed in space but invisible. This phenomenon we call *nutation*. The cone described by the axis of symmetry may be called the cone of nutation (its centre line is the invariable line or axis of momentum). The number of times the cone is described per second may be called the frequency of nutation n_N, the angular velocity of nutation being denoted by ω_N.

Further details about nutation are given in the following section. Here we merely draw one conclusion from experiment which will be of use later: the nutation dies down with the lapse of time. This is a result of the inevitable friction at the bearings, i.e. in our case at the pivot.

10. Nutation of a Top under no Forces; its Axis of Momentum Fixed in Space.

The nutation which we have just observed experimentally is a direct consequence of the principle of the conservation of momentum. Imagine that the plane of the paper in fig. 31 passes through the axis of symmetry of the top (A) and its instantaneous axis (Ω). The top rotates about this instantaneous axis with angular velocity ω, represented by the length of the arrow in the direction of the axis Ω. This angular velocity ω may be resolved into two components ω_1 and ω_2; ω_1 is the angular velocity about A, the axis of greatest moment of inertia (I_1) and ω_2 the angular velocity about C, the axis of least moment of inertia (I_2). The angular momenta are accordingly $G_1 = I_1\omega_1$ in the direction of the axis of symmetry A, and $G_2 = I_2\omega_2$ in the direction of the axis C at right angles to the axis of symmetry.

The two angular momenta are indicated in the figure by the two lines with arrow-heads which are combined graphically so as to form the resultant angular momentum G. The arrow marked G represents the angular momentum in magnitude and direction. That is, the direction of the angular momentum, the axis of momentum, lies between the axis of symmetry and the instantaneous axis in the plane passing through them.

Now by hypothesis the top is " under no forces ". It is supported on a point at its centre of gravity. It is not acted on by any couples.

Hence by the principle of conservation of momentum, its angular momentum must remain unchanged in magnitude and direction. The axis of momentum must always remain in one and the same fixed direction in space (hence the name *invariable line*). Both the axis of

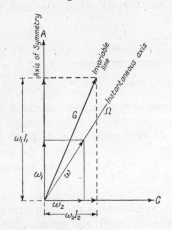

Fig. 31.—The three axes of a top

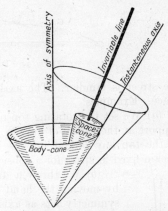

Fig. 32.—The body-cone and the space-cone

symmetry A and the instantaneous axis Ω must revolve round the axis of momentum fixed in space. The axis of symmetry describes a cone, the cone of nutation with which we are already familiar experimentally. The way in which this motion arises is shown very clearly in fig. 32, which represents the whole succession of figures corresponding to fig. 31. The invariable line is surrounded by a cone fixed in space (*space-cone, herpolhode-cone*). A hollow cone encloses this cone fixed in space and rolls on its outer surface. This cone is to be imagined as rigidly fixed to the axis of symmetry of the top (*body-cone, polhode-cone*). The line of contact at any instant of these cones with a common vertex gives the direction of the instantaneous axis of rotation Ω.

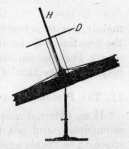

Fig. 33.—Top under no forces with the axis of symmetry remaining fixed in space.

The matters discussed in this section are somewhat troublesome to follow, but the effort is worth while. The root of the matter is contained in the following statement: it is the *invisible* axis of momentum, and *not* the axis of symmetry immediately obvious to the eye, that plays the important part in gyroscopic motion. ,

In special cases the axis of momentum and the axis of symmetry of a top may coincide. This occurs if

(1) the flat top becomes a spherical top, or

(2) the instantaneous axis of a flat top is made to coincide with its axis of symmetry.

This second case may be realized in various ways. We then have the phenomena of the top under no forces, with which the man in

Fig. 34.—Trajectory of a discus

the street is familiar. The axis of symmetry then remains fixed in space. Examples:

(*a*) We set the rotating top of fig. 33 very gingerly on its point of support at the centre of gravity, taking care to avoid giving the axis of the top any lateral impulse. The axis of the top does actually remain fixed in space for a considerable time.

Fig. 35.—
The diabolo

(*b*) We throw a discus, setting it in motion like a top by moving the hand in the well-known way. The axis of symmetry acts as axis of momentum (**G**) and remains fixed in space (fig. 34). In the *downward* part of its trajectory the discus, like the wing of an aeroplane, continues to fly through the air at a fixed angle of incidence (*a*), and is subject to a lifting force, as in the aeroplane case (p. 206). It falls to the ground more slowly than a stone does, and hence its range is greater, as is shown by the dotted parabolic trajectory of an ordinary projectile. In this case, of course, the term " under no forces " only applies approximately, for in actuality the top is acted on by small couples due to the resistance of the air.

(*c*) The diabolo shown in fig. 35: even when thrown to a great height the direction of its axis of symmetry remains fixed.

11. The Precession of the Axis of Momentum.

If no external couples are acting on the body the invisible axis of momentum, and not the visible axis of symmetry, remains fixed in space. The visible axis of symmetry revolves round the axis of momentum on the curved surface of a cone, which we call the cone of nutation. It is only in special cases that the axis of momentum and the axis of symmetry coincide; the fact that in these cases the axis of symmetry remains fixed in space is exceptional. This summarizes the contents of the previous section.

In the present section we shall examine the effect of external couples on a gyrostat. For the sake of clearness we anticipate the general result, which is as follows:

External couples give rise to precession of the axis of momentum; the latter no longer remains fixed in space, but in its turn begins to describe a cone of precession fixed in space; the axis of momentum remains the central line of the cone, as before. The gyrostat now has three characteristic angular velocities:

(1) its angular velocity ω about the axis of symmetry;

(2) the angular velocity ω_N with which the axis of symmetry revolves about the axis of momentum on the cone of *nutation*;

(3) the angular velocity ω_P with which the axis of momentum describes the cone of *precession* fixed in space.

The facts which we have stated in advance here will be established by suitable experiments in this section and the two which follow.

Gyroscopic motions in which both nutation and precession occur simultaneously exhibit very complicated phenomena. Hence for demonstration purposes efforts must be made to separate nutation and precession as completely as possible. To this end it is usual to begin with a gyrostat with no nutation, i.e. a gyrostat of the exceptional type in which the axis of momentum and the axis of symmetry coincide.

Fig. 36 (Plate IV) shows a gyrostat with a horizontal axis. The frame of the gyrostat is supported on a sharp point at the centre of gravity of the whole system. Initially no couple acts on the gyrostat. The axis of symmetry remains fixed in space. The gyrostat at rest is now to be acted on by a couple of moment **M**, with its axis at right angles to the axis of symmetry. This may be effected e.g. by hanging a weight from the frame of the gyrostat. The arrow denoting the couple **M** passes through the point of support and points towards the observer. This couple gives the gyrostat a small angular momentum $d\mathbf{G}$, which could, if we liked, be indicated by an arrow in a direction at right angles to the axis of symmetry. If the gyrostat is at rest, we have (p. 103) the quantitative relationship

$$\frac{d^2\beta}{dt^2} = \frac{\mathbf{M}}{I_g}, \qquad \ldots \ldots \ldots \ldots (15)$$

where I_g is the moment of inertia of the whole apparatus about the centre of gravity and β is the angle of displacement in the *vertical* plane passing through the axis of the gyrostat. Thus, if the gyrostat is *at rest* (fig. 36), the left-hand end of the support descends with angular acceleration $\mathbf{M}/I_g$. If the gyrostat is *in rotation*, however, something quite different takes place: the gyrostat exhibits small nutations. We pay no attention to these, for at the same time something much more striking happens: the axis of the gyrostat begins to revolve *horizontally* with constant angular velocity ω_p: *the axis of the gyrostat does not follow the couple but moves in a direction at right angles to it.* This is the precession of the gyrostat, described without taking account of the co-existing nutations.

How this remarkable precessional motion arises is easily explained. When the gyrostat is moving the angular momentum $d\mathbf{G}$ due to the couple $\mathbf{M}$ has to be combined with the larger angular momentum $\mathbf{G} = I\boldsymbol{\omega}$ of the gyrostat (fig. 37, Plate IV), the resultant angular momentum being in the direction of the point R^*. In time dt the axis of the gyrostat turns through the angle $d\alpha$ not in the vertical plane but in the horizontal plane. By p. 103

$$\mathbf{M} = \frac{d\mathbf{G}}{dt}. \quad \ldots \ldots \ldots (16)$$

The additional angular momentum $d\mathbf{G}$ arising in the interval of time dt has altered the direction of $\mathbf{G}$ only and not its magnitude. By fig. 37

$$d\mathbf{G} = \mathbf{G}\,d\alpha,$$

so that

$$\mathbf{M} = \mathbf{G}\omega_P \quad \ldots \ldots \ldots (17)$$

or

$$\omega_P = \frac{\mathbf{M}}{I\boldsymbol{\omega}}.$$

(The moment $\mathbf{M}$ may be measured e.g. in kg.-force-metres and I in kg.-force-sec.^{2}-m.)   The angular velocity of precession is directly proportional to the acting couple $\mathbf{M}$ and inversely proportional to the previously-existing angular momentum of the gyrostat, $\mathbf{G} = I\boldsymbol{\omega}$.

This statement is confirmed by experiment. Increasing the couple in fig. 37 (by putting on a bigger weight) has the effect of increasing the angular velocity of precession.

Nutation has been ignored in this primitive account of precession, which, however, is sufficient to explain many practical applications of precession. We shall confine ourselves to three examples:

(a) *Riding a Bicycle without touching the Handle-bars.*—Fig. 38 (Plate IV) shows the front wheel of a bicycle. Suppose the rider tilts a little towards the right. The front axle is thereby subject to a couple about the horizontal line of advance B. Simultaneously the front wheel, acting as a gyrostat, describes a precessional motion about the vertical C and curves to the right. The line joining the points of contact of the front and back wheels with the ground is again brought under the centre of gravity of the rider; that is, the point of support is brought back under the centre of gravity. The directions of the rotations and momenta are indicated in fig. 38.

In order to ride a bicycle in this way it is necessary that a certain minimum velocity should be reached; otherwise the necessary velocity of precession of the front wheel is too small and the curvature of the path too slight.

---

* The phenomena are more easily explained in terms of Coriolis forces. See p. 136.

PLATE IV

Chap. VII, Fig. 36.—Tilting of a gyrostat at rest under the influence of a couple.

Chap. VII, Fig. 37.—Precession of a rotating gyrostat under the influence of a couple

Chap. VII, Fig. 38.—To illustrate what happens when one rides a bicycle without touching the handlebars.

Chap. VII, Fig. 5.—Small rotator placed vertically, with a sphere on top. In this apparatus the flexural elasticity of a coiled spring is used. Its torsional rigidity $\mu' = 0 \cdot 0056$ kg.-force-metre /radian $= 0 \cdot 055$ large dyne-metre/radian.

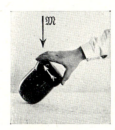

Chap. VII, Fig. 44   Chap. VII, Fig. 45
To illustrate the rising of a top as a result of friction

A demonstration experiment with a small model of a bicycle is very convincing. The wheels are set in rapid rotation by bringing them into brief contact with a rotating disc (fig. 39), and the longitudinal axis of the model is then held horizontally in the air without support. If the bicycle is carefully tilted about this longitudinal axis, say to the right, the front wheel promptly describes a curve to the right; similarly with the left. Set on the floor, the model immediately runs off in a perfectly straight path. A rider is quite unnecessary. All that he does when riding without touching the handle-bars is very little: he has only to learn to avoid disturbing the automatically-occurring precessional motion of the front wheel. The same physical phenomena are obviously at work in the case of the child's hoop.

Fig. 39.—A model bicycle being set agoing by contact with a disc on the shaft of a running electric motor.

Fig. 40.—A piece of cardboard acting as a discus.

(b) *A Piece of Cardboard acting as a Discus.*—A cardboard disc is held almost horizontally in the right hand and is thrown somewhat obliquely upwards. The piece of cardboard only flies like a good discus (fig. 34) at the beginning. The angle it makes with the horizontal soon increases and it rises steeply into the air, instead of ascending gradually as at first. At the same time the disc is raised on the right-hand side and flies to the right; during the steep ascent it loses all its velocity of advance and falls abruptly to the ground from the highest point of its path.

Explanation.—The angular momentum of the cardboard disc is much less than that of the heavy discus with its large moment of inertia. The couple exerted on the rotating disc (gyrostat) by the air flowing past it gives rise to a large precession of the axis of the disc; hence the change in the inclination of the disc.

A non-rotating disc would have its front end tilted up by the couple (see p. 199). That is, the air flowing past the disc gives it angular momentum in the direction of the axis C across the line of flight. The rotating disc already has the angular momentum G. The two angular momenta combine and the axis of the disc executes the precessional motion indicated by the slanting arrow.

(c) *The Boomerang.*—The moment of inertia of the cardboard disc may be increased and the disturbing precession diminished without altering the weight or the " lift " (p. 206). All that we have to do is to shift the greater part of the mass to the edge of the disc, i.e. thicken the outer regions at the expense of the centre.

We may use a cardboard ring about 20 cm. in diameter, 2 cm. broad, and 4 mm. thick, with a sheet of notepaper pasted over the opening.

According to equation (17), a disc with its moment of inertia increased in this way will be subject to only a small precession. The path will again become increasingly steep and the disc will thereby lose its velocity of advance. At the top of its path, however, it will still have an appreciable inclination to the horizontal, which will bring it gliding back, still rotating, to the thrower, so that it exhibits the

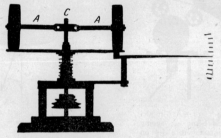

Fig. 41.—Model of a mill   (1/8.)

characteristic property of the *boomerang*. The traditional bent form of the implement is therefore by no means *essential*.

A circular disc, of course, is not a good " aerofoil " (see p. 206). A long, slightly hollowed rectangular shape is considerably better, forming a very good boomerang (it is a non-symmetrical gyrostat). For lecture purposes we use a strip of cardboard about 2·5 × 12 cm. and 0·5 mm. thick.

Small boomerangs are not thrown from the hand. Instead, we lay them on a book held somewhat slanting, in such a way that one end projects, and strike this end with a stick moved parallel to the edge of the book. By tilting the book a little to one side or the other we can make the boomerang fly to the left or to the right as we please, return in the vertical plane in which it set out, oscillate repeatedly about the vertical through the starting-point, and so on. By giving the boomerang the ordinary bent form and twisting the ends like a propeller (fig. 33, Plate VI) we obtain still other paths of flight and can perform a good many other neat tricks.

(d) Fig. 41 shows a model of a mechanism frequently used in mills, which was known even in Roman times. It makes use of precession in order to raise the milling pressure. The horizontal shafts AA of the two millstones are driven by a common vertical shaft C in the middle.

The jointed connexions with the central axis enable the shafts AA to be tilted in the vertical plane common to all three shafts. When revolving each of the millstones forms a gyrostat. Both gyrostats must have the directions of their axes continually changing during their revolution. They are accordingly forced to describe a precessional motion with angular velocity ω_p. This precession requires a couple

$$M = \omega_p \cdot I\omega$$

at right angles to the axes of the gyrostats, where I is the moment of inertia of the millstone, calculated from equation (6), p. 104, and ω the angular velocity of the millstone about the shaft A; M is in large dyne-metres if I is measured in kg.-mass.-m.^{2}

This couple M must press the millstones against the milling-table, leading to an increased milling pressure. In our lecture model the milling-table is supported by a spiral spring; its increased compression makes the rise in the milling pressure due to the forced precessional motion visible at a distance.

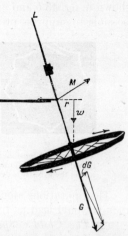

"A couple gives rise to precession of the axis of momentum of a gyrostat. Precession of the axis of momentum of a gyrostat involves the existence of a couple." This is the gist of what we have said above.

Under suitable experimental conditions precession of the axis of momentum as the result of a couple causes a well-defined cone of precession to be exhibited. Examples:

Fig. 42.—A gyrostat suspended so as to vibrate (three degrees of freedom). At the upper end there is a small electric lamp for recording the motion photographically as shown in fig. 43.

(1) *The Gyrostatic Pendulum.*—A gyrostat is suspended in a stable fashion but so that it can be swung in any direction (by means of a "Cardan joint"). It consists of the rim and spokes of a bicycle wheel (possibly loaded with lead). When out of the vertical it is acted on by the couple **M** due to the weight mg acting at the distance r. The line representing it is inserted in the figure and so is the additional angular momentum $d\mathbf{G}$ produced by the couple. Let go in the position shown, the gyrostat begins to describe a well-marked cone of precession with a relatively small angular velocity. At the same time it always exhibits small nutations. The lower end of the axis of symmetry describes not a smooth circle but a wavy one (fig. 43a). The greater the momentum of the gyrostat the smaller the nutation; the latter may become practically inappreciable. The precession is then said to be *pseudo-regular*. The opposite of pseudo-regular precession is genuine

regular precession. In the latter the small nutation set up by the external couple is suppressed. This is done by certain methods of starting the motion. At the instant of release the gyrostat is struck in such a way as to give it a nutation exactly equal and opposite to that which the couple would produce. Further details, however, would lead us too far out of our way here.

Instead of doing this we shall make nutation more and more prominent by diminishing the momentum of the gyrostat; in practice this means diminishing the angular velocity about the axis of symmetry. The tip of the axis of symmetry describes paths like those shown in fig. 43 *b* and *c*. Under suitable initial conditions it is actually possible to suppress precession entirely.

a   b   c

Fig. 43.—(*a*) Small nutations of a suspended gyrostat; approximation to pseudo-regular precession: (*b*) and (*c*) show the increase of nutation as the angular momentum of the gyrostat is decreased (photographic positive).

Then nutations alone persist in spite of the couple; but here again details would lead us too far afield.

(2) *The Child's Spinning-top.*—This is to be considered in the same way as the gyrostatic pendulum, except that its position when at rest is unstable. This, however, does not appreciably affect the gyrostatic phenomena.

In addition to the familiar cone of precession with superposed nutations, however, the spinning-top exhibits a peculiarity of its own. When moving on a smooth surface it rises gradually to the vertical but falls down whenever its tip lands in a hole. This may again be explained by the simple law of precession (equation (17), p. 122).

Case (1).—Fig. 44 (Plate IV) shows a spinning-top on a smooth surface, fig. 45 (Plate IV) a greatly magnified model of its hemispherical tip. The model is made to rotate by hand in the direction in which the axis of the top rotates. The tip rubs against the table and rolls ahead of the hand and centre of gravity of the top. There results a couple in the direction of the arrow M and an additional angular momentum $d\mathbf{G}$. The combination of the two momenta has the effect of making the axis of momentum approach the vertical.

Case (2).—If the top runs against an obstacle the tip is held back while the centre of gravity of the top continues to advance; the top becomes subject to a couple M with its axis vertically upwards. A top with its tip in a hole may be regarded as one which is continually running against an obstacle, so that the top falls down.

(3) *The Earth as a Gyrostat.*—The earth exhibits a very well-known instance of precessional motion. The earth is not a perfect sphere, but is somewhat flattened, the equatorial diameter being about 1/300 part greater than the axis of symmetry, the line joining the north and south poles. Roughly speaking, we may regard the earth as strictly spherical except for a swelling along the equator. The attraction exerted by the sun and moon on this protuberance gives rise to a couple acting on the earth. The result is that the axis NS describes a cone of precession of semi-vertical angle $23\frac{1}{2}°$ once in about 26,000 years. At the same time the couple gives rise to minute nutations, so that at any instant the axis of rotation deviates a little from NS,

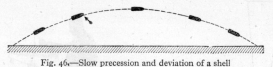

Fig. 46.—Slow precession and deviation of a shell

the axis of symmetry of the earth; the points in which the two axes intersect the surface, however, are only about 10 metres apart.

On these minute nutations in the physical and engineering sense are superposed nutations in the *astronomical* sense. In the physical and engineering sense the latter are *forced* vibrations of the axis of rotation of the earth (§ 11 of Chapter XI, p. 253). These are due to the periodic variations of the effective couple, which, of course, varies with the varying positions of the sun and moon relative to the earth.

(4) *Angular Momentum of a Projectile* (fig. 46).—If projectiles are made long instead of round the mass can be increased without increasing the calibre. With elongated projectiles, however, special precautions have to be taken to prevent them turning over and over in the air. The long axis must lie as nearly as possible along the tangent to the trajectory at each instant, so as to keep down the air resistance. To this end the projectile is either made long and pointed (as in the trench mortar) or is made to rotate about its long axis (by the rifling of the gun). The rotating projectile is a gyrostat, and as such is caused to precess by the resistance of the air. Precession sets in about the point marked by the arrow. The opposing force of the air meets the projectile a little behind its nose, thereby giving rise to a couple whose axis is at right angles to the plane of the paper. The couple is not constant, for the tangent to the path is continually varying in direction. Hence no proper cone of precession results and the nose of the projectile describes cycloidal arcs instead of circles. If its spin is to the right, the nose of the projectile lies successively above and to the right of the tangent to the trajectory, to the side and to

the right, below and to the right, and finally along the tangent. In field-guns the whole process takes about 1 second, i.e. a time which is short compared with the time of flight (about 20 seconds). The nose never deviates far from the tangent to the path, and the projectile reaches its target nose first. A lateral deviation, however, cannot be avoided (" drift "); if the spin is to the right, the drift is to the right. For in the ascending part of the trajectory the resistance of the air always acts on the left-hand side of the precessing projectile.

12. Gyrostat with only Two Degrees of Freedom.

All the gyrostats we have considered hitherto had three degrees of freedom; that is, their axes could take up any position in space. They

Fig. 47.—A gyro-stat suspended so as to vibrate about the axis A (two degrees of freedom).

Fig. 48.—Stabilization by means of negatively-damped precessional vibrations of a gyrostat (mono-rail); between the gyrostat and the man's body there is a guard and on the right below the gyrostat a com-pensating mass.

were at most confined within certain regions of space by accidental external obstacles due to the manner of their construction.

In contradistinction to these we shall now consider a gyrostat with only *two* degrees of freedom. An example of this type is shown in fig. 47, in which a bicycle-wheel is suspended so as to oscillate as a gravity pendulum about the fixed horizontal axis A.

The period of the pendulum is the same no matter whether the gyrostat is rotating or not. In moving the pendulum in its plane of vibration the hand feels *no* effect due to the motion of the gyrostat. All the same a gyrostatic pendulum executing linearly polarized vibrations must be handled very carefully. Vibrations of the gyrostat in the prescribed plane mean forced precessions or changes in direc-tion of the axis of momentum. By equation (17), p. 122, precession never occurs unless couples are present. These couples, however, are taken up by the supports of the axis A and in the bearings of the axis of symmetry of the gyrostat, and under certain circumstances may damage the bearings and journals. We must not be deceived into thinking that the motion and precession of the gyrostat in the direction of the prescribed plane of vibration do not make themselves appreciable at all.

The following remarkable experiment will serve to demonstrate the couples to which the bearings of the pendulum are subject. Fig. 48 shows a horizontal bar supported on ball bearings KK and carrying a seat and a motor-driven gyrostat. The gyrostat is capable of vibrating in a U-shaped frame R in the direction of the length of the bar. A man sits on the seat. The centre of gravity of the whole system (bar, gyrostat, and man) lies far above the bar, i.e. the system is quite unstable. Suppose, for example, that it tilts towards the right. This tilting causes a couple to act on the axis of the gyrostat. The gyrostat reacts by precessing. Suppose it moves in a counterclockwise direction as seen from above. In this case the upper end of the gyrostat moves away from the man. The man then pushes the upper end of the gyrostat a little farther away from him, and feels scarcely any more resistance to the motion than when the gyrostat is at rest. Nevertheless, the precession gives rise to a large couple acting on the bearings of the gyrostat and hence on the bar, which tilts back into its original position. If the original tilt was to the left the phenomena are exactly the same except that the direction of rotation is reversed. The upper end of the gyrostat moves nearer the man and the man pulls it a little nearer him. In this way he can easily keep his balance. The gyrostat executes small oscillations in the plane prescribed by its bearings. The man has only to see that these precessional oscillations are "negatively damped", i.e. he has to *increase* the deflection which exists at any moment.

In a surprisingly short time our body learns how to exert this "negative damping" in a purely automatic way. With gyrostats of certain dimensions there is no time for conscious thought; the muscular sense, however, rapidly grasps the physical situation, and after a few minutes' practice one feels just as secure on this top-heavy horizontal bar as an expert cyclist does on his machine.

This use of negatively damped gyrostatic vibrations was discovered empirically by Chinese rope-dancers long ago. As gyrostat they use an umbrella which they keep in rapid rotation with their fingers. They hold the stick of the umbrella approximately parallel to the rope and balance themselves by tilting the stick slightly in one direction or the other. Most rope-dancers, however, merely utilize the parachute effect of an umbrella held still.

Attempts have been made to apply the gyrostat with two degrees of freedom and negative damping on a large scale in the construction of a *mono-rail car*. The motion of the muscles of the arm is replaced by a suitable auxiliary engine which reverses its motion as the car tilts to left or right.

Gyrostats with only one degree of freedom are more conveniently dealt with by the methods of next chapter.

CHAPTER VIII

Accelerated Systems of Reference

1. Preliminary Remarks: Forces of Inertia.

Hitherto we have considered physical phenomena from the point of view of the fixed ground or floor of the lecture-room. Our system of reference was the earth, regarded as rigid and at rest. Occasional exceptions to this have always been clearly indicated as such.

The passage to another system of reference may in certain cases be of no consequence. In such cases the new system of reference must be moving with *constant* velocity relative to the ground. Its velocity may not alter in magnitude or direction. Experimentally this condition is occasionally realized in a vehicle which is moving very "steadily", e.g. a steamer or railway carriage. In these cases the motion of the system of reference is not "felt" in any way in the interior of the vehicle. Everything goes on in the vehicle just as it would in a lecture-room at rest. In actual experience, however, such cases are quite exceptional.

A vehicle of any kind in general forms an "accelerated" system of reference; its velocity varies both in magnitude and direction. This acceleration of the sytem of reference gives rise to important changes in the physical observations made. In order that the physical phenomena may be simply described from the standpoint of the accelerated system of reference we require to introduce new concepts. An accelerated observer finds that new forces appear. The general name for these is "forces of inertia". Some of them have special names in addition (centrifugal force, Coriolis force). A discussion of these forces of inertia forms the subject of the present chapter.

Hitherto we have always distinguished between two cases of acceleration: acceleration wholly in the line of motion (wholly tangential) and acceleration wholly at right angles to the line of motion (wholly normal), or change of velocity in *magnitude* only and in *direction* only. In the same way we shall now deal with systems accelerated wholly in the line of motion and systems accelerated wholly at right angles to the line of motion as two separate limiting cases.

Systems of reference accelerated wholly *in* the line of motion are in fact commonly met with, e.g. vehicles of all kinds when being *started* or *stopped* in a straight line. The duration of the acceleration, however, is generally very small and the magnitude of the acceleration is constant for a few seconds at most. This limiting case may therefore be dismissed with the comparatively brief treatment given in the next section.

It is quite different in the case of systems of reference accelerated wholly *at right angles* to the line of motion. In a merry-go-round with constant angular velocity the normal acceleration can be kept constant for any length of time. Above all, the earth itself is a big merry-go-round. We must therefore investigate this type of accelerated system thoroughly. This is done in the remaining sections of the chapter.

In order to make our discussion simpler to follow, we shall adopt the device of printing the text in two parallel columns. In the left-hand column the phenomena will be briefly described as hitherto, from the standpoint of a system of reference at rest, i.e. the ground or floor of the lecture-room. In the right-hand column the same phenomena will be described from the standpoint of an accelerated observer. *Both observers preface their descriptions with Newton's second law of motion* ($F = mf$) *as fundamental.*

2. System of Reference Accelerated wholly in the Line of Motion.

We consider the following examples:

(1) One observer sits still on a chair placed on a truck and in front of him there is a ball on a frictionless table-top (fig. 1). The effects of gravity are thus got rid of. The table and chair are screwed to the truck. The truck is accelerated to the left in its longitudinal direction (by means of a push with the foot). The ball and the man on the truck then move closer to one another.

There are now two possible ways of describing what happens. These are given below; in both the terms *left* and *right* apply to the reader.

Observer at rest	*Accelerated observer*
The ball remains at rest. No force acts on it, for it is supported without friction. The truck and the man sitting on it, however, are accelerated towards the left. The man moves nearer the ball.	*The ball moves towards the right with acceleration.* Hence it is acted on by a force towards the right, which is called a *force of inertia.*

In the choice of this name it is assumed that the observer is aware of his own acceleration. A more colourless term or a word coined for the purpose, analogous to the word "weight", would have been more suitable.

Fig. 1

(2) The observer on the truck holds a dynamometer attached to the ball (fig. 2). The truck is again accelerated towards the left. During the acceleration the observer on the truck experiences a sensation of force in the muscles of his hand and arm. The dynamometer records the deflection F.

Observer at rest	*Accelerated observer*
The ball is accelerated towards the left. It is acted on by a force F towards the left. The magnitude of the acceleration is given by $f = F/m$.	*The ball remains at rest*. It is not accelerated. Thus by the second law of motion the sum of the forces acting on it must be zero. The force of inertia pulling it towards the right and the muscular force pushing it towards the left are equal and opposite. Their magnitude is recorded by the dynamometer and we find that $\dot{F} = mf$.

Fig. 2

(3) The truck is accelerated towards the left. The observer standing on the truck must at starting take up the inclined position shown in fig. 3, otherwise he would overbalance.

In the accounts of both observers below the expressions *left* and *right* again apply to the reader.

Observer at rest	*Accelerated observer*
The centre of gravity of the man must have the same *acceleration* in magnitude and direction as the truck. The man produces the force to the left necessary for the acceleration of his centre of gravity by means of his weight, by leaning forwards.	The centre of gravity of the man must remain at *rest*. The line of action of the force acting on his centre of gravity must therefore pass through his point of support (his feet). This force points downwards and to the right, for it is the resultant of two forces acting at G, the centre of gravity of the man: F_2, the weight of the man acting vertically downwards, and $F = mf$, the force of inertia acting horizontally and to the right.

Fig. 3

(4) One observer is in a lift. In front of him there is a table on which is a compression balance with a mass m on it. The deflection of the balance indicates the weight F_2. The lift then begins to move *downwards* with acceleration. The balance now registers the smaller deflection F_1.

Observer at rest	Accelerated observer
The mass is accelerated downwards. It is acted on by two forces which differ in magnitude and direction, the weight F_2 pulling it downwards and the force of the spring F_1 pulling it upwards. The effective force is therefore $F_2 - F_1$ and the acceleration $f = (F_2 - F_1)/m$.	*The mass is at rest.* The deflection of the balance is reduced by the amount $F_2 - F_1$. Hence the downward force F_2 must be opposed by a force of inertia $F_2 - F_1 = mf$ directed upwards.

(5) One observer, holding the compression balance in his hands, jumps from a high table to the ground. The balance carries a mass (a weight). Immediately after he jumps off the deflection of the balance drops from F_2 back to zero (fig. 4).

Observer at rest	Accelerated observer
The mass falls at the same rate as the man, with the gravitational acceleration $g = F_2/m$. The only force acting on it is the weight F_2 pulling it downwards; the muscular force is no longer pushing it upwards.	*The mass is at rest.* The sum of the forces acting on it is zero. The weight F_2 acting on it downwards and the force of inertia acting upwards are equal and opposite, the absolute value of both forces being mg.

These examples may be taken as sufficiently illustrating the term "force of inertia". *The force of inertia exists only for an accelerated observer.* The observer must— at least in thought—participate in the acceleration of his system of reference. *The force of inertia enables the accelerated observer to retain Newton's second law as the fundamental theorem of his dynamics.*

Fig. 4

3. System of Reference Accelerated wholly at Right Angles to the Line of Motion. Centrifugal Force and Coriolis Force.

(1) One observer sits on a rotating stool with a vertical axis and large moment of inertia (fig. 5; cf. also fig. 13). In front of him there is a smooth horizontal table attached to the stool. On this the

observer sitting on the stool places a ball (fig. 5). It flies away from him and off the table.

Observer at rest	*Accelerated observer*
The ball is not accelerated; no force acts on it. Hence it cannot participate in the circular motion. It flies off tangentially with the constant velocity $v = \omega r$ (ω being the angular velocity of the stool and r the distance of the ball from the axis of rotation at the moment when it is laid on the table).	The ball moves away from its position of equilibrium with acceleration; that is, it moves away from the centre about which the table is rotating. Hence the ball at rest is acted on by a *force of inertia* which is given the special name *centrifugal force*. Its magnitude is $F = m\omega^2 r$.

(2) The observer on the rotating stool inserts a dynamometer between the muscles of his hand and the ball, the horizontal longitudinal axis of the dynamometer pointing towards the axis of the stool. During the rotation of the stool the dynamometer registers a force $F = m\omega^2 r$.

Fig. 5

Observer at rest	*Accelerated observer*
The ball moves in a circle of radius r; *it is accelerated*. This involves the existence of a radial force $F = m\omega^2 r$ acting on the ball, directed towards the axis of rotation of the stool.	The ball remains at rest. It is not accelerated. Hence the sum of the two forces acting on it is zero. The centrifugal force pulling it radially outwards and the muscular force pulling it radially inwards are equal and opposite. The absolute value of both is $F = m\omega^2 r$.

(3) The observer on the rotating stool has a gravity pendulum, say a ball on a string, suspended over his table. The pendulum does not hang straight down (fig. 6); it is displaced outwards from the vertical through the angle α in the plane containing the radius vector to the pendulum bob and the axis of rotation. The angle α increases as the rate of rotation does.

Fig. 6

Observer at rest	*Accelerated observer*

Observer at rest

The bob of the pendulum moves in a circular path of radius r; *it is accelerated*. This involves the existence of a radial force $F = m\omega^2 r$ directed horizontally towards the axis of rotation. This force arises from the weight in combination with the elastic force of the stretched string.

Accelerated observer

The bob of the pendulum is at rest. The pendulum is in its position of equilibrium. Hence the string l must be in the direction of the force acting on the pendulum bob. This force points outwards and downwards. It is the resultant of the two forces acting on the bob: the weight F_2 acting vertically downwards and the centrifugal force $F = m\omega^2 r$ acting horizontally outwards.

(4) In the preceding experiments observations have been made on a body *at rest* relative to the rotating stool. The only question was whether the body was *accelerated* out of this position or not. We shall now direct our observations to a body *moving* relative to the rotating stool, namely, a bullet. For this purpose a small pistol is fixed horizontally to the table of the rotating stool; its longitudinal axis may make

Fig. 7

any angle α with the line joining it to the axis of rotation. The pistol is aimed towards a point a on a target at a distance A from the muzzle. The target is supported by bars, and takes part in the rotation of the stool (fig. 7). First a shot is fired with the stool at rest and the point a where it strikes the target determined. The stool is then set in rotation with angular velocity ω. *From now on the stool is to rotate in a counter-clockwise direction as seen from above.* The second shot is then fired. It strikes the target at a point b removed s cm. towards the right.

Numerical Example.—Stool rotates once in 2 sec.; velocity of bullet $(w) = 60$ m./sec. (air pistol); distance of target $(A) = 1·2$ m.; deviation to the right $(s) = 0·075$ m. $= 7·5$ cm.

Observer at rest

When the stool is at rest the bullet strikes the point a aimed at. If the rotating stool is stopped immediately after the shot is fired, the point b where the bullet strikes the target is to the left of the point aimed at. For in this case v, the velocity of the muzzle of the pistol, is combined with u, the velocity of the bullet. Hence the bullet flies through the lecture-room in the direction w.

In the experiment actually carried out the stool continues to rotate after the firing of the shot. The bullet, on the contrary, after leaving the muzzle flies in a straight path in the direction w through the lecture-room under no forces. *Hence the line of sight rotates relative to the line of flight.* At the end of the time of flight Δt the point aimed at lies at a'. Thus b, the point where the bullet strikes the target, is now displaced from the point aimed at towards the right through the distance s. From fig. 8 we infer that

$$s = A\omega \, \Delta t.$$

For both trajectories (i.e. in the directions u and w) the time of flight of the bullet from weapon to target is the same, namely,

$$\Delta t = A/u.$$

Hence $s = \omega u \, (\Delta t)^2$ . . (1)

Accelerated observer

During its flight the bullet is accelerated at right angles to its path. Its path is curved towards the right. During the time of flight Δt the bullet is deflected towards the right through the distance $s = \frac{1}{2} f (\Delta t)^2$. According to the annexed statement for the observer at rest $s = \omega u (\Delta t)^2$ Hence the observed acceleration f is $2\omega u$. It is called the Coriolis acceleration after its discoverer. No acceleration f can exist without a force $F = mf$. Hence the moving bullet is acted on in a direction at right angles to its path by a Coriolis force

$$F = m \cdot 2\omega u. \quad . \quad (2)$$

Or in general: *Suppose a system of reference is rotating with angular velocity ω. Let a body (particle) move within this system with velocity* u. *Then the moving body is acted on by a Coriolis force* F $= 2mu\omega$ *at right angles to its path.* Thus the Coriolis force is a *force of inertia* acting on a moving body. It is at right angles to the arrows representing the angular velocity and the linear velocity.

The equation found by both observers, $s = A \cdot \omega \Delta t = A^2 \cdot \omega / u$, gives a very simple method for measuring the velocity of a bullet.

(5) The above example exhibited the lateral deviation of a body *moving* relatively to an accelerated system of reference for a single initial direction of motion only. The amount of the deviation should be independent of the initial direction (of the pistol). But we de-

liberately refrained from demonstrating this, for it can be done much
more rapidly and simply by slightly altering the experiment.   We

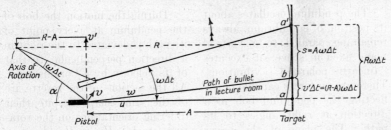

Fig. 8.—Here v' is the component of w, the velocity of the bullet, parallel to the target, and v the
velocity of the muzzle of the pistol. For the sake of clearness the angle $w\Delta t$ has been drawn too
large; this gives rise to the trifling blemish that the line of sight no longer appears to be perpendicular to the target at a'.

replace the bullet by the bob of a pendulum. The pendulum is suspended in the usual way over the table of the rotating stool. To facili-
tate observation the moving bob is made to
trace out its own path.   For this purpose the
bob is made to contain a small ink-pot with a
fine dropping-tube at the bottom.   A sheet of
white blotting-paper is fixed to the table of the
rotating stool.   The observer on the stool at
first holds the bob fixed and keeps the tube
closed (fig. 9), the thread of the pendulum making
any angle with the mean position.   When let

Fig. 9

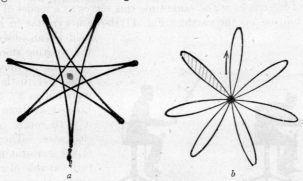

Fig. 10—Rosette-like paths described by a pendulum on a rotating table.
In fig. 10a the pendulum was let go above the ink-blot in a position of
maximum displacement and began by moving towards the right. The end of
the path happens to coincide with the beginning. In fig. 10b the pendulum
was set going by a push out of its position of equilibrium.

go the pendulum oscillates with gradually decreasing amplitude about
its *non*-vertical position of equilibrium (see fig. 6), and describes the
rosette-like path reproduced in fig. 10a.

The two observers now describe their experiences.

Observer at rest	*Accelerated observer*

The pendulum oscillates about its position of equilibrium, always remaining parallel to a vertical plane fixed in space; it executes "linearly polarized" vibrations. There are no forces capable of deflecting the pendulum bob in a direction at right angles to its path. The plane of the paper rotates under the vibrating pendulum.

During the motion the bob of the pendulum at every point of its path is deviated to the right in a direction perpendicular to that of its velocity by a Coriolis force. All the sections of the rosette are of the same shape despite their differing orientations on the rotating stool. *Hence the direction of the path in the accelerated system of reference makes no difference to the magnitude of the Coriolis force.*

The deviation of the mean position of the pendulum from the vertical has already been explained above under (3).

In a lecture experiment the rotating stool should be given only a small angular velocity, otherwise the eye will be unable to recognize the position of the plane of vibration of the pendulum.

The deviation of the mean position of the pendulum from the vertical is a result of the centrifugal force (see above under (3)). *Thus a body moving in an accelerated system is acted on both by the Coriolis force and the centrifugal force.**

(6) We now consider the behaviour of a gyrostat in an accelerated system of reference (at the same time this serves as a model of a gyroscopic compass on the earth). Fig. 11 shows a gyrostat in a frame held by an observer on the rotating stool. For the sake of brevity we shall refer to the rotating stool as a "globe". Looked at from above it is to rotate counterclockwise. The frame of the gyrostat is in its turn capable of rotating about the axis B at right angles to the axis

Fig. 11

* In our deduction of the Coriolis force (fig. 8) we used a body moving very rapidly, namely, a pistol bullet. We were thus dealing with a simplified limiting case: we could neglect changes in the velocity of the bullet due to centrifugal force. If the initial velocity were small, the man on the rotating stool would not observe a small deviation to the right, but a gradually widening spiral path arising from the combined effect of the centrifugal force and the Coriolis force.

of symmetry of the gyrostat. The axis B lies in a meridional plane of the rotating stool or globe, i.e. it points in the direction of its vertical (NS) axis. The axis B can also be set for various latitudes, i.e. it may make any arbitrary angle ϕ between 0° (equator) and 90° (pole) with the horizontal plane of the rotating stool. The observer on his rotating stool sets the gyrostat going by twirling the spokes. He then leaves it to itself. After a few torsional vibrations about the axis B, the axis of symmetry of the gyrostat sets itself in the meridional plane (on the right of fig. 11). In the arrangement chosen here the axes of the gyrostat and of the rotating stool rotate in the same sense. (If, however, the axis of the gyrostat is supported in a different way it is possible to have the two axes rotating in opposite directions.)

For the sake of simplicity both observers start with the axis of symmetry of the gyrostat in the same initial position, in the plane of a " parallel of latitude ".

Observer at rest

The rotation about NS, the axis of the stool or globe, causes the axis of symmetry of the gyrostat to be acted on by the couple $\mathbf{M}$. This has a component $\mathbf{M}_1$ perpendicular to the axis B. The axis of the gyrostat sets itself at right angles to this couple $\mathbf{M}_1$. It executes a *precessional* motion in the horizontal plane of the globe, i.e. in the plane perpendicular to the B axis. At first it swings past the meridian; but the friction at the bearings of the axis B damps down the vibrations very rapidly. The axis of the gyrostat remains pointing along the meridian. For only in this position does the $\mathbf{M}_1$ component of the couple lie along the axis of symmetry of the gyrostat (A). It is only when the gyrostat is in this direction that the component of the couple can give rise to no further precession. The axis of the gyrostat lies along the meridian of the globe just like a compass-needle.

Accelerated observer

The parts of the rim near β are deviated towards the right out of their path by Coriolis forces. The right half of the gyrostat (as seen by the reader) moves out of the plane of the paper towards the reader. This brings the axis of the gyrostat into the meridional plane. Coriolis forces still act on the moving rim but no longer give rise to a couple about the axis B.

So much for our experiments intended to define the ideas of *centri-fugal force* and *Coriolis force*. *Both forces exist only for an observer accelerated in a direction at right angles to that of his motion.* The observer must—at least in thought—participate in the rotation of his system of reference. With these new forces he can still keep Newton's second law of motion, $f = F/m$, as the foundation of his dynamics even in a system of reference accelerated in a direction at right angles to that of its motion.

How does the accelerated observer stand with respect to the law of action and reaction? Answer: he is in the same position as an observer on the earth is relative to the reaction to a weight. An accelerated observer is unable to detect the reactions corresponding to forces of inertia. This brings us back to the question raised in § 1 of Chapter V (p. 65). Does even the weight of a body depend only on an acceleration of the system of reference, i.e. of the earth, the nature of which is still unexplained?

We shall illustrate the meaning of this question by a final obser-vation on the rotating stool. The rotating stool, as usual, rotates counterclockwise as looked at from above. The observer on the rotat-ing stool looks at a spectator A at rest in the lecture-room. There are now two ways of describing the state of affairs:

Observer at rest	Accelerated observer

The spectator is at rest. He is not accelerated. He is not acted on by any forces.

The spectator describes a cir-cular path of radius r with linear velocity $v = \omega r$ in a clockwise direction. We are again dealing with a body moving in an ac-celerated system of reference. Hence the body is acted on both by a centrifugal force and by a Coriolis force. The centrifugal force $F = m\omega^2 r$ is directed radi-ally outwards, the Coriolis force $F_c = 2mv\omega = 2m\omega^2 r$ radially in-wards (in such a direction as to deviate the spectator to the right in his path). As difference there remains a force $m\omega^2 r$ directed radially towards the axis of rota-tion, which enables the spectator to maintain his motion in a cir-cular path.

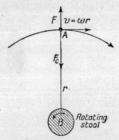

Fig. 12.—To illustrate the text on the right

We see that the occurrence or non-occurrence of forces depends on the way in which the system of reference is chosen. The " reality " of forces and the distinction between "real" and "apparent" forces are not matters which can be settled by physics at all.

4. Vehicles as Accelerated Systems of Reference.

The choice between unaccelerated and accelerated systems is in many cases a matter of taste, e.g. in the case of the circular motion of bodies about shafts supported on bearings. But it is essential that the system of reference used should be clearly stated (cf. the beginning of § 8 of Chapter III (p. 41). In other cases, however, the accelerated system of reference is certainly to be preferred. This applies particularly to any physical phenomena occurring in ordinary vehicles; the acceleration to which these systems of reference are subject is often very complicated, being usually a combination of acceleration in the line of motion (starting and stopping) and acceleration at right angles to the direction of motion (going round curves).

All our everyday experience of forces of inertia in vehicles is already included in the cases of motion described in the last two sections; for example:

(a) the oblique position necessary to avoid overbalancing when a train is starting, stopping, or going round a curve;

(b) the oblique position taken up by a bicycle and its rider, a horse and its rider, and an aeroplane and its pilot when going round a curve;

(c) the lateral displacement, due to Coriolis forces, on the deck of a steamer altering her course; one cannot walk straight to one's destination without making one foot cross over the other;

(d) the Coriolis forces which are "felt" particularly strongly on a rotating stool with a high moment of inertia and hence a very constant angular velocity. If one tries to move a weight (say of 2 kg.) *rapidly* in any direction (fig. 13), the result takes one aback. One feels as if one's arm were embedded in a current of viscous liquid. This is a particularly important experiment.

Numerical Example.—One revolution in 2 sec., i.e. $n = 0.5$ per second; $\omega = 2\pi n = 3.14$; mass of metal block 2 kg.; $v = 2$ metres per second; Coriolis force $= 2mv\omega = 2 \cdot 2 \cdot 2 \cdot 3.14 = 25$ large dynes $= 2\frac{1}{2}$ kg.-force, i.e. greater than the weight of the metal block moved!

Now, too, we are at last in a position to understand the action of the forces in the experiment shown in fig. 23, p. 14. The muscular force had to move the metal blocks in opposition to the Coriolis forces. The rotating stool had only a *small* moment of inertia; consequently it reacted to the force opposing the muscular force by large changes in its angular velocity.

These qualitative experiments may be multiplied to a considerable extent. It is more instructive, however, to discuss quantitatively what at first sight appears a peculiar special case, *a horizontal torsion balance on a rotating table.* The rotating table is shown in profile in fig. 14. On it is placed a torsion balance with a rod-like " pendulum ". The axis of the pendulum is at a distance R from the shaft of the rotating table. The long axis of the pendulum is to point in the direction of the axis of rotation of the table, *independently of any acceleration of the table.*

If the rotating table has constant angular velocity ω the pendulum remains in the position of equilibrium: for the wholly *normal* acceleration in this circular motion is exactly in the direction of the long axis of the pendulum. Such accelerations can never give rise to a couple.

In order to test this the axis of the pendulum may be made to slide along a rail and the direction of its length set parallel to the rail. Then the pendulum does not react to any accelerations in the direction of the rail.

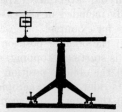

Fig. 13.—A rotating stool with a high moment of inertia, for demonstrating Coriolis forces. The additional masses shown here may also be used with advantage in the experiments shown in figs. 5, 7, 9, and 11 of this chapter.

Fig. 14.—A torsion balance on a rotating table. The torsion balance consists of a wooden rod on the small rotator of fig. 5, Plate IV. (1/40.)

On the other hand, every change in the angular velocity ω, i.e. every angular acceleration $\dot{\omega}$ of the rotating table throws the pendulum out of its mean position, the deflections being quite considerable. For now the accelerations are *at right angles* to the long axis of the pendulum. At first sight the fulfilment of the condition stated above seems entirely out of the question. Nevertheless, it may be attained quite easily. *Merely by suitable choice of its moment of inertia* (I), it is possible to make the pendulum entirely insensitive to any angular acceleration. We must have

$$I_0 = mRs, \quad \ldots \ldots \ldots \ldots \quad (3)$$

or

$$I_g = ms(R - s), \quad \ldots \ldots \ldots \quad (4)$$

a formula which is more convenient for numerical calculations, I_0 being the moment of inertia of the pendulum about its axis of rotation, I_g that about a parallel axis through its centre of gravity, m the mass of the pendulum, s the distance between the centre of gravity and the axis of rotation, and R the distance between the axis of the pendulum and the axis of the table.

The value of the torsional rigidity μ' for the coiled-up spring of the pendulum is of no importance; it does not enter into the calculation at all. In the silhouette of fig. 14 we see a pendulum of this kind in the form of a rod (for dimensions see p. 144). This pendulum actually does remain at rest, no matter how great the angular acceleration of the rotating table. The experiment is very striking. Small alterations in R or s restore the former sensitiveness to angular accelerations.

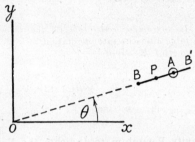

Fig. 15

Proof.—In fig. 15, BB' is the rod, carried on the spindle A, which is attached to the turning table. The rod has a torsional spring control about A and initially the rod points towards O—the equilibrium position. Ox, Oy are a set of fixed axes. Let r, θ be the polar co-ordinates of any particle, P, of the rod.

Due to the rotation of the table, P moves relative to the fixed axes and has an absolute acceleration relative to these axes. This acceleration is f_r, along OP, f_t, tangential to OP. As P is, initially (at $t = o$), in the line OA, the radial force, mf_r (m being the mass of the particle at P), has no moment about A and hence cannot disturb the rod.

It is otherwise with mf_t—this force gives rise to a torque $mf_t \times (PA)$, about A.

Now $$f_t = 2\dot{r}\dot{\theta} + r\ddot{\theta}.$$

Initially, $\theta = 0$, but $\ddot{\theta} \neq 0$, $\therefore$ at the beginning of the motion of the table, $f_t = r\ddot{\theta}$ only. $\therefore$ the torque about A, due to the mass, m, at P is

$$mr\ddot{\theta} \times (R - r), \quad \ldots \ldots \ldots \ldots (1)$$

where $R = OA$.

This formula shows that, *initially*, the transverse acceleration of the particle, P, produces a torque about A which is proportional to $r(R - r)$. *This quantity changes sign for particles lying between A and B'*—hence it is possible that the whole torque, due to *all* the particles composing the rod, may be zero.

From (1), the whole torque is, initially,

$$\Sigma[\ddot{\theta}\{mr(R - r)\}].$$

If x is the distance of P from A, $x = R - r$ and $r = R - x$,

i.e.   whole torque $= \ddot{\theta}\Sigma[m(R - x)x] = \ddot{\theta}\Sigma[mRx - mx^2].$

Now $\Sigma(mx) = M \cdot s$, where s is the distance of the centre of gravity of the rod from A, and $\Sigma mx^2 = I_a$, the moment of inertia of the rod about A.

$$\therefore \text{Whole torque} = \ddot{\theta}[MRs - I_a].$$

This is zero (whatever value $\ddot{\theta}$ has), if

$$MRs = I_a.$$

But

$$I_a = I_g + Ms^2 = Mk^2 + Ms^2.$$

$$\therefore Rs = k^2 + s^2$$

or

$$k^2 = (R - s)s.$$

Consequently, if the radius of gyration of the rod is such that

$$k^2 = (R - s)s,$$

there will be no torque due to traverse acceleration, i.e. the rod will *continue* to point towards O, when the table alters its rate of rotation.

Now for a rod of length l, $k^2 = \tfrac{1}{12}l^2$,

$$\therefore \tfrac{1}{12}l^2 = (R - s)s,$$

or

$$l^2 = 18Rs - 12s^2.$$

Numerical example (fig. 14).—$R = 50$ cm., $s = 5$ cm., $l = 52$ cm. approx.

5. Use of the Gravity Pendulum to Indicate the Vertical in Vehicles Subject to Acceleration.

In order to navigate an aeroplane over great distances when the earth's surface may be hidden by clouds or mist (as e.g. in transatlantic flights), it is necessary that the *vertical* direction should at all times be known with certainty. Otherwise a pilot cannot even tell whether he is flying in a straight line or in a curve when the ground is invisible. His muscular sense and the position of his body are absolutely unreliable; they merely indicate the direction of the resultant of his weight and the centrifugal force, never the true vertical, i.e. the direction of the centre of the earth.

An observer at rest on the earth ascertains the vertical by using the gravity pendulum as a plummet. At first sight it seems futile to use the gravity pendulum in this way in an accelerated vehicle. Everyone is familiar with the behaviour of pendulums in ordinary vehicles; for example, a strap hanging from the rack in a railway carriage swings about freely, making no resistance to the forces of inertia. Nevertheless, it is *theoretically* possible to use a gravity pendulum to indicate the vertical in a vehicle accelerated in any way. The reason is as follows: any acceleration of the motion of a vehicle may be resolved into a vertical component and a horizontal one. Vertical accelerations merely have the effect of accelerating the point of suspension of the pendulum in the direction of the length of the pendulum; that is, they have no effect on a pendulum in its mean position. There remains the *horizontal* component of the acceleration.

Now comes the decisive point. A horizontal motion, i.e. a motion parallel to the surface of the earth, really takes place *not* in a *straight* path, but in a *circular* path about the centre of the earth. This statement is quite independent of the rotation of the earth on its axis and would be true even for a stationary earth. For every horizontal motion is in a line parallel to a great circle of the earth, i.e. in the last resort is a motion on a rotating table even with a stationary earth. Hence we can now go straight back to the remarkable experiment discussed in last section.

It is only necessary to give the pendulum the moment of inertia required by equation (4), where R is put equal to the earth's radius ($6.4 \cdot 10^8$ cm.). That is, we refer back to the diagram of fig. 15: R denotes the radius of the earth, so that s is obviously negligible compared with R. We have

$$I_0 = msR. \qquad \ldots \ldots \ldots (4b)$$

In a gravity pendulum, in contradistinction to an oscillating spring, the moment of inertia is definitely connected with the constant μ'. The choice of the latter is no longer unrestricted; in a gravity pendulum its value depends on the weight. By p. 107

$$\mu' = mgs.$$

Hence by equation (11), p. 107, the period of oscillation of this pendulum is given by

$$\tau = 2\pi \sqrt{\frac{I_0}{\mu'}} = 2\pi \sqrt{\frac{msR}{mgs}} = 2\pi \sqrt{\frac{R}{g}}:$$

$\tau = 84$ minutes, corresponding to a simple pendulum (p. 66) of length equal to the radius of the earth!

Unfortunately the construction of an actual pendulum with as long a period of oscillation has hitherto proved impossible. Even with the use of suspended gyrostats the greatest period of oscillation (of precession) so far attained is 15 minutes. Such a pendulum, it is true, represents a fair approximation to the theoretically required ideal, but it is only an approximation. With its present period of oscillation even the slowest gyrostatic pendulum is but a makeshift. In unfavourable circumstances, especially with certain periodically-recurring accelerations, it may fail to act. The real problem of the " artificial horizon " will only be solved by the construction of a gravity pendulum with a period of 84 minutes (M. Schuler).

6. The Earth as an Accelerated System of Reference: Centrifugal Forces on Bodies at Rest.

As a final example of an accelerated system of reference we shall discuss the earth as a merry-go-round, taking account of the daily rotation of the earth relative to the fixed stars. A complete rotation takes place in 86,140 seconds. The angular velocity of the earth is therefore small; it is given by

$$\omega = \frac{2\pi}{86,140} = 7 \cdot 3 \cdot 10^{-5}. \quad \ldots \ldots \quad (6)$$

This angular velocity ω gives rise to a centrifugal force $F = m f_z$ or centrifugal acceleration f_z acting on every body at rest on the surface of the earth. Let the body be in geographical latitude ϕ (fig. 16). Let $r = R \cos\phi$ be the radius of the corresponding circle of latitude. Then the centrifugal acceleration is

$$f_z = \omega^2 r = \omega^2 R \cos\phi = 0 \cdot 03 \cos\phi \; \text{m./sec.}^2 \; (\text{approx.}). \quad (7)$$

This centrifugal acceleration is directed outwards along the radius of the circle of latitude (r). Along the vertical, i.e. the direction of the radius of the earth (R), there is only a component of this centrifugal acceleration, namely,

$$f_R = f_z \cos\phi = 0 \cdot 03 \cos^2\phi \; \text{m./sec.}^2. \quad \ldots \quad (8)$$

It is directed outwards from the centre of the earth, being opposed to the " gravitational acceleration " g_0 wholly due to the weight. Hence on the *rotating* earth the gravitational acceleration in latitude ϕ must be a little less than it would be if the earth were at rest. We obtain

$$g_\phi = g_0 - 0 \cdot 03 \cos^2\phi \; \text{m./sec.}^2 \quad \ldots \quad (9)$$

Here g_0 is the value of the gravitational acceleration for the earth *at rest*.

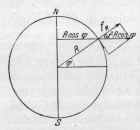

Fig. 16.—Weight and centrifugal force at the earth's surface in latitude ϕ.

There is, however, an additional complication. The action of the centrifugal force is by no means confined to bodies on the surface of the earth. Every particle of which the earth is composed is subject to a centrifugal force acting radially outwards in the plane of its parallel of latitude. The net effect of all these forces is an elastic deformation of the earth's surface. The earth is slightly flattened at the poles, the polar diameter being shorter by about 1/300 part than the equatorial diameter. Owing to this flattening the change in the gravitational acceleration g_ϕ with latitude (ϕ) is greater than that calculated from equation (9). Empirical observations lead to the equation

$$g_\phi = (9 \cdot 832 - 0 \cdot 052 \cos^2 \phi) \text{ m./sec.}^2. \quad . \quad . \quad (9a)$$

At sea-level in latitude 45° we find that $g = 9 \cdot 806$ m./sec.². The correction term attains its maximum value for $\phi = 0°$, i.e. at the equator. The correction then amounts to five parts in a thousand; that is, in many experiments it may be neglected without serious error. Nevertheless, a pendulum clock at the equator loses as much as 3·5 minutes a day as compared with a similar clock at the pole.

The flattening by 1/300 applies to the solid parts of the earth. The deformation of its liquid covering, the oceans, by the centrifugal forces is much greater. This deformation, however, is never manifested by itself, but has superposed on it the attractions of the moon and the sun on the water, which vary periodically during the day. The oceans are also much more affected by these attractions (cf. p. 68) than the solid parts of the earth. The superposition of centrifugal forces and attractions gives the complicated phenomena of the ebb and flow of the tides. The problem is one of " forced vibrations " (Chap. XI, § 11, p. 253), which meanwhile we can only refer to in passing.

7. The Earth as an Accelerated System of Reference: Coriolis Forces on Moving Bodies.

In fig. 17 *HH* denotes the horizontal plane of an observer at a point in geographical latitude ϕ. This plane revolves about the vertical R with angular velocity equal to the vertical component of the angular velocity (ω) of the polar axis; that is,

$$\omega_v = \omega \sin \phi = 7 \cdot 3 \cdot 10^{-5} \sin \phi. \quad . \quad . \quad . \quad (10)$$

By equation (2), p. 136, a body moving in this horizontal plane with linear velocity v is subject to a Coriolis acceleration

$$f = 2v\omega \sin \phi.$$

For an observer looking towards the north pole the earth appears to rotate counterclockwise; that is, the direction of rotation is that of the rotating stool in § 3 (p. 135).

Among the many and varied experimental arrangements for demonstrating the existence of the Coriolis acceleration in the *horizontal* plane, the Foucault pendulum is distinguished by its simplicity. The principle on which it depends has already been illustrated on p. 137 for the case of a rotating table. The pendulum describes the rosette shown in fig. 10.

Any gravity pendulum consisting of a ball on a thread and placed at the earth's surface will describe a similar rosette. Except at the poles, however, a day is not sufficient for the rosette to close, a somewhat longer period, 24 hours/sinϕ, being required. In London ($\phi = 51°$ 30') the time taken is about 30½ hours.

There is no difficulty in verifying the occurrence of these phenomena (rosette-shaped path described by the pendulum, deviation to the right, time taken to describe the complete rosette) in any lecture-room. A common form of the experiment is shown in fig. 18. The essential part of the apparatus is a good astronomical objective, which produces a greatly magnified image of the thin suspending thread when

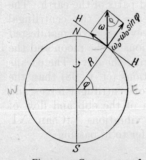

Fig. 17.—Components of the Coriolis force at the surface of the earth.

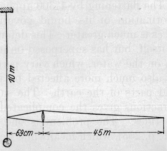

Fig. 18.—Apparatus for demonstrating the rosette-shaped path described by a long gravity pendulum at the surface of the earth (Foucault's pendulum).

it passes through the extreme points of the rosette loops. The figure contains the necessary numerical data. If these dimensions are adopted the extreme points of the successive loops of the rosette follow one another at intervals of about 2 cm. in the magnified image. The experiment is quite a simple one.

We shall further mention some qualitative examples of the deviation to the *right* due to the horizontal component of the Coriolis acceleration in our (*northern*) hemisphere.

(*a*) In all double-line railways the right-hand rail of each track wears away more rapidly than the left, because the rims of the wheels are pressed hard against it by the Coriolis force.

(*b*) Right banks of rivers are more undermined than left.

(*c*) Projectiles are always deviated to the right, even apart from the phenomenon illustrated in fig. 46, p. 127.

(*d*) Air flows into barometric depressions along curved paths and hence vortices are formed (cyclones).

It is equally easy to demonstrate the vertical component of the Coriolis acceleration. The corresponding component of the angular velocity of rotation of the earth (ω) is $\omega \cos \phi$ about the tangent to the meridian of the point of observation.

Among quantitative experiments we may mention the rotating Eötvös balance. A sensitive balance is loaded and placed on a table rotating about an accurately vertical axis. The mass which at any time is moving eastwards appears lighter. In order to obtain large deflections the time of rotation of the vertical axis is made equal to the period of vibration of the balance beam. The deflections of the balance are thereby greatly increased owing to the well-known phenomenon of "resonance" (Chapter XI, § 11, p. 253).

Among qualitative experiments we may mention the deviation to the right of a falling stone. For this experiment, however, a large drop is required and it is best carried out in the shaft of a mine.

8. The Gyroscopic Compass in Ships and Aircraft and its Inevitable Theoretical Error.

We shall conclude our chapter on accelerated systems of reference with an account of the very important practical application of the Coriolis forces in the gyroscopic compass. The principle of the gyroscopic compass is easily understood from the experiment of fig. 11, p. 138. In practical forms of the apparatus a gyrostat with a horizontal axis is suspended to form a gravity pendulum. The gyroscopic compass is primarily intended to be used in ships and aircraft, and this fact gives rise to questions which are interesting from the physical point of view. The difficulties are at bottom the same as those

Fig. 19.—Error of the gyroscopic compass in a moving vehicle. An electric motor with reduction gear Z is placed on a rotating table; the disc G is thereby made to rotate with a small angular velocity ω_2. This disc is meant to represent a great circle of the "earth" indicated by dotted lines. The gyrostat has a ring-shaped frame (instead of the fork-shaped one in fig. 11, p. 138), which is free to rotate about the axis A, which lies along the radius of the great circle G, and is therefore normal to the "horizon" of the point where the gyrostat is placed. The leads to the motor are shown at S; the gyrostat is that shown in figs. 36 and 37, Plate IV. (1/22.)

involved in the use of an ordinary pendulum to indicate the vertical. *Vertical* accelerations of the vehicle are without effect, but any *horizontal* acceleration of the vehicle gives rise to forces of inertia which throw the axis of the gyrostat out of its mean position (near the meridian). These disturbances, though in themselves very considerable, could be got rid of by the device which would make an ordinary pendulum capable of indicating the vertical even in a vehicle subject to horizontal acceleration. The gyroscopic compass would have to be constructed with a period of oscillation of 84 minutes; it would then be entirely insensitive to all accelerations of the ship or aircraft (starting, stopping, going round curves). A further device (simultaneous use of three gyroscopes) would eliminate disturbances due to rolling.

Nevertheless, a *theoretical* error in the indications of the gyroscopic compass is inevitable even with a perfectly constructed instrument. The velocity of the ship itself gives rise to an error in the indication of the compass. The way in which this error arises may be illustrated by fig. 19. The ship is supposed to be moving along a great circle of the earth.* The "earth" is indicated by dotted lines, the edge of the metal disc G being taken as the great circle. The motion of the ship

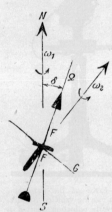

Fig. 20.—Figure to illustrate the preceding after a rotation of 90° about the axis OZ (this axis is shown broken off above Z).

in this great circle is a rotation about the axis ZO which passes through the centre of the earth O and is at right angles to the plane of the circle G. This rotation of the ship (with angular velocity ω_2) combines with the rotation of the earth on its axis (with angular velocity ω_1) to give a resultant rotation (fig. 19). The resultant rotation has angular velocity Ω, and its axis—here is the important point—always deviates from the polar axis of the earth, except for the special case where the ship is moving exactly along the equator. In short, the compass is on a merry-go-round whose axis of rotation is the arrow Ω. Hence, FF, the axis of symmetry of the gyrostat or the arrow parallel to it, takes up a position in the plane passing through the arrow Ω. In fig. 19 this plane does pass through the polar axis of the earth. For here the ship and its gyrostat are at a point of nearest approach to the pole. At all other points of the great circle G, e.g. in fig. 20, this is not the case, and the gyrostat shows a considerable error (δ). To demonstrate this error we "stop the ship" by stopping the electric motor (making ω_2 zero). The axis of the circle then drops back into the meridian plane, moving through the angle δ. It is only at the two points of nearest approach to the poles (fig. 19 shows that nearest to the north pole) that δ is zero. In modern liners this error rarely exceeds 3°. In aeroplanes, with their much higher velocity, it is correspondingly greater. The error can of course be taken account of only by calculation, correction tables being used, just as with the old magnetic compass. These tables give the magnitude of the error for various

---

* In this connexion the student should re-read the third paragraph of § 5 (p. 144).

localities on the earth's surface and for various velocities and directions of motion.   In spite of this inevitable theoretical error the modern gyroscopic compass represents a tremendous practical advance, for it is absolutely free from disturbance due to neighbouring iron parts. It also has a higher torsional rigidity than the magnetic compass.   It can easily be used to drive "daughter" compasses in large numbers and even the steering-gear of the ship.

CHAPTER IX

Liquids and Gases at Rest

1. Preliminary Remarks.

Our next subject is the motion of liquids and gases. As a preliminary we require to have a general idea of the physical properties of liquids and gases at rest. These we shall discuss in a fairly summary manner, as most of the details (vessels in communication, Hero's fountain, &c.) are dealt with very fully in school science courses. In an attempt to bring out the connexions between the various isolated facts more clearly, we shall be guided by one or two ideas derived from atomic theory.

We subdivide the material according to the following scheme:

(a) The nature of liquids:

    (1) The freedom to move and close packing of the molecules of a liquid (§ 2, pp. 152–156).

    (2) Consequences of these (§§ 3–8, pp. 156–166).

    (3) The internal friction, "tenacity", and surface tension of liquids (§§ 9–10, pp. 166–175).

(b) The nature of gases:

    Gases as fluids with a low density and no surface (§§ 11–18, pp. 175–191).

In selecting the material we have chiefly considered what light the individual facts may throw on the phenomena of the motion of liquids and gases. A number of points which are merely touched on or omitted altogether in this chapter are more appropriately dealt with in the subject of Heat.

2. The Free Movement of Molecules in a Liquid: the Brownian Movement.

The idealized form of a solid body is the *crystal*. The characteristic feature of the crystal is the arrangement of its elementary constituents (atoms, molecules, or ions) in a space lattice with a definite geometrical structure. In a rough formal way a two-dimensional crys-

tal model may be sketched as in fig. 1, the elementary constituents
being symbolically represented by balls and the forces between them
by springs. We consider a crystal as a lattice structure in equilibrium,
formed of a large number of our well-known spring pendulums (p. 50).
The masses of these elementary pendulums or oscillators are not in
equilibrium but vibrate continually, moving in the complicated paths
of Lissajous figures in space (p. 62). This is the motion to which our
sense of touch reacts with the sensation of " heat ". According to the
atomic conception of matter the well-
known loss of kinetic energy in all
mechanical processes is only apparent.
Part of the kinetic energy of the large-
scale visible bodies passes over to the
invisible elementary pendulums; " the
body becomes hot as a result of friction".

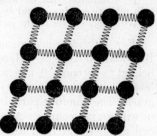

The heating-up of a crystal means
an increase in the kinetic energy of its
elementary pendulums or oscillators.
The amplitudes of the vibrations increase as the temperature of the crystal

Fig. 1.—Two-dimensional crystal model

rises.   If the amplitude increases beyond a certain value the limits of
stability are exceeded. The complicated lattice structure collapses
like a house of cards. The crystal melts and becomes a liquid. Sharp-
ness of melting point is an important property characteristic of a
crystal.

According to this conception of the melting process, the main
features of which have a firm experimental basis, we have *a priori*
to reckon with a considerable stock of kinetic energy in the molecules
of the liquid. We must assume that the molecules of the liquid are
continually in rapid motion. The oscillatory motions about a position
of equilibrium in the lattice must be replaced by motions of translation and rotation on the part of the individual molecules.

We have available a reproduction of these thermal motions in
liquids, which, though on a large scale, is no doubt to a great extent
true to nature, namely, the phenomenon known as the Brownian movement.

The principle may be illustrated by an analogy which is positively
childish in its simplicity. Imagine a dish full of live ants, and suppose
the onlooker is short-sighted or at a considerable distance. He cannot
distinguish the individual insects crawling about, but merely sees a
featureless brownish-black surface. A simple device, however, greatly
assists the observer. A number of larger and lighter bodies of a con-
spicuous colour, say down or bits of paper, are thrown into the dish.
These " particles " do not remain at rest. Pulled and pushed about
by invisible individuals, they move about and rotate irregularly. The

onlooker sees a large-scale picture of the movements in the seething mass of insects.

The Brownian movement is observed in a similar way, but a fairly powerful microscope is necessary. Between slide and cover-glass we interpose a drop of liquid, in the simplest case water, to which a fine insoluble powder has previously been added; it is convenient to use e.g. a minute amount of Indian ink, which consists of very finely ground carbon.

To demonstrate the Brownian movement to a large audience a powder of high refractive index should be used, e.g. the mineral rutile (TiO_2). The high refractive index gives bright pictures with strong contrast.

Few physical phenomena are so fascinating to the observer as the Brownian movement. Here the observer is for once granted a peep behind the scenes. A new world opens out before him, full of the restless and bewildering activity of a multitude of individuals too great to grasp. The minutest particles shoot across the field of view like arrows, in wild zigzag courses. A larger number advance in a slow and dignified manner, but their directions are also changing continually. The majority merely reel about in one neighbourhood. Their abrupt changes in direction clearly indicate the existence of rotatory motions about axes whose direction is constantly changing. Never does a trace of system or order manifest itself. The rule of blind, lawless chance—such is the overwhelming impression stamped on the mind of the unsophisticated observer. The Brownian movement is indeed one of the fundamental phenomena in the realm of modern observational technique. No description in words can even approximately replace the effect of personal observation.

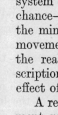

Fig. 2. — Advance of the boundary between two liquids as a result of diffusion.

A really effective presentation of the Brownian movement requires a magnification of several hundred. This magnification readily leads one to over - estimate the observed velocities. This error may be avoided by another method of observation, in which the particles suspended in the liquid are made visible as a *swarm* or *cloud* but individual particles can no longer be distinguished. Fig. 2 shows water containing fine particles, e.g. an extremely dilute solution of Indian ink, with a layer of clean water superposed, the boundary between the two liquids being sharply defined.

There are numerous ways of attaining this. One way is to pour in the lower liquid first, and to place a thin flat disc of cork on top, and then to pour the clean water carefully on to this in a fine stream.

At the same time we set up a parallel experiment. The average diameter of the fine particles in Indian ink is about $0{\cdot}5\,\mu$.* For the

* [μ is the symbol for a *micron*, i.e. 10^{-3} millimetre.]

parallel experiment we take much smaller floating particles, namely, the molecules of some dye *soluble* in water. In both cases the boundary between the coloured water and the uncoloured becomes blurred in course of time. The Brownian or thermal motions of the particles of carbon or dye carry them into the water which was previously clean. The colouring matter diffuses into the clean water. That is, *diffusion* and *Brownian movement* are two names for the same thing. Both refer to the turmoil of the thermal motions of the elementary particles of a liquid. If we use the term Brownian movement we imply *microscopic* observation of certain individual particles distinguished by their great size. In the case of macroscopic observation we speak of *diffusion*, quite independently of the size of the individual particles. That is, the particles visible as a swarm or cloud may be particles of dust or tiny molecules of colouring matter beyond the reach of any microscope.

In our connexion the essential point is the *velocity* with which this diffusion process takes place. The boundary of the swarm advances surprisingly slowly. It takes days or even weeks, according to the size of the particles of colouring matter, to move through a measurable distance.

The slowness of the diffusion process is due to the close packing of the seething mass of individual molecules in the liquid. In liquids the mean distance between the molecules is of the same order as for the corresponding (crystalline) solid. This follows from two facts:

(*a*) The *density* of the liquid usually agrees with that of the solid to within a few per cent. By density we mean the mass of unit volume, as a rule the mass in grammes of a cubic centimetre of the substance. Any method of measuring mass (weighing) combined with any method of measuring volume gives a method for measuring density. It is not necessary to recount all these methods and give special names to them. We shall confine ourselves to one or two numerical results.

Substance	Temperature	Density	
		Solid, gm. (mass)/cm.^{3}	Liquid, gm. (mass)/cm.^{3}
Mercury	$-38.9°$	14·2	13·7
Sodium	$+97.6°$	0·95	0·93
Sodium chloride ..	$+800°$	—	1·55
Water	$0°$	0·92	1·00

If the units kg.-mass and metre are used the numbers given have to be multiplied by 10^3, e.g. water has the density 1000 kg.-mass/m.^{3}

(*b*) Liquids are only very slightly compressible. Numerical data vouching for the truth of this statement will be found on p. 161.

Bearing these considerations in mind we can easily replace an actual liquid by a *model* liquid, by means of which we may study the characteristic properties of liquids. An excellent model would be a vessel full of live ants or rounded beetles with hard wing-cases. But a vessel full of small smooth steel balls will suffice, except that the proper motions (thermal motions) of these model molecules must be reproduced rather clumsily by shaking the whole vessel. In future we shall not always mention this shaking.

The chief characteristic of this model liquid is the freedom of all the liquid molecules to move relative to one another. This freedom of movement immediately explains three important facts:

(1) the shape of the surface of a liquid (§ 3, pp. 156–157);

(2) the distribution of the pressure in a liquid (§§ 4, 5, pp. 157–161);

(3) the distribution of pressure in a gravitational field, and the upward thrust of a liquid (§§ 6–8, pp. 162–166).

3. The Shape of the Surface of a Liquid.

The surface of a liquid is always at right angles to the direction of the force acting on its molecules. Examples:

(*a*) In the case of a liquid in a broad shallow dish the *weights* of the individual molecules of liquid are the only forces acting, and the surface takes the form of a *horizontal* plane. When the model liquid is poured into a dish the molecules are free to slide over one another and only come to rest when the centre of gravity of the whole reaches its lowest possible position.

(*b*) In the basins of lakes and seas the directions of the weights at different points can no longer be regarded as parallel, the weight being always directed radially towards the centre of the earth. Hence their surfaces form parts of a spherical surface.

(*c*) In a vessel rotating about a vertical axis the surface of the liquid assumes the form of a *paraboloid*. We shall consider the motion of a molecule of the liquid from the standpoint of an accelerated system of axes. Each individual particle (molecule) is acted on by two forces: mg, the weight of the particle, acting downwards, and $m\omega^2 r$, the centrifugal force, acting outwards. The two forces combine to give the resultant R. The surface takes up a position at right angles to this resultant. By fig. 3, we have

$$\frac{dr}{dz} = \tan\alpha = \frac{mg}{m\omega^2 r}, \quad \ldots \ldots \quad (1)$$

or
$$\frac{g}{\omega^2} dz = r\, dr,$$

whence   const. $z = r^2$, $\quad \ldots \ldots \ldots \ldots \quad (2)$

which is the equation of a parabola.

For the sake of clearness we shall also describe the same phenomenon from the standpoint of the system of reference we used formerly (the earth or the floor of the lecture-room).

Every particle of liquid must move in a circular path of definite radius r. This requires the existence of a radial force $m\omega^2 r$ acting radially on the particle *towards* the axis of rotation. This is supplied by the weight with the co-existence of a " slope ". This " slipping down the slope " is very clearly shown in the snapshot of the model liquid in fig. 4. For a given angular velocity ω the steepness of the slope must increase with the distance r between the particle and the axis of rotation. The paraboloidal " slope " merely provides a new solution of a problem we have already met with, namely the production of a radial force proportional to the radius r, a solu-

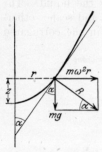

Fig. 3.—Paraboloidal surface of a rotating liquid.

Fig. 4.—Parabolic section of a rotating model liquid consisting of steel balls (instantaneous photograph).

tion which takes the place of our previous solution by means of a spring and the linear law of force on p. 43. This new solution may be demonstrated very simply with a spherical bead on a wire bent into the shape of a parabola. If the wire rotates with a suitable angular velocity the bead when struck backwards or forwards with a stick remains in equilibrium at any height.

4. Pressure in a Liquid: Manometers.

In general we define *pressure* as the quotient force/area. The chief pressure units are as follows (see also p. 163):

1 atmosphere $= 1 \cdot 0333$ kg.-force/cm.$^2 = 1 \cdot 013 \cdot 10^5$ large dynes/m.^{2} $= 1 \cdot 013$ bar $= 14 \cdot 7$ lb.-force/sq. in.

1 lb.-force/sq. in. $= 0 \cdot 068$ atmosphere $= 6 \cdot 9 \cdot 10^3$ large dynes/m.$^2 =$ $6 \cdot 9 \cdot 10^4$ dynes/cm.$^2 = 6 \cdot 9 \cdot 10^{-2}$ bar.

1 bar $* = 10^3$ millibars $= 10^6$ dynes/cm.$^2 = 10^5$ large dynes/m.^{2}

The order of magnitude of the pressures which occur in ordinary life is usually underestimated. The finger when shielded by a thimble can easily exert a force of a kilogram on a needle. Suppose the point of the needle has a diameter of about 0·2 mm., i.e. a cross-section of the order of 4.10^{-4} sq. cm. Then the point of the needle forces its way through the material being sewn with a pressure of about

$$\frac{1 \text{ kg.-force}}{4.10^{-4} \text{ sq. cm.}} = 2500 \text{ kg.-force/sq. cm. or 2500 atmospheres.}$$

* In American and Continental usage 1 bar = 1 dyne/cm.^{2}, and the terms *megabar, kilobar* are accordingly used in place of the British *bar, millibar*.

Again, a razor-blade exerts a pressure of about 10^4 atmospheres on the hairs. What one has to pay attention to is not the pressure but the product of pressure and area.

Having defined pressure as above, we may now proceed to the construction of *pressure gauges* or *manometers* for liquids. Fig. 5 shows a piston arranged so as to move with very little friction in a hollow cylinder which forms part of the vessel containing the liquid. The piston is attached to a spring balance with pointer and scale; otherwise, the piston and spring may be combined, as in the corrugated

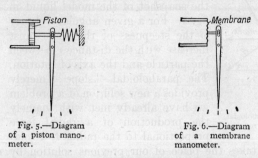

Fig. 5.—Diagram of a piston mano-meter.

Fig. 6.—Diagram of a membrane manometer.

membrane in fig. 6 (or even a flat membrane will suffice). The pressure makes the membrane bulge, and this movement actuates the pointer. Uncalibrated, these instruments only enable us to tell whether pressures at different places or times are equal or not. A method for calibrating them will be described on p. 159.

Now that we have manometers available (even if still uncalibrated), we shall proceed to investigate the distribution of pressure in liquids. For the sake of simplicity we distinguish between two limiting cases:

(1) The pressure is due entirely to the weight of the liquid itself (which we may call *gravity pressure*).

(2) The liquid is in a vessel which is entirely closed; a cylinder and piston attached to it are used to produce a pressure compared to which the gravity pressure may be neglected as insignificant. (We may refer to this as *piston pressure.*)

We shall begin with the second limiting case (§ 5).

5. Equality of (Piston) Pressure in all Directions.

Fig. 7 shows an iron vessel of complicated shape, entirely full of water and provided with four ordinary pressure gauges all exactly the same. On the right a piston is pressed into the vessel by means of a screw. All the (uncalibrated) pressure gauges show the same deflection and accordingly demonstrate that the pressure is the same in all directions.

In order to understand this, we may imagine the model liquid (steel balls) filled into a sack and a piston pushed in through a suitable hole. The sack is blown out in all directions. Owing to the free movement of the steel balls no preference for any particular direction can arise.

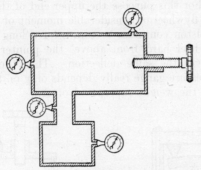

Fig. 7.—Distribution of pressure in a liquid when piston pressure predominates (1/6.)

We shall now give three important applications of the fact that piston pressure is the same in all directions.

(1) *Calibration of a Pressure Gauge.*—The pressure gauge considered is of a type very much used in practice, which we have not mentioned hitherto. In fig. 8 the flexible membrane is replaced by a flat tube R of elliptical cross-section. When subjected to the pressure

Fig. 8.—Calibration of a pressure gauge by means of a rotating piston K (1/10.)

of a liquid the tube expands and thereby causes a pointer to move (compare the well-known children's toy consisting of a flat " elephant's trunk " of paper which in the state of equilibrium is rolled up in a coil). The pressure gauge is joined by tubing of some kind to the cylinder Z, into which a piston K fits. All the hollow space is filled with some liquid, e.g. an oil. Pressure is the product of force and area. Thus the pressure exerted by the piston is equal to the weight of the piston itself plus the weight superposed on it, divided by the cross-section of the piston. Now here comes the essential point: the friction between the cylinder and the piston must be got rid of. This is done by letting the piston be continually bathed in a thin skin of liquid, a state of affairs attained by making the piston revolve uniformly about its vertical axis.* For this purpose the upper end of the piston is made in the form of a flywheel of considerable moment of inertia; once set in motion, the piston continues to rotate for a long time. If we hit the moving flywheel hard from above, the pointer of the pressure gauge always gives the same deflection. Thus the position of the pointer of the pressure gauge really depends only on the weight of the piston itself and the weight superposed on it.

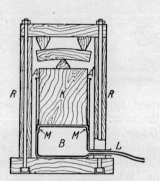

Fig. 9.—Improvised hydraulic press (1/20.)

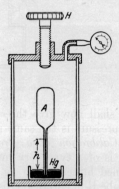

Fig. 10.—To illustrate the compressibility of water (1/8)

(2) *The Hydraulic Press.*—This machine is of great practical importance, and is used to produce large forces with the help of quite small pressures. An improvised form is shown in fig. 9. Its essential parts are a cylindrical cooking-pot A, a thin-walled rubber balloon B, a wooden piston K, and a solidly-made rectangular frame R. The filling-tube of the rubber balloon emerges through a side hole L, and is connected to the water mains by a thick rubber tube. A leather collar M round the edge of the piston prevents the formation of culs-de-sac between the piston and the wall of the pot.

Numerical Example.—The pressure of the water in the lecture-room is about 4 atm. (58·7 lb.-force/sq. in.). The pot used has an inside diameter of about 12 in., i.e. the cross-section of the piston is about 113 sq. in. Hence the press gives a force of about 6630 lb., and will snap oak blocks 1·5 in. by 2 in. and of length 16 in.

(3) *The Compressibility of Water.*—The small compressibility of liquids was mentioned on p. 155 merely as a hypothesis. The equality of liquid pressure in all directions enables us to verify this hypothesis

---

* This experiment at the same time illustrates the fact that the lubrication of bearings depends on "creeping" motion (see § 2 of next chapter, p. 192).

quantitatively. The principle adopted is as follows: a liquid is forced at high pressure into a measuring vessel, care being taken to prevent the measuring vessel bulging out like a balloon by having the vessel surrounded by a liquid at the same pressure. We thus have the arrangement shown in fig. 10.

A is the measuring vessel filled with the liquid under investigation. As an example we choose water. The vessel ends in a narrow tube of known cross-section q, open at the bottom and dipping into a small dish of mercury which acts as a barrier liquid. The measuring vessel and the dish of mercury are enclosed in a wide glass cylinder, which is entirely closed and filled with water also. A screw worked by the knob H enables us to push a piston into the water and so to exert pressure on the water, the amount of the pressure being read off on a pressure gauge in pounds per square inch or atmospheres. As the pressure rises the barrier liquid rises in the glass tube. A rise of h cm. means a decrease of $\Delta V = hq$ sq. cm. in the volume of the water contained in the measuring vessel. Actual measurements of this kind show that water can only be compressed to a very trifling extent. The decrease of volume ($\Delta V/V$) for an increase of pressure of one atmosphere is only about 5.10^{-3} per cent. Only at a pressure of 1000 atmospheres does the decrease in volume of water under pressure amount to as much as 5 per cent. This very slight compressibility of water gives rise to a variety of striking experiments, in all of which large forces and pressures are produced by only a trifling compression. We may mention one of these.

Suppose we have a fairly watertight rectangular box without a lid, filled with water, the surface of the water being free to the air. A shot is fired through the side of the box. The water is thereby compressed by an amount corresponding to the volume of the bullet, for there is no time for the water to get out of the way by rising. Considerable pressures result, and the box is reduced to fragments.

The following variant of this experiment is easier to perform. A Rupert's drop is made to burst in a glass beaker. Rupert's drops are made by dropping molten glass into water. They are solid glass drops which, owing to rapid cooling, are in a state of great internal strain (fig. 11). Rupert's drops are very insensitive to blows, and will stand being knocked about with a hammer. On the other hand, they cannot stand the least damage to their thread-like tails. If the tip of the tail is broken off the drop bursts with a loud report. A Rupert's drop may be exploded in this way in the closed fist; then one distinctly feels the

Fig. 11.—Rupert's drops (1/4.)

particles flying apart, although they cause no pain or injury. The harmlessness of this experiment in the hand is in surprising contrast to the complete destruction of the beaker filled with water.

6. Distribution of Pressure in a Gravitational Field.

Suppose we have an upright cylindrical vessel of cross-section A (fig. 12) filled to a depth h with a liquid of density ρ. That is, the mass of liquid in the vessel is $m = Ah\rho$. In dynamical units the weight of this column of liquid is

$$W = mg = Ahg\rho. \quad . \quad . \quad . \quad . \quad . \quad (3)$$

Dividing the weight by the area of the cross-section, we obtain the pressure P acting uniformly over the bottom of the vessel:

$$P = \frac{W}{A} = hg\rho. \quad . \quad . \quad . \quad . \quad . \quad (4)$$

This argument involves a definite *assumption*, which, however, is satisfactorily confirmed by experiment: i.e. that the density of the

Fig. 12. — To illustrate gravity pressure in a liquid.

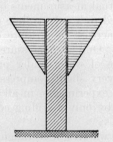

Fig. 13.—To illustrate pressure on the base of a vessel.

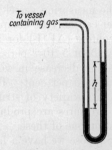

Fig. 14.—Pressure gauge depending on the use of a liquid.

liquid is the same throughout the column. Owing to the minute compressibility of liquids the density of the bottom layers is not noticeably increased by the pressure of the upper layers resting on them. The density of water is $\rho = 1000$ kg.-mass/m.^{3}, and $g = 9.81$ m./sec.^{2}, so that, by equation (4), the pressure of a column of water 1000 metres high on its base is $1000 . 1000 . 9.81 =$ about 10^7 large dynes/m.$^2 =$ about 100 atmospheres. But this pressure only compresses the lowest layer of water by $\frac{1}{2}$ per cent (see p. 161).

The shape and cross-section of the vessel do not enter into equation (4). This, as we may easily see, is still true for non-cylindrical vessels of any shape. Fig. 13 shows a cylindrical wooden block and a cone with a hole through the middle, which can be slid up and down. The weight of the cone cannot possibly have any effect below the bottom of the cylinder. The case of a funnel-shaped vessel filled with liquid is quite analogous. The molecules of liquid are free to move relative to one another, and thus slide along the imaginary boundary surface. The independence of the shape of the vessel may be similarly proved in all other cases. We thus obtain the important theorem: the gravity pressure at any point of a liquid depends only on the *vertical depth* of the point below the surface of the liquid, and is given quantitatively by equation (4).

Among the many important applications of this theorem we may mention the well-known type of manometer depending on a liquid, which is used to measure the pressure of a gas or vapour. The simplest practical form consists of a glass U-tube containing a liquid of suitable density (fig. 14). Water or mercury is most commonly used as the barrier liquid. Manometers of this type may, of course, be graduated in terms of the ordinary pressure units, e.g. pounds per square inch. As a rule, however, we content ourselves with the difference of level of the liquid in the two limbs of the tube, and speak e.g. of a pressure of 4 in. of water. The conversion factors are immediately obtained from equation (4). It is merely necessary to know the density of the liquid. The following data are frequently used:

1 atmosphere = 14·7 lb.-force/sq. in. = 760 mm. (30 in.) of mercury.

1 bar = 10^5 large dynes/m.^{2} = 10^6 dynes/cm.^{2} = 750·06 mm. of mercury.

7. The Upward Thrust of Liquids on Solid Bodies immersed in them.

The most familiar consequence of the way in which the pressure is distributed in a gravitational field is the *statical upward thrust* on bodies in a liquid. We shall first consider the upward thrust on a *solid* body immersed in the liquid. For the sake of simplicity we assume that it has the form of a flat cylinder (fig. 15). The pressure of the liquid is the same in all directions. This is a consequence of the freedom of the molecules of the liquid to move relative to one another. Hence there is an *upward* pressure $p_1 = h_1 g \rho$ on the bottom of the cylinder and a *downward* pressure $p_2 = h_2 g \rho$ on the top. All the pressures on the curved surface of the cylinder cancel one another in pairs, so that there remains only the difference of the two pressures, $(p_1 - p_2)$. This, multiplied by A, the area of the base of the cylinder, gives an upward force F acting on the body, which is called the *upward thrust*. We have

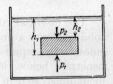

Fig. 15.—To show how the upward thrust of a liquid arises.

$$F = g \rho A (h_1 - h_2). \quad . \quad . \quad . \quad . \quad (5)$$

But the product on the right-hand side is the weight of a mass of liquid having a volume equal to that of the immersed body. In this way and in other ways we obtain the general result that *the upward thrust on a body immersed in liquid is equal to the weight of the liquid displaced.*

The liquid is subject to a downward force, the reaction to the upward thrust. This is shown by means of the apparatus of fig. 16, which requires no further explanation.

Many quantitative experiments on upward thrust or buoyancy have been devised. Instead of describing these we shall make use of our model liquid to discuss how the upward thrust arises. Fig. 17 shows a silhouette of a glass dish containing steel balls, among which two large balls, one of wood and one of stone, have previously been buried. As before we

Fig. 16.—The reaction to the upward thrust

replace the missing thermal motions in our model liquid by shaking the dish. The upward thrust immediately brings the two large balls to the surface, on which they "float", the wooden ball sticking high up and the stone one about half immersed.

Of course no quantitative verification of the value of the upward thrust can be expected from this experiment; the substitution of shaking for the thermal motions is too primitive for that.

Fig. 17.—Upward thrust in a model liquid consisting of steel balls (1/5.)

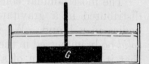

Fig. 18.—Prevention of the upward thrust

The essential condition for the occurrence of an upward thrust on solid bodies is a pressure on the *bottom surface* of the body. The molecules of the liquid must, thanks to their freedom to move, be able to penetrate below the body. This is shown by the well-known experiment illustrated in fig. 18. The dish used has a smooth bottom made of plate glass. On this we lay a glass block G, likewise with a smooth bottom. Holding it firmly against the bottom of the dish, we fill up the dish with mercury. The mercury cannot penetrate into the space between the block and the dish. The liquid can only press vertically on the upper side of the block. Hence the block remains sticking to the bottom of the dish as if by suction, and can readily be slid about on this plane, as may be shown if the block has a handle fixed to it.

If the handle and block are tilted the mercury is enabled to penetrate under the block, which immediately rises to the surface of the mercury.

The weight of a body and the upward thrust of a liquid on it oppose one another. If the *weight* predominates the body sinks to the bottom of the liquid: if the *upward thrust* predominates, it rises to the surface. Between these two possibilities there is an intermediate case. The body and the volume of liquid displaced by it may have the same mass. In this particular case the body can float at any height in the liquid. This may be realized in a large number of ways: as a single example we may mention an amber ball suspended in a zinc sulphate solution of suitable concentration.

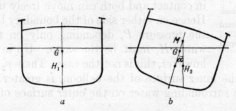

Fig. 19.—The metacentre

If the upward thrust predominates, part of the body rises out of the liquid. The body comes to rest when the mass of liquid displaced is equal to its own mass. We then speak of a body floating in liquid. For practical purposes (ships) it is of the greatest importance that the floating position should be *stable*. This is determined by the position of the *metacentre*. Imagine a ship listed through an angle a from its equilibrium position (fig. 19b). Let H_2 be the centre of gravity of the volume of water displaced by it in the *oblique* position, i.e. the point of application of the upward thrust in the oblique position. Through this point H_2 we draw a vertical. The point where it intersects the centre line of the steamer is called the *metacentre*. The metacentre must not come below G, the centre of gravity of the ship, in any oblique position. Otherwise the couple acting on the ship would not bring her back to the position of equilibrium. A ship will not float in stable equilibrium unless the metacentre is *above* the centre of gravity.

8. The Upward Thrust of one Liquid on Another.

Hitherto we have considered the statical upward thrust on a *solid* body immersed in a liquid. We shall now consider the upward thrust on a *liquid* immersed in another liquid. We shall choose a particular case which will be of use later on, that of a liquid contained in a vessel open at the bottom and immersed in another liquid of greater density. Fig. 20 shows a glass bulb open at the bottom and filled with oil, placed in a larger vessel filled with water. The upward thrust on the oil and its glass envelope are greater than its weight, but a small wire ring (not shown in the figure) prevents it from reaching and penetrating the surface of the water.

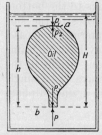

Fig. 20. — Upward thrust on a bulb filled with liquid in a liquid of greater density.

From the purely formal point of view we may apply the theorem deduced on p. 164. The upward thrust on the oil and its envelope is equal to the weight of water displaced. All the same it is worth while to investigate the distribution of pressure within the "balloon", as this throws light on the whole process. The balloon is open at the bottom. The oil and the water are in contact and both can move freely up and down. Hence on either side of the boundary between them the pressure P, depending only on the depth of water H, must be the same. Up in the balloon, however, this is not the case. There p_2, the pressure of the oil on the inner surface of the balloon, is greater than p_1, the pressure of the surrounding water on the outer surface of the balloon: for

$$p_1 = P - hg\rho_1,$$
$$p_2 = P - hg\rho_2,$$

where ρ_1 is the density of the water and ρ_2 the density of the oil. The density of the oil is *less* than that of the water: hence $p_1 < p_2$, and we have

$$p_2 - p_1 = hg\,(\rho_1 - \rho_2). \quad . \quad . \quad . \quad . \quad . \quad (6)$$

This difference of pressure gives rise to the upward force acting on the envelope of the balloon. For a cylindrical balloon of cross-section A the force is $(p_2 - p_1)A$.

In order to demonstrate this distribution of pressure in practice the balloon is made with a small hole at a, which is initially closed with a cork. When this cork is removed the oil shoots out into the water in the form of a free jet: that is, p_2, the pressure in the oil, was greater than p_1, the pressure in the water. We shall come back to this experiment in § 16, p. 186.

9. Viscosity and Internal Friction in Liquids.

In the model liquid we have hitherto used (steel balls) three important properties of real liquids are missing: (1) viscosity, (2) the adhesion of liquids to solid bodies, (3) the obvious cohesion between the molecules of liquids. When an actual liquid is poured out, the individual molecules do not scatter separately in all directions like the steel balls of the model liquid; an actual liquid always *hangs together in drops* of differing size and shape.

We shall discuss the viscosity of liquids and their adhesion to solid bodies in this section, and their internal cohesion in the following one.

There is no doubt that the principal characteristic of all actual liquids is the complete *freedom of movement* of the individual molecules. In liquids there is no " internal hooking-together ". Two solids in contact, in fact, are always " hooked together " along their surfaces. The best polishing processes merely diminish the hooking together of minute irregularities in the surfaces; they can never remove it entirely. Two bodies with smooth horizontal faces in contact *cannot* be set in motion relative to one another by an indefinitely small force parallel to these faces (fig. 21). In order to initiate motion the force must always exceed a certain limiting value. In a liquid,

Fig. 21.—Apparatus to illustrate external friction

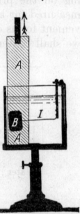

Fig. 22.—Internal friction in a liquid (1/6)

on the other hand, there is no hooking together and no limiting value of the force required to cause motion. In all liquids the particles are completely free to move in all directions.

In spite of this free movement of the particles many liquids possess a considerable degree of viscosity. This viscosity gives rise to the force in moving liquids known as " internal friction ", which we shall have to investigate more closely.

If a force is referred to as frictional, the existence of motion of one body relative to another is thereby implied. We have already learnt to distinguish two cases of this:

(1) Two solid bodies are *in contact* along their boundary surfaces and slide on one another under a greater or smaller pressure. *During* this motion each of the bodies is acted on by a force (the action being equal to the reaction). This force is called the *external friction*.

(2) Two solid bodies are separated by liquids or gases. Then each of the bodies is still acted on by a force during the motion. This force is called *internal friction*.

Internal friction is measured by an apparatus on the principle illustrated by fig. 22. A large glass plate A is moved upwards in some way (e.g. by the hand) with constant velocity v. A second plate B of cross-section a is at a distance x in front of it. The plate B can be moved in the direction of its length and is held by a dynamometer I (leaf spring) on the right. The distance between the plates is small compared with the diameter of B. Both plates are surrounded by liquid. The force indicated by the dynamometer is the internal friction. Its magnitude is found to be directly proportional to a, the area of the plate, and v, the relative velocity of the two plates, and inversely proportional to x, the distance between the two plates. Thus for the magnitude of the force known as internal friction we obtain

$$F = \eta a \frac{v}{x}. \quad . \quad . \quad . \quad . \quad . \quad . \quad . \quad (7)$$

The factor of proportionality η is called the *coefficient of viscosity*. Numerical values of this constant are obtained by a method depending on the principle illustrated in fig. 22. That is, the force F is measured for known values of v, a, and x. Technical details and convenient forms of apparatus are a matter for practical textbooks. Here we shall content ourselves with a list of numerical values.

Substance	Temperature	Coefficient of Viscosity in large dyne-sec./m.² (1 large dyne = 0·102 kg.-force)
Liquid carbon dioxide ..	20°	$7\cdot0.10^{-5}$
Benzene	20°	$6\cdot4.10^{-4}$
Water	0°	1·8
	20°	1·0
	98°	0·3 $\Big\}.10^{-3}$
	−21·4°	1·9
Mercury	0°	1·6
	100°	1·2
	300°	1·0
Castor oil	20°	1·02
	30°	.0·45
	40°	0·22
Pitch	20°	10^7

For gm.-mass and centimetres the above numerical values have to be multiplied by 10: thus e.g. the coefficient of viscosity of water at 20° C. is 0·01 dyne-sec./cm.²

The next experiment, illustrated in fig. 23, is intended to throw further light on the phenomena connected with the occurrence of internal friction. It really gives us a side-view of the experiment of fig. 22. This time the lower half of the liquid is coloured violet. The long plate A is moved upwards with constant velocity v. During this motion the liquid behaves as shown in the instantaneous photograph reproduced in fig. 23: the liquid *sticks* to the moving body. The boundary layer of the liquid next the moving body moves with the same velocity as this body. The neighbouring layers of liquid likewise receive a velocity in the direction of the plate as the latter moves verti-

cally upwards, but this velocity decreases as the distance from the
moving plate (x) increases. The velocity in the liquid falls off at rght
angles to the direction of the velocity at the rate
$\partial v/\partial x$. In the special case of the body with plane
parallel sides used here the fall of velocity is approximately linear. For on the right of A we find that
the boundary layer between the coloured liquid and the
uncoloured liquid in the region between the plate and
the vessel departs very little from a straight line.
Hence we may write v/x for $\partial v/\partial x$. The quotient v/x
in our experimental equation (7) thereby acquires a
physical meaning; it stands for the *fall of velocity* within
the liquid. Equation (7) must be rewritten in a more
general form, and we thus obtain

Fig. 23.— To
illustrate how
internal friction
arises.

$$F = \eta a \frac{\partial v}{\partial x} \quad . \quad . \quad . \quad . \quad . \quad . \quad . \quad (8)$$

for the internal friction.

According to this extended conception the force which we call
internal friction is not confined in its action to the solid surfaces forming the boundary of a liquid. On the contrary, internal friction exists
even between neighbouring particles of liquid which differ in velocity.
It is this internal friction that enables moving particles of liquid to
drag along neighbouring particles previously at rest. In viscous liquids
a constant velocity of motion can only be maintained by the continued
action of external forces.

10. The Cohesion, Tenacity, and Surface Tension of Liquids.

Liquids stick to solid bodies and possess viscosity. The latter
gives rise to the force known as "internal friction". These characteristic features have hitherto not been reproduced in our model
liquid. Such was the gist of the previous section.

Hitherto our model liquid has also failed to reproduce a third
feature of actual liquids, namely, the cohesion between individual
molecules. This forms the subject of the present section.

Actual liquids, in short, behave like a model liquid consisting of
magnetized steel balls. Between the individual molecules of the liquid
there exist forces. In the case of actual liquids these are formally
referred to as "molecular forces". According to our present-day
knowledge they are exclusively of electromagnetic origin. Molecular
forces are also responsible for liquids adhering to solid bodies. They
act between the molecules of liquid on the one hand and the molecules
of the solid body on the other. This adhesion may go so far as to
cause *wetting* of the solid surface. In such cases the cohesion between

the molecules of the liquid and those of the solid body is *greater* than that between the molecules of the liquid.

Unfortunately it is not easy to experiment with a model liquid consisting of magnetized steel balls. A heap of magnetized steel balls, it is true, will cohere very nicely to form a " drop ", which

Fig. 24. —
Tenacity of a
model liquid.

may then be made to slide about on a glass plate. But it is no longer possible to produce "thermal motion" in this model liquid by shaking, for the "liquid" is now as viscous as thick honey. In addition to this, ordinary steel balls do not readily retain their magnetization, and it has to be renewed before every experiment with the help of a strong electromagnet. Nevertheless, this model liquid is capable of doing us good service.

Fig. 24 shows a longitudinal section of a vertical iron tube closed at the top and filled with the model liquid. The magnetic "molecules" stick to the iron walls. We have a model of a coherent column of liquid. The latter supports its own weight, i.e. possesses considerable "tenacity".*

Fig. 25.—Tenacity of
a column of water

Fig. 25 shows the same experiment carried out with a real liquid, namely, with a column of water. The water is previously freed thoroughly from air by boiling in a vacuum. The wide limb B is pumped free of air. Columns of water many metres high may be held up in this way. Their tenacity is often surprising to the beginner. It is convenient to mount the long glass tube on a board. The board may be bumped hard against the ground, i.e. the column of water may be subjected to strong downward forces of inertia (the board being regarded as an accelerated system of reference). Frequently the column is not broken until after many fruitless attempts.

Fig. 26.—A drop of oil
held between two finger-
tips and the correspond-
ing model experiment.

The essential condition for demonstrating the tenacity of the water is obviously this: lateral constriction of the column or formation of a " waist " must be prevented. Adhesion to the walls tends to prevent this contraction. Hence the absence of the minutest air bubbles is essential, for these immediately form the starting-point for the formation of a " waist ".

In a second series of experiments we deliberately allow constriction to occur. On the left of fig. 26 we see a drop of our model liquid

* Ger., *Zerreissfestigkeit.*

between two "wetted" steel bodies a and b, for which it is simplest to use the poles of two bar magnets. On the right we see the same experiment with an actual liquid, namely, a drop of oil between two finger-tips. If the distance between them is slowly increased, two things happen:

(1) formation of a " waist ";

(2) increase in the surface area of the liquid.

The experiment with the model liquid shows very prettily how new molecules continually pass from the interior of the drop into the outermost layer of molecules, or surface of the liquid.

This is particularly well shown in projection. The image should be projected in a direction *parallel* to the broader side of the drop as it slowly degenerates into a thin film.

Within the liquid all the molecules are subject to symmetrical forces on all sides; hence their free movement. When they enter the surface layer, however, this all-round symmetry no longer exists. For every new molecule entering the surface layer a place must be made in spite of the opposing effect of the forces of attraction between

Fig. 27.—Apparatus for measuring the surface tension of water. R a ring of diameter 8 cm., T a shallow dish of water, M a sliding micrometer, G a counterpoise; $F =$ 3·8 gm.-force. The difference in weight corresponding to the observed distance between the " duck's bills " is ascertained subsequently.

neighbouring molecules. Some molecules which were originally neighbours must be separated from one another. Hence work is done in increasing the surface area. The work required to produce one unit of additional surface is called the *surface tension* (σ). Its dimensions are work/area or force/length (for work = force $\times$ length). The product of the circumference of a drop and the surface tension gives a force. This force is at once felt by the muscles in the case of the magnetic model liquid and may easily be measured in the case of actual liquids. Many methods are available, most of them involving the confinement of the drop of liquid to the region between two solid bodies, one of which is connected with a dynamometer. The word " drop " is not to be taken too literally; usually it is in the form of a ring or flat film. To produce a shallow ring of liquid we may use the apparatus shown in fig. 27. Its essential part is a light metal ring R suspended from a dynamometer. The pull on the dynamometer is increased until the film gives way. The value of the force which it can just sustain, divided by the area of the liquid surface, gives the required surface tension. The following numerical values are the results of similar experiments.

Substance	Temperature	Surface Tension (in Air), in watt-sec./m.² or large dyne/m. (1 large dyne = 0·102 kg.-force)
Mercury	18°	500
Water .. {	0°	75·5
	20°	72·5
	80°	62·3
Glycerine .. .	18°	64
Castor oil .. .	18°	36·4
Benzene	18°	29·2
Liquid air	−190°	12
Liquid hydrogen ..	−254°	2·5

(All values $\times 10^{-3}$)

For gm.-mass and centimetres the factor 10^{-3} must be omitted; the surface tension is then obtained in dyne/cm.

In order to form a liquid film we use the apparatus shown in fig. 28. Lateral constriction of the film towards the right and the left is prevented by two wetted wires. The only constriction which can take place is one at right angles to the plane of the paper. The liquid practically always used for demonstrations of this kind is soap solution.

Fig. 28.— A soap film, bounded at the foot by a sliding wire.

This apparatus illustrates an important point: the surface of a liquid is often compared to a rubber membrane. This analogy has to be used with caution. In fact, the tension of a rubber membrane *increases* as its surface is increased, which is by no means true of the surface tension of a liquid. This follows at once from the molecular conception illustrated by fig. 26. The extension of the surface of a liquid means that gaps are made in order that new molecules may enter the surface. The force required for this depends merely on the attractions of neighbouring molecules in a surface and not on the extent of the surface. This independence of the surface tension and the increase of area may readily be demonstrated by the apparatus of fig. 28. The lower boundary wire sliding on the side bars may be made to remain at any height by suitable loading. The weight required to load it is equal to $2l\sigma$, where l is the length of the loop; the factor 2 comes in because the film has two surfaces.

By what we have said above the concept of surface tension loses all meaning in the case of a layer consisting of only one layer of molecules (" monomolecular " film). This case is represented diagrammatically in fig. 29, but cannot be realized in practice (owing to evaporation). Now all the molecules are already in the surface and there are no more to come in. Tearing of the film for a definite load would, in contradistinction to fig. 28, give not the surface tension but the tenacity of a "monomolecular" film of liquid. This imaginary experiment brings out the importance of molecular forces both for tenacity and for surface tension. The numerical connexion between the two cases may be calculated theoretically. For this, however, it is necessary to know how the molecular forces depend on the distances between the molecules.

Here we confine ourselves to recounting some further demonstrations of the effects of surface tension. These, of course, are only a very small selection from a great variety of possible experiments.

In the first three examples the surface of the liquid behaves like a readily expansible skin or membrane.

Fig. 29. — Rough diagram of a monomolecular film of liquid.

(1) A body which is not wetted by a liquid can rest on the surface as on a loosely-stuffed cushion, say an air cushion; the surface is hollowed out under it. For example, a needle which is not quite free of grease can be made to float on water without any difficulty (cf. the legs of water-beetles).

(2) A drop of liquid will lie on an unwetted sieve just as a rubber balloon full of water will lie on wide-meshed wire-metting, with a curved portion of the surface protruding from each mesh.

(3) A coating of powder which is not wetted (e.g. lycopodium) prevents one's finger from being wetted when dipped in water. The surface of the liquid is upheld by the minute dust particles in the same way as a tent is by its pole. The parts of the surface which hang down between the supports cannot reach the skin.

In the remaining examples surface tension is found to bring about the maximum diminution of surface compatible with the conditions of the experiment.

(4) A fine jet of mercury is directed into a shallow clock-glass filled with acidulated water. At first the mercury collects at the bottom of the glass in numerous fine drops about 1 mm. in diameter (fig. 30). That is, the total surface area of the mercury is very large. But the drops soon proceed to unite by fits and starts. Here and there a small drop is engulfed in a larger one. The necessary communicating " bridges " arise at the boundaries of the drops as a result of the random variations of the thermal motions. After a few minutes only one big drop is left. As a result of surface tension the surface of the

mercury has contracted to the least possible value. This is a very pretty experiment.

(5) A large coloured drop of castor oil is made to float on water. Owing to its own weight it has a flattened form. A stick is thrust vertically into the blob from above. The blob clings to the stick like a wet rag and is pushed down below the surface. The stick is then held still. In the course of a few minutes the thick oily covering of the stick forms itself into a spherical ball as a result of surface tension. This ball crawls slowly up the stick.

When it reaches the surface the spherical drop again forms a flattened blob owing to its weight, for then the upward thrust which is exerted on it by the water and which largely equalizes the weight is again wanting.

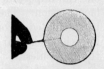

Fig. 30. — Drops of mercury of various sizes becoming united.

Fig. 31.—Soap film with a loop of thread

(6) A knotted loop of thread is thrown on to a soap film. The latter is then pierced somewhere within the loop, preferably with a match dipped in alcohol. The hole in the film bounded by the loop of thread is circular (fig. 31). The maximum diminution of the surface of the film compatible with the length of the thread is thus attained.

The match used to pierce the soap film had just been wetted with alcohol. This is our first example of the great changes in surface tension brought about by the introduction of foreign molecules. According to the mechanism of surface tension illustrated in fig. 26 even minute quantities of foreign molecules must have a considerable effect.

We give a few further examples of the effect of foreign molecules on surface tension.

(7) The behaviour of drops in glasses of heavy wine ("tears of wine"). A drop flowing slowly down the glass loses alcohol by evaporation. The surface tension of the drop is less than that of the wine in the glass. When the drop reaches the surface of the wine some molecules of alcohol pass into the surface of the drop. The surface tension of the drop rises and the drop suddenly contracts.

(8) Another example is the behaviour of a grain of camphor on water. Different parts of its surface go into solution at different rates. Hence the surface tension differs in different directions and the grain of camphor dances about on the surface of the water. Motions of this kind play an important part in the organic world. We may mention the modes of motion of many small organisms and above all the contraction of animal muscle.

(9) "Pouring oil on the waves." This transforms "breakers" with their overhanging tops of foam into a low swell. In order to bring about the necessary change in surface tension it is only necessary to drop a very trifling quantity of oil overboard in the form of small drops.

If foreign molecules are present the phenomena of surface tension lose their simplicity. The surface tension becomes anomalous; that is, its value, like that of the tension of a rubber membrane, comes to depend on the previous expansion of the surface. Expansion of the surface also gives rise to heat. Kinetic energy is destroyed and re-appears as heat. These matters, which in some respects are very interesting, belong to the subject of Heat.

11. Gases and Vapours as Fluids with a Low Density and no Surface.

As compared with liquids, gases have an extremely low density. As an example we shall measure the density of air. Fig. 32 shows a glass bulb of capacity 7 litres on the left-hand scale-pan of an ordinary pair of scales. This bulb has previously been exhausted of air by some

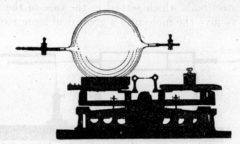

Fig. 32.—Measurement of the density of air

form of pump. The exhausted bulb is counterpoised by means of a few weights. We then open the top of the bulb and let the air of the room rush into it. The pan with the bulb then sinks. We have to take 9 gm. from the left-hand pan in order to bring the scales back to equilibrium. That is, the mass of the 7 litres of air is about 9 gm. The mass of a cubic centimetre (the density) is 0·0013 gm./cm.³ That is, the density of the air of the room is about 1/800 that of water.

The molecules of a gas and of the corresponding liquid are the same. Hence the small density of gases can only arise if the distances between the individual molecules are great. Further, the following facts point to the molecules in gases and vapours being far apart.

(1) Gases, in contradistinction to liquids, are very highly compressible, as we see every time we use a bicycle pump.

(2) The Brownian movement may be observed with a much lower magnification in gases than in liquids. For the visible dust particles we may very conveniently use the products of combustion of which tobacco smoke consists. A small glass bulb containing smoky air is illuminated from the side and observed by quite a primitive type of microscope.

(3) The molecules of a gas or vapour move about in all directions

quite independently of one another.   They distribute themselves
throughout any space which is made available to them (cf. an escape
of gas in a room, or the fragrant gaseous particles of perfumes).   In
contradistinction to liquids, no cohesion between the molecules is
observed unless refined methods are used.   In any case, in gases a surface is no longer formed.   The forces of attraction between individual
molecules obviously do not attain their fullest effects when the molecules are far apart.

12. Model of a Gas: Gaseous Pressure a Consequence of Thermal Motions: the Equation $pv = \text{const.}$

The above facts enable us to replace an actual gas by a model and
hence to study further properties of gases.   As our molecules we shall
again use the steel balls which served in the case of the model liquid,
but this time we give the molecules a great deal more room by placing

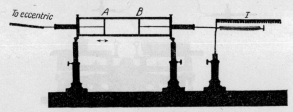

Fig. 33.—Gas-holder for a model gas consisting of steel balls. The balls
are filled in between A and B; B is the piston of the dynamometer I (1/10.)

them in a wide " gas-holder ", consisting of a shallow box with large
glass windows (fig. 33).   We also ensure that the molecules shall have
continual and lively " thermal motions ", instead of the occasional
shaking-up which we gave to the model liquid, by making a rapidly
vibrating steel piston A one of the side walls of the gas-holder.   A
second side wall B likewise consists of an easily movable piston, which
is connected with a dynamometer I (a spring balance) by means of a
rod.

When the apparatus is working all the steel balls fly about rapidly.
The molecules continually collide with one another or with the walls.
These collisions are elastic.   The velocity of any molecule is continually changing in magnitude and direction.   We have a good picture
of *really* " random " thermal motions.

These thermal motions give rise to a pressure exerted by the model
gas against the walls of the container.   We first establish the existence
of this pressure experimentally by means of the dynamometer I.   *The
pressure of a gas on the walls of a vessel thus arises in quite a different
way from that of a liquid.*   In the case of a liquid the pressure on the
walls arises either from the weight of the liquid (gravity pressure) or

from the driving in of a piston into a closed vessel of liquid (piston pressure). In liquids there is no question of a pressure on the walls of the containing vessel due to *thermal motions*. Here gases and vapours exhibit a completely *new* phenomenon, due to the absence of a surface and of cohesion between the individual molecules.

We can easily see *qualitatively* how the pressure arises from the thermal motions. The molecules are continually impinging on the walls of the vessel. Every time a molecule is reflected the wall is subject to an impulse $\int F\,dt$. Together all these impulses have the effect of a continuously-acting force of magnitude pA (where A is the area of the wall).

We can, however, also give a quantitative account of how this gaseous pressure arises, provided we assume that all the N molecules have the same average velocity v independent of V, the volume of the containing vessel, and hence the same kinetic energy $\frac{1}{2}mv^2$.

Let the box in fig. 34 represent a vessel containing gas. There are N/V molecules of gas in each cubic centimetre of its volume; that is, the density of the model gas within it is

$$\rho = \frac{Nm}{V}. \qquad \cdots \cdots \cdots \quad (9)$$

We proceed to calculate the pressure on the left-hand wall of the vessel (the surface A).

In time t a molecule of velocity v describes a path of length $s = vt$. Hence during the time t only those molecules which are within the shaded part of the vessel can reach the left-hand wall. The volume of this shaded part is $As = Avt$. In *one* cubic centimetre there are N/V molecules. Hence the number of molecules in the shaded volume As is $NAvt/V$. The molecules are flying at random and have no particular preference for any of the six directions of space, so that only one-sixth of the molecules are flying towards A. Hence during the time t only one-sixth of the above-mentioned number of molecules impinge on the surface A, i.e. $NAvt/6V$ molecules. For the sake of simplicity we shall assume that all these molecules impinge normally on the wall. Then each molecule gives the wall momentum $\int F\,dt = 2mv$ (p. 89), for the collisions are elastic. The sum of all the impulses occurring during the time t is

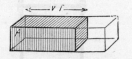

Fig. 34.—To illustrate the calculation of the pressure of a model gas.

$$2mv \cdot \tfrac{1}{6}\frac{N}{V}Avt,$$

or

$$\tfrac{1}{3}\frac{mN}{V}v^2At. \qquad \cdots \cdots \cdots \quad (10)$$

The sum of these unequal and irregular impulses in rapid succession may be replaced by *one* impulse Ft, the *constant* force F being supposed to act throughout the time t. This force divided by the area A gives the required pressure p. We have

$$Ft = \tfrac{1}{3}\frac{mN}{V}v^2 At;$$

hence
$$p = \frac{F}{A} = \tfrac{1}{3}\frac{mN}{V}v^2, \quad \ldots \ldots \quad (11)$$

or, if we introduce the density of the gas, $\rho = mN/V$,

$$p = \tfrac{1}{3}\rho v^2. \quad \ldots \ldots \quad (12)$$

In equations (11) and (12) not only are m and N constant, but also, by *hypothesis*, v. Hence these equations deduced for a model gas may be expressed in the very simple forms

$$pV = \text{const.}, \quad \ldots \ldots \quad (13)$$

$$\rho = \text{const.}\, p. \quad \ldots \ldots \quad (14)$$

That is, for our model gas,

(1) *the product of the volume V and the pressure p is constant*;
(2) *the density ρ is proportional to the pressure p.*

The sole assumption we made was that the velocity v is constant and hence also the kinetic energy $(\tfrac{1}{2}mv^2)$ of all the molecules of equal mass m contained in the vessel.

Good qualitative confirmation of these results is obtained by the use of the model apparatus. Exact quantitative agreement is out of the question. The assumed constancy of the velocity of the balls can be realized only approximately with the simple apparatus we used.

The behaviour of actual gases, on the other hand, agrees remarkably well with that calculated for the model gas. To demonstrate this we use the apparatus shown in fig. 35. A glass cylinder is subdivided by a gas-tight partition W. In the lower part a piston consisting of liquid is free to move, and may be made to compress the gas (e.g. air) enclosed in the upper part of the vessel. A hole is bored in the partition and a pressure gauge (one of the instruments we have already used with liquids: see figs. 5 and 6, p. 158) is fixed in the opening. The upper part of the glass tube is exhausted by an air-pump. The volume of the gas enclosed below the partition at any time is pro-

Fig. 35.—Apparatus for investigating the relationship between the pressure and the volume of a gas.

portional to the distance between the piston and the partition. In making the observations it is essential that the temperature of the enclosed gas should be kept as constant as possible. The following table gives some typical results.

Pressure p at 0° C. in Atmospheres	Product pV in Atmospheres . m.³		
	1·293 kg.-mass Air	0·0899 kg.-mass Hydrogen	1·977 kg.-mass Carbon Dioxide
0·01	1·000	1·000	1·007
0·1	1·000	1·000	1·006
1·0	1·000	1·000	1·000
10	0·995	1·006	—
50	0·975	1·031	0·105
100	0·968	1·069	0·202
300	1·097	1·209	0·560
1000	1·992	1·776	1·656

Hence the densities ρ at 0° C. for a pressure of one atmosphere in kg.-mass/m.³ are

$$| \quad 1·293 \quad | \quad 8·99.10^{-2} \quad | \quad 1·977 \quad |$$

or in gm.-mass per c.c.

$$| \quad 1·293.10^{-3} \quad | \quad 8·99.10^{-5} \quad | \quad 1·977.10^{-3} \quad |$$

We find that equation (13), $pV =$ const., is a good approximation in the case of air and hydrogen up to pressures of about 100 atmospheres. These gases are therefore called " ideal ". Other gases, such as carbon dioxide, exhibit constancy of the product pV only for small pressures. Such gases are sometimes called *vapours*. A vapour is a gas which obeys the " ideal gas law ",

$$pV = \text{const.}, \quad \ldots \ldots \ldots \quad (13)$$

imperfectly or not at all.

The " ideal gas law " is accordingly a typical *limiting* law for actual gases. It gives excellent approximations, especially in the region of low pressures and densities, but no gas obeys it with perfect accuracy.

13. The Velocities of Gaseous Molecules.

According to what we have said in the previous section, our idealized model gas is capable of reproducing the behaviour of actual gases to a considerable extent. The molecules of our model gas are elastic balls incapable of penetrating into one another, moving with a velocity whose time-average is constant. There is a simple relationship between this velocity v and the pressure p and the density ρ of the gas; by equation (12)

$$p = \tfrac{1}{3}\rho v^2.$$

From the above table we extract an arbitrary pair of values of p and ρ for air at room temperature, e.g.

$$p = 1 \text{ atmosphere} = 10^5 \text{ large dynes/m.}^2; \quad \rho = 1 \cdot 3 \text{ kg.-mass/m}^3.$$

Inserting these values into equation (12) we obtain

$$\mathbf{v} = 480 \text{ m./sec.}$$

for the velocity of air molecules at room temperature. Similarly we find that the molecular velocity for hydrogen at room temperature is about 2 km./sec.

This calculation is certainly trustworthy as regards order of magnitude, but, of course, only gives average values about which the actual velocities are grouped, but from which they may differ widely.

14. The Atmosphere: Experiments on Atmospheric Pressure.

Like our model gas, the air distributes itself throughout any space made available for it. It does not possess the cohesion arising from the existence of a surface. How, then, does the earth's atmosphere remain in place? Why do the air molecules not fly out into space? The answer is that like all bodies the air molecules are attracted towards the centre of the earth by their *weight*. The same is true for air molecules as for any projectile (p. 72); in order to leave the earth a velocity of at least 11·2 km./sec. is necessary. Now the *mean* velocity of air molecules remains far below this limiting value of 11·2 km./sec. Hence the vast majority of the molecules of the air are chained to the earth by their weight.

Were it not for their *thermal motions* all the air molecules would fall to the ground like stones * and—as we may mention in passing— would form a layer about 10 metres thick. Were it not for their *weight*, they would immediately desert the earth for ever. But the competition between thermal motions and weight keeps the air molecules floating about, hence the formation of the atmosphere. The solid surface of the earth prevents the atmosphere from reaching the centre of the earth. Hence the surface of the earth has to bear the whole *weight* of the masses of air contained in the atmosphere. The weight on one square centimetre gives the normal atmospheric pressure of " one atmosphere " or " 76 cm. (30 in.) of mercury ".

" We human beings live a deep-sea life at the bottom of the huge ocean of air." Nowadays this is known to every schoolboy. The experiments to prove the existence of " atmospheric pressure ", which a few hundred years ago were sensational, nowadays form part of the most elementary school courses on physics. Nevertheless, for reasons of historical sentiment we shall mention two classical experiments.

* This may easily be illustrated by the model experiment described on p. 185.

Both are due to the Magdeburg burgomaster Otto von Guericke * (1602–86).

(1) A small box is closed by an air-tight membrane and exhausted by a pump through a side-tube. The membrane is forced inwards and in many cases bursts with a loud report.

This deformation of a membrane on an exhausted vessel may be transferred to a pointer and scale. This is the principle of the well-known aneroid barometer.

(2) Guericke fixed two copper hemispheres 42 cm. in diameter tightly together by means of a greased leather washer and exhausted the air through a side-tube. The hemispheres were then tightly pressed together as a result of atmospheric pressure, the force being the product of the cross-section of the sphere ($A \sim 1400$ sq. cm.) and the

Fig. 36.—Two Magdeburg hemispheres being pulled apart by horses

atmospheric pressure ($p \sim 1$ kg.-force/sq. cm.), i.e. 1400 kg.-force. Hence Guericke required eight horses to separate the two hemispheres. The much reduced woodcut reproduced in fig. 36 shows this celebrated experiment being performed. The picture actually shows sixteen horses instead of eight. This, however, was no doubt a bluff intended to impress ignorant spectators. Eight of the horses could quite well have been replaced by a solid wall, for even then the law of action and reaction held!

Nowadays a degenerate form of the Magdeburg hemispheres plays a modest but useful part in everyday life, namely, the well-known utensils for bottling fruit, consisting of a glass jar, a rubber ring, and a glass lid. These are not exhausted by means of a pump; the air is replaced by steam. When the steam cools and condenses a " vacuum " is left.

* A good extract from Guericke's principal work, *Nova Experimenta (ut vocantur) Magdeburgica*, was published by R. Voigtländer, Leipzig, in 1912. Ingenuity in experiment and clearness of exposition are typical of Guericke's work.

(3) In elementary physics courses the well-known "siphon" is also brought forward as an effect due to atmospheric pressure. This statement, however, is applicable only with very considerable restrictions. The principle of the "siphon" has *nothing* to do with atmospheric pressure; it is illustrated by fig. 37. A chain hangs over a frictionless pulley. Each end hangs rolled up in a glass. If one of the glasses is made to rise or fall the chain runs down into the lower glass. It is pulled downwards by the weight of the overhanging end H.

Fig. 37. —
Chain " siphon "

Exactly the same is true for liquids, for even liquids possess tenacity like solid bodies (p. 170). The liquid, however, must be free from bubbles of gas. Hence a siphon works beautifully in a vacuum. An apparatus of this kind is shown in fig. 38. The overhanging end of the column of water is indicated by the length H. That is, in theory a siphon works entirely independently of atmospheric pressure.

Liquids in everyday life, and water in particular, however, are never free from small bubbles of air. These greatly diminish the tenacity of the water. This was abundantly shown in § 10, pp. 169–170. Hence columns of ordinary water containing air bubbles readily break apart.

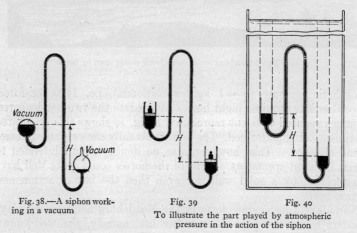

Fig. 38.—A siphon working in a vacuum

Fig. 39

Fig. 40

To illustrate the part played by atmospheric pressure in the action of the siphon

This quite secondary difficulty may be avoided in many ways. A possible method is shown in fig. 39. The liquid in the siphon is subjected to equal pressures on both sides, due to two equally-loaded pistons. This pressure prevents the column of liquid containing air from breaking under its own weight. The liquid flows from the left-hand vessel to the right-hand one; it is pulled over by the weight of the column H indicated by dotted lines. In practice, of course, the pistons cannot be made sufficiently frictionless. It is much simpler to replace the piston pressure by the gravity pressure of a liquid of lower density (e.g. oil) as shown in fig. 40; that is, the solid piston in fig. 39 is replaced by the columns of liquid bounded by dotted lines in fig. 40. In fig. 40 only one thing is essential, namely, that the auxiliary liquid should have a lower density than the liquid used in the siphon. Hence we may at once replace the oil used in fig. 40 by the atmosphere. That is, *atmospheric pressure plays only a very subsidiary part in the action of the siphon*; it prevents the formation of bubbles which so readily occurs in liquids containing air, and thus prevents the columns of liquid from breaking apart.

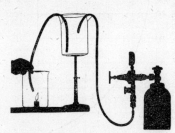

Fig. 41.—A gas siphon: on the right a cylinder of carbon dioxide with reducing valve and rubber tubing for filling the beaker.

Fig. 42.—A stream of ether vapour made visible in shadow form (an extremely simple case of the so-called " Schlieren " method).

The same is *not* true of siphons used with a gas. Gases have no tenacity. In contradistinction to liquids, gases by themselves can never form a column. Thus, in order to siphon a gas, it is always necessary to use an auxiliary pressure as in figs. 39, 40. Fig. 41 shows a gas siphon working. The invisible gas carbon dioxide is made to flow through a rubber siphon from the upper beaker to the lower. The arrival of the gas in the lower beaker is visibly indicated by its extinguishing of a candle flame.

The gas siphon leads us to mention a subsidiary part played by the atmosphere which is useful in many experiments. Gases possess no surface, but the presence of the atmosphere compensates for this in certain respects. The absent surface is replaced by the boundary between the gas or vapour and the surrounding air. Hence, for example, we may manipulate ether vapour just as we do a liquid. We tilt a bottle containing some ether, but not enough to let the liquid flow out; yet we see the ether vapour flowing out like a stream of liquid. This is particularly well shown by projection on a screen.

We may catch the ether vapour in a beaker counterpoised on a balance (fig. 42). As the beaker fills, the scale-pan on which it is placed sinks, for ether vapour has a greater density than the air it displaces from the beaker. At the close of the experiment we empty the beaker by turning it upside down and again see the ether vapour flowing out like a broad stream of liquid and falling to the ground.

15. Distribution of Pressure in Gases in a Gravitational Field: the Barometric Formula.

Hitherto we have only considered the atmospheric pressure at the ground-level. Apart from small changes associated with the state of the weather, it has a practically constant value of about 1 kg.-force/ sq. cm. (14·7 lb./sq. in). This is equal to the pressure of the water at the bottom of a pond about 10 metres deep.

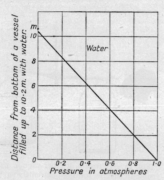

Fig. 43.—Pressure distribution in water   Fig. 44.—Pressure distribution in air

In every fluid the pressure decreases as we rise upwards from the bottom. For liquids this decrease of pressure is a linear one. In water, for example, the pressure falls by $\frac{1}{10}$ atmosphere or about 0·1 kg.-force/sq. cm. for a rise of 1 metre (see fig. 43). In water each layer of thickness dh contributes an *equal* amount $dp = g\rho\,dh$ to the total pressure. For liquids are almost incompressible and the density ρ is practically the same at all levels, the lower layers not being noticeably compressed by the weight of the upper layers superimposed on them. The phenomena are quite different in gases. Gases are very compressible. The lower layers are compressed by the pressure of the superimposed layers. The density (ρ) of any layer is proportional to the pressure p there. We have

$$\frac{\rho}{\rho_0} = \frac{p}{p_0} \text{ or } \rho = \rho_0 \frac{p}{p_0},$$

where ρ_0 is the density of the gas at normal atmospheric pressure p_0. The contribution of each layer of vertical height dh to the pressure is accordingly

$$dp = -g\rho_0 \frac{p}{p_0} dh. \qquad \ldots \ldots (15)$$

Integrated up to the height h this gives

$$p_h = p_0 e^{-g\rho_0 h/p_0}, \qquad \ldots \ldots (16)$$

where p_h is the pressure h m. above the ground in large dynes/m.², p_0 the normal atmospheric pressure at the ground-level in large dynes/m.², ρ_0 the density of air at pressure p_0 and temperature 0° C. in kg.-mass/m.³, and g the acceleration due to gravity (9·81 m./sec.²).

Inserting the numerical values for temperature 0° C. we obtain

$$p_h = p_0 e^{-0·127h}$$

for the atmospheric pressure at a height of h kilometres above sea-level, p_h and p_0 being measured in terms of any pressure unit so long as the same is used for both. This " barometric formula " is represented graphically in fig. 44, as a contrast to the distribution of pressure in water shown in fig. 43.

Fig. 45.—Two snapshots of a model gas consisting of steel balls, to illustrate the barometric formula: time of exposure 8 . 10⁻⁴ sec.

The meaning of this barometric formula may be illustrated very clearly by means of our model gas consisting of steel balls. To do this we set the apparatus shown in fig. 33, p. 176, vertical, and observe it by intermittent light. We then obtain on the screen varying instantaneous pictures of the type shown in fig. 45. We see that the molecules are crowded together in the lower layers and that they rapidly decrease in number upwards. We see the competition between thermal motions and weight. Only 2 metres above the vibrating piston (i.e. as seen on the screen) molecules occur but rarely, although stray molecules may wander up to a height of 3 metres. Our " artificial atmosphere " has *no definite upper boundary*.

The phenomena of our atmosphere are to be thought of as quite analogous to this, except that the range of height is very much greater. An upper boundary cannot be assigned to the atmosphere any more than it can to our artificial atmosphere. At a height of 5·4 km. above sea-level the density of the air has fallen to half its value ($e^{-0·69} = 0·5$), at a height of 11 km. to a quarter, and so on (fig. 44). Even several hundred kilometres above the ground, however, molecules of our atmosphere are still to be found darting about, for meteors are observed to begin being luminous even at these heights (they begin to glow on reaching our atmosphere, owing to friction). Auroras are also observed at similar heights; these are due to beams of electric corpuscles penetrating into our atmosphere.

Finally, we add some larger bodies, e.g. splinters of wood, to our artificial atmosphere. These are to represent dust in the air. We see the "dust" dancing about in vigorous "Brownian movements". But it always keeps near the ground, for the weight of a splinter of wood is much greater than that of a steel ball "molecule". (The dust behaves like a gas of high molecular weight.)

16. Upward Thrust or Buoyancy in Gases.

According to the results of last section, the distribution of pressure in a gravitational field is qualitatively very similar in liquids and gases. In this respect gases and in particular the atmosphere may

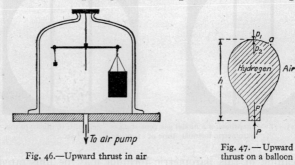

Fig. 46.—Upward thrust in air

Fig. 47.—Upward thrust on a balloon

safely be regarded as liquids. The deviations are only of a quantitative nature. They are easily seen by comparing figs. 43 and 44. Hence the upward thrust in gases may readily be explained by analogy with that in liquids. To begin with, fig. 46 shows an example of the upward thrust of a gas on a solid body. This experiment also goes back to Otto von Guericke. A gas-tight cylinder is counterpoised on a balance enclosed in a large bell-glass. The arrangement is diagrammatically as shown in fig. 15, p. 163. As the bell-glass is exhausted the cylinder sinks, showing that it is now "heavier"; for as the density of the air decreases the upward thrust is diminished.

We shall now consider the upward thrust on a gas surrounded by another gas, and discuss the action of the ordinary balloon. A balloon is shown diagrammatically in fig. 47, which is closely related to fig. 20, p. 166, the water being replaced by air and the oil by a gas of low density, usually hydrogen or coal-gas.

Formally we may again apply the theorem deduced on p. 164: the upward thrust on the balloon is equal to the weight of air displaced by the balloon. As in the corresponding case of two liquids, however, it is again advisable to consider the distribution of pressure within the balloon, which makes the process easier to follow.

A balloon is open at the foot. At the boundary between the air and the gas filling the balloon there is no difference of pressure. This boundary, of course, is not a sharp one, being merely a region where one gas diffuses into another. The effective difference of pressure may be observed in the upper half of the balloon. There the pressure of the gas inside the balloon on the inner surface of the envelope is greater than the pressure of the atmosphere on the outer surface of the envelope. The valve for emptying the balloon is also at the top (*a* in fig. 47).

The upward force acting on the envelope of the balloon is proportional to the difference of density between the air and the gas in the balloon. As the height increases both densities decrease. As regards the gas in the balloon this decrease results, in the case of the "slack" balloon, from gradual expansion of the lower parts of the balloon. If the limit of "tautness" is exceeded the gas in the balloon escapes from the lower opening. That is, as the absolute values of the densities diminish, so does the value of their *difference*. For a definite limiting value of the density the upward force becomes equal to the weight, and in this case the balloon floats at a constant height. A further rise requires a decrease in weight, i.e. ballast must be thrown out.

A distribution of pressure analogous to that in the balloon occurs in the gas-pipes in our houses. The piping in a house of more than one story may be thought of as being represented by a cylinder full of coal-gas and closed at the top. This cylinder, like the pipes in the house, is surrounded by air. Normally the gas in the mains should be subject to a definite pressure. Often, however, this pressure

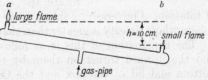

Fig. 48.—Decrease of pressure with height (Behn's tube)

is too feeble. Then the gas will not flow from a tap in the cellar. On the fourth story, however, there is no trouble; if a tap is opened there a powerful jet of gas streams out. These phenomena may be illustrated by means of a very pretty experiment. In fig. 48 the piping system is shown as a glass tube with a small burner at each end, the right-hand burner being 10 cm. lower than the left-hand. Coal-gas is supplied to the tube from the mains by means of a side-tube, the supply being partly cut off by a tap. Then it is easy to light the gas at the upper opening *a*, but not at the lower opening *b* of the same size. At *b* there is no difference of pressure between the air and the gas, whereas only 10 cm. higher up there is a noticeable difference of pressure and a bright flame is obtained. If the tube is set horizontal, flames of equal height are obtained at both openings. If it is tilted the other way a flame can only be obtained at b.

This apparatus is remarkably sensitive. It does not, as might be imagined, show how the pressure of the air diminishes with height, but only the *difference* in the decreases of the pressure in atmospheres of air and of coal-gas.

In this connexion we finally mention the chimneys of houses and factories. They contain warm air of lower density than the surrounding atmosphere. The higher the chimney the greater is the difference of pressure at its upper end between the inside and the outside, and hence the better does the chimney " draw ".

17. Gases and Liquids in Accelerated Systems of Reference.

After the detailed explanations of Chapter VIII (pp. 130–151), a short summary will suffice here. We shall first consider a few examples of a system of reference accelerated at right angles to the line of motion. That is, throughout this section we have a description of phenomena as they are observed by a person on a merry-go-round or rotating stool.

Fig. 49.— To illustrate the principle of centrifugal machines (1/40.)

(1) *Upward thrust arising from centrifugal force: the principle of centrifugal machines.* A closed water-tight box full of water is set along the radius of a horizontal rotating table (fig. 49). Just under the cover there floats a sphere whose density is accordingly less than that of the water. When the table is made to rotate the sphere moves towards the axis of rotation. Conversely, a sphere lying on the bottom of the box, i.e. a sphere whose density is greater than that of the water, moves outwards.

The explanation of this is as follows: The weight of the spheres and the upward thrust on them by the water are balanced by the bottom and lid of the box and the Coriolis forces by its side walls. There remain only the centrifugal forces. Inside the *horizontal* box these have the same effect as the weight would have inside a *vertical* box; for the centrifugal forces the axis of rotation is " top " and the outer edge of the table " bottom ". A body in the liquid is subjected to an " upward thrust ", i.e. a force acting towards the axis of rotation. This force may be greater or smaller than the centrifugal force acting on the body. If the latter preponderates the body moves outwards, i.e. figuratively speaking, " falls to the bottom ". The reverse is true if the " upward thrust " preponderates.

This upward thrust in rotating liquids forms the basic principle of practical centrifugal machines, e.g. for separating butter fat from milk; the butter fat, owing to its low density, collects near the axis of rotation.

(2) *Curvature of a rubber tube as a result of Coriolis forces.* We

again make the rotating stool rotate counterclockwise and place a piece of rubber tubing on it. It is pulled outwards by the centrifugal forces. The man now sends a jet of water into the tube by means of a rubber bulb. The moving water is acted on by Coriolis forces and the jet of water is deviated towards the right. The tube enclosing the water is curved to the right.

(3) *Deviation and curvature of a candle flame as a result of centrifugal forces and Coriolis forces.* A lighted candle enclosed in a large glass box and carefully shielded from draughts is placed

Fig. 50.—A flame subjected to forces of inertia (1/40.)

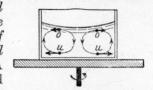

Fig. 51.—Radial circulation in a shallow dish of liquid

on the rotating table. The flame is bent towards the axis of rotation (fig. 50); it also becomes curved to the right looking from above.

Explanation: the resultant of the weight and the centrifugal force is directed obliquely outwards and downwards. The flame gases are lighter than air, so that the upward thrust of the air drives them obliquely inwards and upwards. This upward thrust gives the flame gases a *velocity*, and hence Coriolis forces arise as well as the centrifugal forces; the former have the effect of curving the flame to the right.

(4) *Radial circulation in liquids where different layers have different angular velocities.* At the middle of the rotating table we put a shallow dish full of water (fig. 51). We then make the table rotate with constant angular velocity. The water gradually acquires an angular velocity as a result of friction; this is greatest for the bottom layers and decreases upwards. The uppermost layer of water does not take up the full angular velocity of the table until several minutes have elapsed. This may be seen by strewing bits of paper over the water. Before this steady state of affairs is reached symmetrically-placed particles of water o and u are subject to unequal centrifugal forces radially outwards, the force on u being greater than that at o. Hence the liquid begins to rotate radially as shown by the dotted lines. Bodies strewn on the bottom of the dish (bits of paper are the most convenient) move along the radius to the outer edge of the bottom of the dish. A converse experiment is well known to all. If a cup of tea is stirred, the whole contents eventually have the same angular velocity. But as soon as we stop stirring, the angular velocity of the lower layers is at once diminished by contact with the bottom of the cup, which is at rest. A radial circulation is set up, but this time in the direction opposite to that in fig. 51, so that the tea-leaves are heaped up at the middle of the bottom of the cup.

The meandering of rivers and streams may be explained in exactly the same way. Fig. 52 shows a profile on an exaggerated scale of the river-bed at ab in the curve. At 1 the water is flowing more rapidly in the direction of the river-flow than at 2, for at 2 it is hindered by internal friction from the bottom upwards. Hence there is a greater centrifugal effect outwards at 1 than at 2, and a circulation is set up in the direction indicated by the arrow. The right-hand bank is undermined, and the sand carried away is deposited in the neighbourhood of a by the circulation. Hence the river-bed moves in the direction of b, and the bend becomes more pronounced.

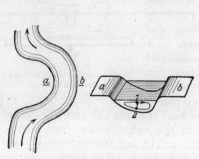

Fig. 52.—To illustrate how river-beds are formed

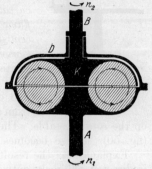

Fig. 53.—The Föttinger hydraulic transmitter (1/4.)

(5) *Utilization of the Coriolis forces in radial circulation*: *Hydraulic transmission.* Hitherto we have only spoken of a radial circulation of the water; in reality the paths of the particles in the horizontal plane are curved to the right, owing to Coriolis forces acting on the radially-moving particles.

These Coriolis forces are utilized in the design of an instructive type of hydraulic transmission gear. For this purpose the lower half of the vessel of fig. 51 is subdivided by radial partitions, which are shown shaded in fig. 53, and are attached to the shaft like the gills of a mushroom. A " coupling disc " K with similarly constructed radial partitions is introduced through the cover D. The partitions are brought to within a few millimetres of one another. The lower shaft A is to represent the shaft of a motor and the shaft of the upper disc leads to the " working shaft ". The shaft of the motor when running has an angular velocity somewhat greater than that of the working shaft (" slip "). Hence there is a continual circulation, towards the shaft above and away from the shaft below. The Coriolis forces on these moving masses of water press against the radial partitions and force the coupling disc to rotate almost as fast as the working shaft. Hydraulic couplings have been built on this principle up to thousands of kilowatts. They are outstandingly useful, as with their use it is possible to couple together steam turbines, which give a constant torque, and reciprocating engines, which give a variable torque. This is extremely important in the case of a steamer with only a single screw, for the steam which has escaped total utilization in the low-pressure cylinder can still be used to drive a low-pressure turbine, and the total efficiency may thereby be raised by about 25 per cent.

Fig. 54 shows a silhouette of a hydraulic transmitter suitable for demonstration experiments, fixed to the vertical shaft of an electric motor ($\frac{1}{3}$ kilowatt). To demonstrate the "slip" a small bell G is fixed to the casing. Its clapper is actuated by the cam N on the working shaft. The number of times the bell rings per second at once gives the difference of the number of revolutions of the shaft of the motor and the working shaft, i.e. the slip. As the load increases (e.g. if we clasp the shaft more tightly) the slip increases (this is a good analogy to the induction motor; see *Physical Principles of Electricity and Magnetism*, pp. 158, 166).

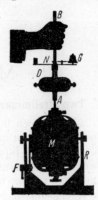

Fig. 54.—Model of a hydraulic transmitter attached to an electric motor M. The motor is supported in the frame R so that it can turn about a horizontal axis, and can be fixed at F ($1/13$.)

In conclusion we shall mention a case of gases in a system of reference accelerated in the direction of motion. We let a burning candle fall freely in a box free from draughts (a stable-lantern), catching it on a cushion at the bottom. During its fall the candle goes out. The reason is that there is no upward force to carry away the flame gases, as the forces of inertia are equal and opposite to the weights of the gas molecules (cf. pp. 133, 186, 189).

18. Concluding Remark.

With our discussion of circulatory motions we have passed away from the realm of liquids and gases at rest. Such motions form a connecting link with the subject of the next chapter, namely, the motion of liquids and gases.

CHAPTER X

The Motion of Liquids and Gases

1. Two Preliminary Remarks.

(1) Between liquids and gases there is an essential difference due to the existence in the former case of a surface. In spite of this the phenomena in liquids and gases *at rest* may be discussed formally in very much the same way. In the case of the *motion* of liquids and gases the joint treatment may be carried much farther. For example, up to a velocity of 50 m./sec. air may safely be treated as an *incompressible* liquid. For this velocity is still small as compared with the velocity of sound in air (340 m./sec., see p. 247). In this chapter we shall for brevity use the word *fluid* as meaning both fluids with a surface and fluids without a surface, i.e. in ordinary language, both liquids and gases.

(2) In dealing with the mechanics of solid bodies we have always endeavoured as far as possible to eliminate the forces which we call friction. In the case of the motion of fluids we shall at first do the exact opposite. The motions which we shall begin by describing are those which occur when the effects of *friction* predominate, the so-called " creeping motion " of fluids. This unusual procedure has the great advantage that it very soon puts us in possession of a very convenient method which will subsequently save a great deal of calculation.

2. The Creeping Motion of Fluids when Internal Friction Predominates.

Fig. 1 (Plate V) shows the front view of a very shallow glass cell placed vertical. In the corresponding cross-section (fig. 2) we see that the glass plates are continued upwards by two metal plates, each of which forms the inner wall of a vessel containing fluid. These vessels communicate by holes with the interior of the shallow cell, which is only about 1 mm. across. The holes in the left-hand vessel as compared with those in the right-hand one are displaced through half the distance between successive holes. To begin with both vessels are filled with water, the left-hand one higher than the right-hand one. Some ink is then added to the right-hand one. The fluid flows out from both vessels through the holes into the glass cell and runs away slowly at the bottom through a fine dropping tube. While the

192

PLATE V

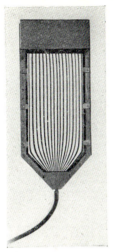

Chap. X, Fig. 1.—Stream-tube apparatus (front view). In projection it is convenient to rotate the image through 90° (by means of two rectangular prisms); the flow is then horizontal (cf. figs. 3, 7, &c.).

Chap. X, Fig. 4.—Distribution of velocity in creeping flow through a tube.

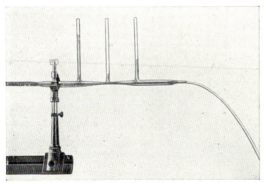

Chap. X, Fig. 8.—Distribution of static pressure in fluid flowing past a constriction. The three vertical glass tubes act as manometers.

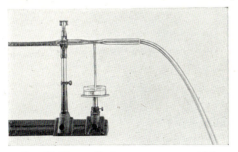

Chap. X, Fig. 9.—Static pressure at a constriction; it is less than the atmospheric pressure. A mercury column acts as a manometer.

fluid is in motion the cell exhibits the phenomenon shown in fig. 1, namely, a system of equidistant stream-tubes of a " creeping motion ".

In another experiment a circular disc (of celluloid) is inserted between the glass plates. It is meant to represent the cross-section of a sphere. The stream-tubes of the fluid (water) are as shown in

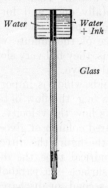

Water   Water
                     + Ink

Glass

Fig.   2.— Stream-tube
apparatus in longitudinal
section (1/6.)

Fig. 3.—Creeping flow round
a sphere or cylinder

fig. 3. We see the velocity of the alternate coloured and uncoloured bands relative to one another. The stream-tubes nearer the sphere have a longer path to cover. The velocity falls off radially from the centre of the sphere but at differing rates in different directions. By equation (8), p. 169, this fall of the velocity gives rise to the force which we call internal friction, which as the " resistance of the sphere in viscous fluid " acts in opposition to the relative motion of the sphere and the fluid. Calculation of this resistance gives

$$R = 6\eta\pi rv, \quad \ldots \ldots \ldots \ldots \quad (1)$$

where the units are as in the table on p. 168, e.g. η is measured in large dyne-sec./m.^{2}, r in metres, and v in m./sec. That is, as in any case of internal friction, the resistance of the sphere increases proportionally to v, the relative velocity of the sphere and the fluid.

A sphere placed in a sufficiently viscous fluid is at first *accelerated* by its weight, but after it has fallen through a short distance this acceleration ceases, for when a certain velocity is attained the resistance R and the weight W are equal and opposite. The sum of the forces acting on the sphere is zero. Hence its velocity is thenceforth constant.

The constant velocity of fall of spheres in fluids where viscosity predominates has many important practical bearings. For example:

(1) It may be used to measure the coefficient of viscosity η.

(2) It enables us to obtain the radii of small spheres (e.g. drops of oil) floating in the air. This method is often more convenient than direct measurement by means of a microscope, especially in the case of swarms or clouds consisting of large numbers of separate droplets.

(3) It may be used for sounding. The time of fall of *small* pear-shaped bombs is observed by means of a stop-watch. Their velocity of fall in sea-water is determined experimentally once and for all; usually it amounts to 2 metres (about a fathom) per second.

(4) Were it not for the frictional resistance of the minute water drops the clouds would fall on our heads. As it is, they fall very slowly, evaporating below and usually being re-formed above.

Another special case of creeping flow in fluids which is important in practice is the flow of fluids in narrow tubes (e.g. the flow of blood through the capillaries). Fig. 4 (Plate V) shows how the velocity falls off in a fluid flowing in a square tube. A coloured column of glycerine was placed on top of an uncoloured one. In the figure the direction of flow is downwards. In the case of a cylindrical tube the distribution of velocity is such that the boundary surface is a paraboloid. For this case the quantity of fluid flowing per second, or discharge, is given by the equation

$$i = \frac{\pi r^4}{8\eta} \frac{p_1 - p_2}{l}, \quad \ldots \ldots \quad (2)$$

where r is the radius of the tube, l its length, and p_1, p_2 the pressures at the two ends (the units being as on pp. 37 and 157). The factor $(p_1 - p_2)/l$ is called the resistance of the tube.

The examples we have given furnish a purely experimental demonstration of the existence of creeping motion, which is by no means confined to viscous fluids, like glycerine, but may occur in fluids of very small viscosity, such as water or even air. For the sake of comparison we give the values of the coefficient of viscosity for these three fluids:

	Temperature	Coefficient of Viscosity in large dyne-sec./m.² (1 large dyne = 0·102 kg.-force)
Glycerine .. {	0°	4·6
	20°	$8 \cdot 5 . 10^{-1}$
Water .. {	20°	$1 \cdot 0 . 10^{-3}$
	100°	$3 \cdot 0 . 10^{-4}$
Air	20°	$1 \cdot 7 . 10^{-5}$

For gm.-mass and centimetres the above numerical values have to be multiplied by 10. Thus e.g. the coefficient of viscosity of air at 20° C. is $1 \cdot 7 . 10^{-4}$ dyne-sec./cm.²

The measurements again depend on the principle illustrated in fig. 22, p. 167. No further details need be given here; cf. also fig. 3, p. 26.

Thus the magnitude of the *coefficient of viscosity* is by no means the only factor which determines whether the effects of internal friction on the motion shall predominate. No less important are the *linear dimensions* (e.g. the diameter of a sphere, the width of a pipe) and the *velocity* of flow. Further details of the relationships will be found in § 9, pp. 215–217. Meanwhile we are merely concerned with the *existence* of creeping motion, especially in connexion with the stream-tube apparatus shown in fig. 1.

3. Ideal Motion of a Fluid uninfluenced by Friction: Bernoulli's Equation.

From now on we shall use the method adopted in studying the mechanics of solid bodies; we shall attempt to *eliminate* effects due to *friction* as far as possible. We shall use the flow apparatus shown in fig. 5. It consists of a glass cell 1 cm. wide and 30 cm. long. Bodies with differing profiles may be moved about inside it just in contact with the glass. In fig. 5 the body is circular in outline, whereas in fig. 6 there are two bodies, *a* and *b*, which are supported by invisible bars, and together give an hourglass-shaped outline. For photographic purposes the glass cell is made free to slide along vertical grooves with constant velocity. For observation by projection on a screen the fixed arrangement of the cell shown in fig. 5 suffices. The cell is filled with water, and the next thing we have to do is to make the motion of the water visible. This is achieved by the addition of commercial aluminium-bronze powder (carefully purified from grease) to the water. The specks of metal indicate on the screen the magnitude and direction of the velocities of the individual particles of water throughout the cell at any moment. A photograph taken with an exposure of about 0·1 second shows the path of each speck of metal as a short line. These lines are practically all straight. Each line, briefly speaking, represents a velocity vector, indicating the magnitude and direction of the velocity of an individual particle of the water. The whole picture shows the velocity vectors which co-exist at a given moment. If the exposure is of suitable length (as e.g. in fig. 27, p. 205), the lines join up to form " stream-lines ". These give a remarkably clear picture of the flow of the fluid. They show us all the different directions of the velocities which co-exist in the fluid at a definite instant of time. In the case of *steady* motion the stream-lines also show us the entire paths described by individual particles *as time goes on*, i.e. stream-tubes.

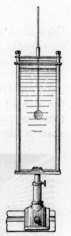

Fig. 5. — Flow apparatus. (1/11.) Here again it is often convenient to rotate the image through 90° in the course of projection, e.g. in figs. 6, 27, 28, and 31 of this chapter.

The photograph gives the configuration of the stream-lines in the clear form shown in fig. 6. More vivid, however, is the impression derived from direct observation by projection on a screen. But often a field with little detail and only a few clear lines is all we obtain. In such cases a peculiar circumstance comes to our assistance. The stream-

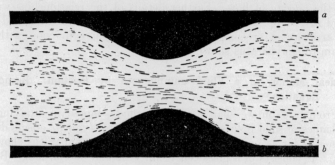

Fig. 6.—Stream-lines of a fluid at a constriction; photographic negative taken with the flow apparatus (fig. 5), using dark-ground illumination

lines of a steady flow practically uninfluenced by friction may be excellently imitated by a model experiment in which the stream-line apparatus of fig. 1 with its creeping flow is used. In spite of the totally different conditions giving rise to the flow, the stream-tubes in creeping flow are in complete formal agreement with the stream-lines in the flow of ideal frictionless fluid. Fig. 7 shows a picture obtained in this way, corresponding to fig. 6. But unlike fig. 6, it is, as we shall again emphasize, merely obtained from a *model* experiment.

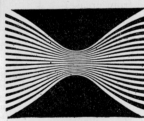

Fig. 7.—Stream-lines obtained from a model experiment (photographic positive with bright-ground illumination: similarly in figs. 11, 14–16, 18, 19, and 29 of the present chapter and in fig. 54 of Chapter XI (p. 249))

Formally, however, the picture is correct, and its simplicity makes it clear and impressive. The practically frictionless flow illustrated in this way can only be maintained for quite a short time. It bears some resemblance to the case of a sphere moving with constant velocity under no external forces in the mechanics of solid bodies. It is an idealized limiting case, for which, however, an important theorem of fundamental importance for all that follows is true. This theorem relates to the "static" pressure, i.e. the pressure of the fluid against a surface parallel to its stream-lines. Stated at first in qualitative form, it is as follows:

In regions where the stream-lines are crowded together or the velocity of flow is increased, the "static" pressure of the fluid is less than in the surrounding regions.

This theorem may be illustrated by the experiments shown in figs. 8, 9 (Plate V). Fig. 8 shows the pressure of the moving fluid *in front oj, at,* and *behind* a constriction. There is nothing diagrammatic about this figure. As a result of unavoidable frictional losses the pressure behind the constriction does not attain quite the same value as in front of it. In fig. 9 the velocity of flow used is considerably higher. In this case the pressure of the water at the constriction is less than the surrounding atmospheric pressure in the room. The water is capable of "sucking up" mercury in a U-tube manometer, and the mercury column thus raised may reach a length of several centimetres.

Quantitatively the relationship between the pressure and the velocity of a fluid is expressed by "Bernoulli's theorem". This theorem is readily obtained by applying the principle of the conservation of energy (p. 81) to a frictionless fluid in motion. We suppose that a mass of fluid m has volume V and density ρ and that a manometer moving with the fluid indicates the "static" pressure p. Then by the principle of energy the *sum* of the three following items must be constant:

I. The work done by the mass of fluid m, e.g., by moving a piston in a cylinder (hydraulic engine). This work amounts to pV, for the pressure p is equal to the force divided by the cross-section of the piston, whereas the volume is equal to the product of the cross-section of the piston and the distance through which the piston is displaced; that is, the product pV is equal to force × distance or work.

II. The potential energy still contained in the mass of fluid, which generally depends on h, a height above the ground; it is given by the product of weight by height, i.e. mgh in dynamical units.

III. The kinetic energy ($\frac{1}{2}mv^2$) still contained in the mass of fluid.

Hence

$$pV + mgh + \tfrac{1}{2}mv^2 = \text{const.,} \qquad \ldots \quad (3)$$

or

$$p + \rho gh + \tfrac{1}{2}\rho v^2 = \text{const.,} \qquad \ldots \quad (3a)$$

where p is measured in large dynes/m.^{2}, h in metres, ρ in kg.-mass./m.^{3}, and v in m./sec., and $g = 9.81$ m./sec.^{2}.

If the fluid is at a constant height above the ground the middle item, the potential energy, also remains constant, and we may write

$$p + \tfrac{1}{2}\rho v^2 = \text{const.} = p_1. \qquad \ldots \quad (4)$$

This is Bernoulli's theorem.

As we mentioned above, p is called the *static pressure,* for in theory it is measured by a manometer moving along with the fluid, that is, *at rest* relative to the fluid: $\frac{1}{2}\rho v^2$ is called the *dynamic pressure* or *velocity pressure.*

Bernoulli's equation therefore expresses the fact that *the sum of the static pressure and the dynamic pressure is constant*. The constant sum of these two pressures is called the *total pressure* p_1.

The apparatus illustrated in fig. 9 serves to measure the *static pressure p* in the moving fluid; the opening leading to the manometer lies *parallel* to the stream-lines. For measurements in the *interior* of broad channels the opening is made in the side of a " static tube ", usually in the shape of a slit or a number of sieve-like holes, the static tube being connected with a manometer by rubber tubing (fig. 10).

The total pressure p_1 is found at a stagnant region, such as is shown

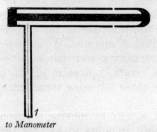

to Manometer

Fig. 10.—Section through a static tube with a ring-shaped slit, used for measuring the static pressure in the interior of a moving fluid.

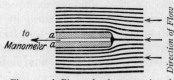

Fig. 11.—A Pitot tube for measuring the total pressure p_1 in a stagnant region. The tubes actually used are made of copper, with right-angled bends, and usually with an external diameter of only 2–3 mm. This figure was obtained from a model experiment with the stream-tube apparatus (fig. 1, Plate V), the outlines of the tube being indicated by subsequent shading

in the model experiment illustrated in fig. 11. In the middle of the stagnant region a stream-line impinges at right angles on the obstacle, and a tube leading to the manometer is brought from here (the Pitot tube in fig. 11). At this point the fluid is at rest, i.e. $v = 0$. By equation (4) the static pressure is equal to the total pressure p_1. The manometer then indicates the total pressure p_1.

The dynamic pressure is measured as the difference of the total pressure p_1 and the static pressure p. By equation (4), the dynamic pressure is

$$\tfrac{1}{2}\rho v^2 = p_1 - p \quad \ldots \ldots \ldots \quad (5)$$

(units as in (3a)). That is, to obtain the dynamic pressure we have to combine a pressure measurement of the type illustrated in fig. 10 with one of the type illustrated in fig. 11. To measure the static pressure p we use a static tube, to measure the total pressure p_1 we use a Pitot tube. For practical purposes it is convenient to combine the two instruments as shown diagrammatically in fig. 12 and in silhouette in fig. 13. Making observations with a combined Pitot and static tube is a favourite way of measuring the velocity of a moving fluid.

Figs. 8, 9 (Plate V) show how the static pressure decreases as the velocity of flow increases. This may also be illustrated by a variety of other lecture experiments, one or two of which we shall now describe.

(1) A smooth plate is set at an acute angle to the direction of flow. The course of the stream-lines (fig. 14) is obtained from a model experiment with creeping flow. We see two regions in which the stream-lines are crowded together. Hence there arises a couple in a

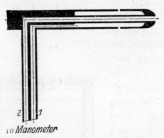

Fig. 12.—Section through a combined Pitot and static tube. A manometer is connected with the two limbs of the tube and gives the dynamic pressure directly as the difference of the total pressure p_1 and the static pressure p.

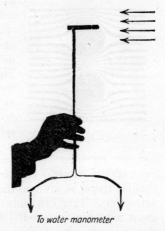

Fig. 13.—Silhouette of a combined Pitot and static tube (actual size)

clockwise direction, which tends to set the plate *at right angles* to the direction of flow. In this position a model experiment (fig. 15) exhibits a completely symmetrical configuration of the stream-lines. No further couple arises. Any semi-stiff sheet of paper falling to the ground exhibits the existence of this couple due to the stream-lines in the air. The sheet merely oscillates vigorously about the symmetrical position.

Fig. 14.—Stream lines of the flow round a plate set obliquely to the direction of flow (model experiment).

Fig. 15.—Stream-lines of the flow round a plate set at right angles to the direction of flow (model experiment; cf. fig. 20).

Subsequently we shall make use of this phenomenon for purposes of measurement.

(2) Two spheres are moving in a fluid. The line joining their centres is at right angles to the direction of the undisturbed stream-lines. The model experiment illustrated in fig. 16 shows that between the

spheres the velocity of flow is increased; as a result the static pressure between the spheres is diminished and the spheres "attract" one another. Fig. 17 shows a slight variation of this experiment. A wooden sphere is suspended, free to turn about a hinge, in a trough of water

Fig. 16.—Stream-lines between two spheres or two cylinders (model experiment).

as a reversed gravity pendulum. A second sphere is moved past it at a small distance by means of a sliding bar. The attraction of the wooden sphere is visible at a distance. A stop prevents the two spheres coming into contact. On the large scale we have the

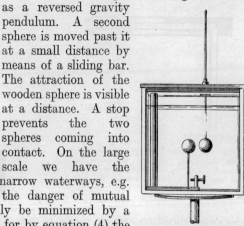

Fig. 17.—Attraction between a sphere at rest and a moving sphere.

case of two ships. In narrow waterways, e.g. canals, there is always the danger of mutual attraction. This can only be minimized by a great reduction in speed; for by equation (4) the dynamic pressure $\frac{1}{2}\rho v^2$ increases as the square of the velocity.

(3) Fig. 18 illustrates the course of the stream-lines at the hood of a ventilator, as obtained by a model experiment. About the points marked a there are regions of diminished pressure, to which the air flows from the shaft attached at b. For a lecture experiment the hood and shaft are made of glass. A bit of cottonwool is put in the shaft

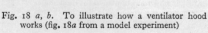

a   b

Fig. 18 a, b. To illustrate how a ventilator hood works (fig. 18a from a model experiment)

Fig. 19.—Stream-lines of the flow about a cylindrical bar (model experiment).

and someone blows from the right against the point of the hood. The cottonwool is sucked up the shaft and flies out at the opening, thus making the motion of the air visible to a large audience.

We now summarize the contents of this section and add a further important statement.

We have sought to observe motions in fluids where internal friction has no appreciable effect. We were supposed to be studying the behaviour of an " ideal " frictionless fluid. As a means for representing the phenomena we used the stream-lines. These stream-lines were really observed by means of the flow apparatus of fig. 5, p. 195. As a convenient way of avoiding laborious drawing they were imitated by model experiments carried out with the apparatus of fig. 1, Plate V. The stream-line diagrams, e.g. for the flow round a sphere and round a plate (figs. 3, p. 193, and 15, p. 199) were obtained in this way. A third figure is added (fig. 19) representing the flow round a long cylinder. This picture is likewise a model drawn from the stream-tubes of a creeping motion.

All these " stream-line figures " of " ideal " fluid have one feature in common. In no region of the fluid whatever do we observe any rotation of the particles. Every region of the fluid is *free from rotation*. This circumstance has determined the nomenclature: the motions of an ideal fluid which we have hitherto considered are said to be *irrotational motions*. The meaning of this term will be explained in the next section.

4. Irrotational Motion. Definition of Circulation (System of Reference Moving with the Fluid).

The form taken by the stream-lines definitely depends on the system of reference we choose. As usual, we referred the motion to the ground or floor of the lecture-room and from this point of view the bodies round which the fluid moved were at rest. We now pass to *another* system of reference, this time one moving with the fluid with a velocity equal to v_0, the velocity of the fluid in the region in front of or behind the body where the stream-lines are parallel to one another. This system of reference moving with the fluid may also be used with the stream-line apparatus. The phenomena of the flow are reproduced by means of a projection apparatus or by photography. All that we have to do is to fix the apparatus firmly to the ascending and descending cell, and again make short exposures. The stream-line pictures obtained in this way are quite different from those previously obtained, e.g. fig. 20 is replaced by fig. 21. Corresponding pictures for the flow round a sphere and round a cylindrical bar are to be found in figs. 22, 23.

The pictures taken with the system of reference moving with the fluid agree formally with the lines of force diagrams in electricity. Fig. 23, for example, resembles the lines of magnetic force of a long coil carrying an electric current, fig. 21 the stray field of a plate condenser (cf. *Physical Principles of Electricity and Magnetism*, fig. 4, p. 88, and fig. 4, p. 22).

Physically speaking, the stream-lines of moving fluid and the lines of force of a coil carrying a current have quite different meanings. In the first case their tangents represent the directions of the velocities of the particles of the water, whereas in the second they represent the direction of the intensity of a magnetic field. From the *formal*

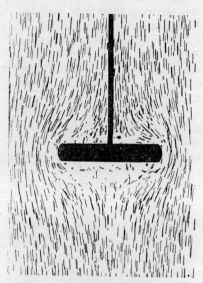

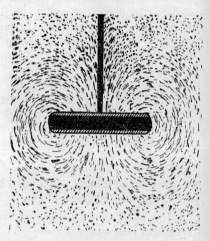

Fig. 20.—Stream-lines of the flow about a plate at right angles to the direction of flow, immediately after the start of the flow. Photographic negative with dark-ground illumination, using the stream-line apparatus of fig. 5, p. 195. This picture corresponds to the model experiment represented in fig. 15.

Fig. 21.—Stream-lines of the flow about a plate at right angles to the direction of flow, the system of reference moving with the undisturbed flow. Photographic negative with dark-ground illumination, using the stream-line apparatus of fig. 5, p. 195. The half-tone outlines of the body, which were blurred by the motion, have been subsequently replaced by shading: otherwise no retouching has been done. The same is true of figs. 22–24.

mathematical point of view, however, the diagrams may be dealt with in exactly the same way in both cases. It becomes necessary to bring in the mathematical symbolism of the " potential theory ", which is used in many other connexions. For this reason the flow in an " ideal " frictionless fluid is said to have a potential.

Physically, all the cases of flow we have hitherto discussed are distinguished by a definite feature. The fluid is free from rotation in all its regions, or free from *circulation*. Here the word *circulation* has the following meaning: imagine a line drawn within a region of the fluid with Indian ink (" fluid line ") and the product $v_s ds$ formed for every longitudinal element ds of this line, v_s being the component of velocity of the fluid in the direction of the line element ds. The sum $\Sigma v_s ds$ may be called the line sum of the velocity. In the "limit" we replace the line sum by the line integral $\int v_s\,ds$. This line integral taken round a closed line is written

$$\oint v_s\,ds,$$

and is called the *circulation* Γ.

"In every element of the region the circulation of the fluid is zero."

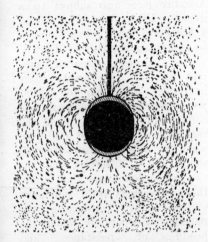

Fig. 22.—Stream-lines of the flow about a sphere, the system of reference moving with the undisturbed flow. The reader should compare this picture with the lines of force diagram for a magnetic sphere (see e.g. fig. 10, p. 93, in *Physical Principles of Electricity and Magnetism*).

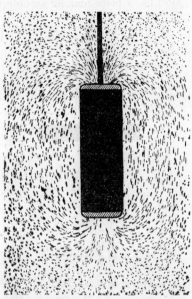

Fig. 23.—Stream-lines of the flow about a cylinder, the system of reference moving with the undisturbed flow.

This is the mathematical criterion for all the irrotational motions we have hitherto discussed, e.g. for those in figs. 20, 21. Or, expressing the same idea in quite a crude physical way: hitherto we have observed no rotations of the fluid particles taking place in any region of our ideal frictionless fluid in motion.

5. The Resistance of Bodies moving relative to Fluid.

From our previous stream-line pictures we derive a very remarkable result. As an example we consider a plate at right angles to the stream-lines (figs. 15 and 21). The configuration of the stream-lines in front and in the rear of the plate is completely symmetrical. By equation (4), p. 197, this complete symmetry of the stream-lines means

complete symmetry of the pressures and forces acting on the front
and rear sides of the plate. The sum of the forces acting on the plate
must be zero. That is, the plate must not be carried along with the
fluid. If we tow a plate along in an ideal fluid our arm should feel
no force. The motion of a plate in an ideal fluid must take place
without any " resistance ". This, of course, is in flat contradiction to
all observation; consider e.g. rowing, the behaviour of a child's
kite, i.e. a plate making an oblique angle with the wind as in the dia-

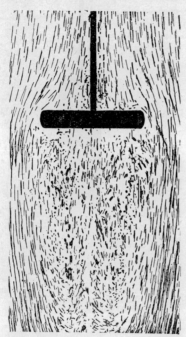

gram of fig. 14. According to this
idealized scheme the kite should
be acted on by a *couple* only.
Actually it is also subject to a
dynamic *lifting force* (" lift ") and
a *backward force* (" drag ") taken
by the string holding it. That is,
we have pushed the idealizing process too far. In reality it is
impossible for fluids to exhibit
perfect symmetry of the stream-line
configuration in front and in the
rear of a body immersed in it, even
when the body has a symmetrical
shape. In this section and the
two which follow we shall study
these deviations from symmetry
experimentally, considering the
phenomena separately under the
headings of " resistance " and
" lift ".

We do not need any new
apparatus; the flow apparatus of
fig. 5, p. 195, will amply suffice for
our experiments. We have only to
lengthen the time of observation
and increase the velocity of flow.

Fig. 26.—Distortion of eddies when
looked at from the point of view of a system
of reference at rest (the floor of the lecture-
room.)

As system of reference we shall use that moving with the fluid (i.e. the
camera moves with the velocity of the undisturbed fluid). We now
obtain the picture reproduced in fig. 24 (Plate VI). The irrotational
motion is only maintained in front of the body (above the body in the
figure). We see that *two* large eddies or vortices are formed at the rear
of the body. These begin to appear symmetrically at the right and at
the left, revolving in opposite directions. (The angular momentum,
initially zero, is therefore conserved.) Soon, however, the eddies become
detached from the body and move off with the fluid, while their place is
taken by fresh eddies. The latter, however, no longer arise symmetri-

PLATE VI

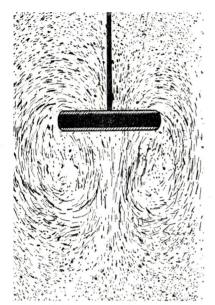

Chap. X, Fig. 24.—Formation of eddies when fluid flows past a plate set at right angles to the direction of flow (system of reference moving with the undisturbed fluid).

Chap. X, Fig. 25.—One of the eddies of fig. 24 after it has detached itself from the body (photographic positive with bright-ground illumination, again taken with the stream-line apparatus of fig. 5, p 195).

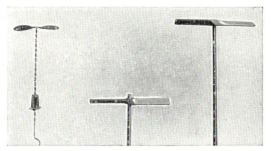

Chap. X, Fig. 33.—Three " propellers " with differing moments of inertia; *a* flies freely as a flat double aerofoil apart from the twisted wire used to set it in rotation, while *b* and *c* have the shafts attached to them.

cally on both sides, but large eddies follow one another on the left and on the right alternately. A photograph of one eddy of this kind is shown in fig. 25 (Plate VI).

With a system of reference *at rest* (the camera or projection apparatus being at rest relative to the body in the fluid), the progressive motion of the fluid is superposed on the eddying motion. The eddies are therefore distorted in the photograph; we obtain e.g. fig. 26, instead of fig. 24. In the case of subjective observation on a screen our eyes automatically eliminate this source of disturbance, and we see a distinct eddying motion superimposed on the slowly-moving background. Compared with subjective observation on a screen, even good photographs taken with the camera moving (e.g. fig. 25) seem very inadequate.

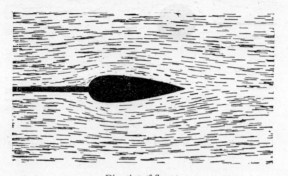

Direction of flow→

Fig. 27.—Drop-shaped profile (photographic negative with dark-ground illumination, using the stream-line apparatus of fig. 5, p. 195).

We now see in a flash how the resistance of bodies immersed in actual fluids originates; it is due to *rotation* in the fluid in the rear of the body. New regions of the fluid are continually being set in rotation. The working up of these eddies, the production of their kinetic energy, necessitates work being done. *The resistance of a body in moving fluid is due to rotations or eddy motion in its rear.* This is the surprising result of observation. Meanwhile we shall not touch on the question how these rotations originate; we shall see in § 8 (p. 212) that they are due to internal friction.

We shall next try to eliminate this troublesome eddy formation by giving the body a suitable shape. Nature has supplied us with many solutions of this problem. Their common feature is the " drop-shaped " profile shown in fig. 27. We may make the water in our glass cell flow past a body of this shape very rapidly without the objectionable eddy formation taking place. With the same fluid velocity a sphere of practically the same diameter at once gives rise to pronounced eddy formation (see fig. 28). The experimental advance obtained with the drop-shaped profile is extraordinary. This is shown by a numerical example: a half-crown in air exhibits the same resistance, owing to eddy formation in its rear, as a drop-shaped body nearly 1 metre long and 30 cm. in diameter. In engineering the drop-shaped profile plays an extremely important part; we mention only the shape of airships, submarines, torpedoes, struts and wires in aeroplanes, and so on.

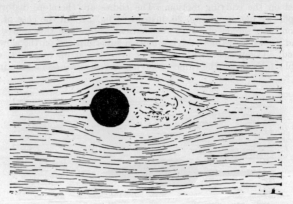

Direction of flow→

Fig. 28.—Formation of eddies behind a sphere (taken in the same way as fig. 26)

6. The Dynamic Lift on an Aerofoil a Consequence of Circulation.

In last section we regarded rotatory motion or circulation as a troublesome source of disturbance. Its occurrence in the rear of a body in moving fluid gives rise to the resistance of the body, which, however, may be greatly diminished by giving the body a suitable shape (drop-shaped profile) but cannot be eliminated altogether.

Fig 29.

In other cases, however, circulation in a fluid may prove extraordinarily useful. It gives rise to the " dynamic lift " on aerofoils and wings of all kinds, both natural and artificial. In these cases, to be sure, we have a particular case of rotatory motion or circulation. It is no longer confined to isolated patches of fluid; on the contrary, the fluid is subject to rotatory motion as a whole.

Fig. 29 shows a very simple type of " aerofoil ", namely, a plate set obliquely to air flowing in a horizontal direction (arrow 1). The air is deflected obliquely downwards and to the right by the aerofoil as roughly indicated by arrow 2. As a result of this change in the direction of motion of the air the aerofoil receives momentum in a direction obliquely upwards and to the right (arrow 3). Its vertical component (arrow 4) gives rise to the *lift* and its horizontal component (arrow 5) give rise to the *drag*. In a good aerofoil the lift considerably exceeds the drag. This is brought about firstly by setting the aerofoil at a small angle to the horizon (small " angle of incidence " or " angle of attack "), and secondly by giving the aerofoil a well-known special form. So much for a rough description of the phenomena which we shall now improve and extend by observation of actual cases of flow.

To begin with we obtain a rough idea of irrotational flow about an aerofoil profile by means of a model experiment on creeping motion. The resulting figure is reproduced in fig. 30. The characteristic feature of this figure is the flow round the sharp rear edge of the aerofoil.

We then make fluid containing specks of metal shavings flow round an aerofoil profile in our flow apparatus. We notice three things:

(1) It is only at the very beginning that the fluid flows round the sharp rear edge (" trailing edge ") of the aerofoil.

(2) Immediately thereafter an eddy is formed at the rear edge (fig. 31). Here, in contradistinction to fig. 24, we have only *one* eddy. In the figure it is ro- tating counter- clockwise and

Fig. 31—Formation of an eddy when an aerofoil begins to move.

Fig. 30.—Irrotational flow about an aerofoil (model experiment)

moves off with the fluid. It can be seen beautifully on a screen but is difficult to photograph.

(3) Simultaneously with the formation of this initial eddy the fluid begins to move more rapidly past the upper surface of the aerofoil and more slowly past the under surface (fig. 32). This may be demonstrated very clearly in a lecture experiment.

That is, experiment shows that the velocity of flow is increased on the upper side of the aerofoil or wing, and decreased on the under side. By Bernoulli's theorem (p. 197), this means decreased pressure above, diminished pressure below. This is how the dynamic " lift " on an aerofoil arises. The aerofoil is chiefly " sucked " upwards, but is also pushed up from below.

But how is it that this increase of velocity above and decrease of velocity below comes about? The answer is contained in our second observation. We see a counterclockwise rotation arising in the fluid. By the principle of the conservation of angular momentum, a second rotation of equal and opposite angular momentum must simultaneously arise in the fluid. Otherwise the sum of the angular momenta (initially zero) would not be conserved. This rotation consists of a *circulation of the fluid about the aerofoil*. In fig. 31 it would be in a clockwise direction with the aerofoil as centre, but it is not shown in the figure. The superposition of this circulation and the original irrotational motion

gives the actual flow about the aerofoil (fig. 32).   Above the aerofoil
the circulation is in the same direction as the progressive motion of
the fluid, and the two velocities combine to give a large resultant
velocity.   Below the aerofoil the two motions oppose one another and
the resultant velocity is small.

Here we have confined ourselves to a qualitative explanation of
the lift on an aerofoil.   If we consider the matter more accurately we
find that the arrow representing the force acting at the centre of
gravity of the aerofoil is always directed somewhat obliquely to the

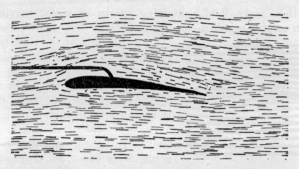

Fig. 32.—Flow about an aerofoil. The velocity of the fluid (shown by
the lengths of the lines) is greater above than below. (Photographic
negative with dark-ground illumination, using the flow apparatus of
fig. 5, p. 195).

rear, i.e. in addition to the lift there is always a disturbing backward
force or drag.

We now return to fig. 29 and inquire where the deflection of masses
of air in a downward direction comes into the representation by stream-
lines.   For the aerofoil cannot receive an upward impulse without other
masses being simultaneously accelerated downwards, by the principle
of the conservation of momentum.   The answer is that at the two
lateral edges of the aerofoil the high-pressure region below the aerofoil
and the low-pressure region above the aerofoil meet and air flows from
below the aerofoil to above.   As a result eddies or trails of eddies are
formed at the two edges of the aerofoil, and these as a whole possess
momentum directed downwards.   The change in direction of the air
indicated in fig. 29, therefore, is only a rough approximation to the
actual facts.

Wings or aerofoils are made use of both in nature and in engineering
in a great variety of ways.   In order that the dynamic lifting force may
arise, it is essential that there should be a flow about the wing-profile.
The aerofoil and the fluid must be in relative motion.   For lack of
space we shall have to confine ourselves to a few examples only:

(1) A child's kite is held in the wind by a string from the ground.

Then the air can flow round the kite without carrying it along horizontally. The approximation of the kite's profile to that of a good aerofoil is, to be sure, only middling, but quite sufficient for practical purposes.

(2) The thrust of the propeller gives an aeroplane a velocity relative to the air in a horizontal direction. The flow of air round its wings gives rise to the vertical lifting force. In ideal frictionless air the pilot could switch off his engine at any height and then continue to fly on at the same height. In reality, however, he loses kinetic energy after switching off the engine, and the aeroplane slows down. The loss of energy must be made good from its store of potential energy, so that the aeroplane slowly glides to earth.

(3) Sea-gulls following a steamer also exhibit gliding flight. They utilize the slight current which rises *obliquely* from the stern. Their downward gliding takes place not in air at rest but in air which is flowing *obliquely upwards*, and their flying height is thus maintained. Their losses of kinetic energy are not made good from the potential energy stored up as the result of muscular exertion, but are supplied by the oblique upward current of air, i.e. in the last resort at the expense of the engines of the steamer.

(4) Motorless aeroplanes for purposes of sport (" gliders ") usually practise gliding flight in upward currents of air. In order that they may rise it is necessary that the upward vertical component of the wind velocity shall exceed the downward vertical component of the velocity of gliding flight.

Gliders, however, are also capable of " sailing " like many birds. In sailing flight the necessary velocity of the wings relative to the surrounding air is produced by *forces of inertia*. For sailing flight a *change* of wind velocity is essential. The air is to be regarded as an accelerated system of reference. The pilot has merely to vary the angle of incidence and the direction of flight in a suitable way (and fill in the periods of constant wind velocity with downward gliding flight). The energy relationships are readily illustrated by model experiments. For horizontal variations of velocity, for example, we take a steel ball in a zigzag shaped glass tube held upright. If the tube is moved backwards and forwards with acceleration in a horizontal direction the ball rises in the tube as a result of forces of inertia.

(5) The sails of old-fashioned windmills are also typical examples of aerofoils, and were developed empirically to a high degree of efficiency; a knowledge of the theoretical principles underlying their action has not led to any appreciable advance.

(6) The propeller of an aeroplane or steamer bores its way into the air or water like a corkscrew. Its blades are nothing more or less than rotating aerofoils. We cover a wooden frame about 20 cm. by 80 cm. loosely with fabric as in a child's kite and set it revolving while held at arm's length. If the surface makes the proper angle of incidence

with the direction of motion of the air the forward thrust of this improvised propeller blade is clearly felt.

(7) The " propellers " of fig. 33 (Plate VI) form very nice children's toys. The moment of inertia may readily be altered by using shafts of various lengths. The two free axes of rotation are then exhibited very nicely. The axes of least and greatest moment of inertia (see fig. 33a (propeller without shaft) and fig. 33c) give stable flight, whereas the axis of mean moment of inertia (fig. 33b) again turns out to be completely unstable; the " propeller " wobbles about and falls to the ground (cf. Chapter VII, § 7, p. 115).

(8) Any rotation may be replaced by a periodic motion to and fro. An oar used as a scull at the stern of a small boat has the same effect as the rotating screw of a steamer. The sculling action may be illustrated equally clearly using a wing-shaped body consisting of a rigid wooden frame.

(9) In the flight of birds and of insects the wings have a double function. Firstly, they have to act as aerofoils and supply the vertical lifting force. Secondly, they have to act as propellers and supply the horizontal forward thrust, thus producing the necessary velocity of the creature relative to the air. This is done by means of sculling motions, some of which are very complicated.

The rotating propeller of the aero-engineer is nowhere realized in the animal world. The supply of blood-vessels and nerves from a bearing to a rotating shaft would appear to be incompatible with the elementary structure of living material.

7. Further Details of the Lift on an Aerofoil.   The Potential of a Circulation.

In our experimental investigation of the lift on an aerofoil described in last section, the essential point is the occurrence of circulation about the aerofoil. This circular motion of a fluid about a solid core is by no means confined to *aerofoil* profiles. Such profiles merely have the advantage of being capable of giving rise to the circulation necessary for the occurrence of the dynamic lifting force by themselves, without the use of any further device.

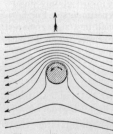

It is possible, for example, to produce a completely symmetrical circulation about a *cylindrical* core. The cylinder in this case must be made to *rotate* about its longitudinal axis. The circulation arises in the same way as in the case of the aerofoil. At first we see an eddy arise in the rear of the cylinder and move off with the fluid. In the end we have the state of affairs shown in fig. 34. If the motions of the fluid and of the cylinder are in the directions shown, the cylinder is

Fig. 34.—Stream-lines of flow about a rotating cylinder.

subject to an upward force in the direction of the feathered arrow. In order to demonstrate this phenomenon we use a light cardboard spool the size of a rolled-up table-napkin (fig. 35), the ends being closed by slightly projecting circular discs. A smooth piece of tape is rolled round the spool and the other end is attached to a stick, the whole forming a kind of whip. The whip-handle is moved smartly to one side in a horizontal direction. The cylinder thereby acquires velocity in a horizontal direction, and the unwinding of the tape simultaneously sets it in rotation. The conditions illustrated in fig. 34 are thus realized. Instead of flying in the ordinary parabolic path of a projectile, starting at the horizontal level, the cylinder flies up steeply and " loops the loop ".

Fig. 35.—Dynamic lift on a rotating cylinder (Magnus effect)

Fig. 36.—Windmill with symmetrically constructed sails

This upward force on rotating cylinders has been used to propel a new type of sailing ship (" rotor ship "), but the commercial results do not appear to have justified the exaggerated claims originally made. On the other hand, similar phenomena are utilized in sport in a great variety of ways. A " cut " tennis or golf-ball, i.e. a ball struck by a racquet or club moving obliquely, has a greater range than a ball moving in an ordinary parabolic path. The path of a " cut " ball resembles that shown in fig. 34, p. 120.

Another case where a dynamic lifting force arises, although the profile is not that of an aerofoil, is that of a small toy windmill with a pair of perfectly symmetrical sails of semicircular section. This merely consists of a stick of semicircular section attached to a shaft so as to rotate freely, as in fig. 36. When brought into a current of air this " windmill " at first remains at rest. Owing to the complete symmetry of the two sides there is no formation and detachment of an eddy at starting as in fig. 31, and hence there is no circulation about the " aerofoil ". Detachment of an eddy at starting, however, may be brought about by artificial means. All that is necessary is to give the mill angular momentum by means of a blow. This immediately sets up a circulation about the aerofoil, and the mill goes on rotating indefinitely. The direction of rotation is determined merely by the direction of the original angular momentum, and may be reversed at will by a blow in the opposite direction. This is a very instructive little toy.

The circulation which is utilized in producing a dynamic lifting force has two important characteristics: (1) the circulation takes place about a solid body; (2) it is *not* confined to a particular region of the fluid, but (in the typical case, at least) involves the whole mass of fluid. The velocity of the circulatory motion, it is true, falls rapidly as the distance from the solid core increases, but does not disappear completely even at a great distance.

The core of a circulatory flow, however, need not be a solid body. A familiar case of circulation in the absence of a solid body may often be observed when a bath is emptying. The core in this case consists of an air tube.

In the mathematical discussion all circulations of the whole mass of fluid about a core have a potential. The simplest case is that of symmetrical circulation about a rotating cylinder. The mathematics is to a great extent analogous to that of electric and magnetic fields. The reader need only compare fig. 34 with fig. 18, p. 144, in *Physical Principles of Electricity and Magnetism* to assure himself of the formal connexion between them.

The rotating cylinder, for example, corresponds to a straight conductor carrying a current. In hydrodynamics (p. 203) we have

$$\oint v\,ds = n\Gamma,$$

and in electricity

$$\oint \mathbf{H}\,ds = ni,$$

when the path of integration encircles the core or wire n times.

8. The Formation and Shape of Eddies: the Formation of Jets.

Rotations of limited regions of fluid we call eddies, or vortices. We have already observed these eddies occurring in actual fluid in two cases:

(1) In the rear of bodies in moving fluid, giving rise to resistance;

(2) At the rear or "trailing" edge of an aerofoil during the setting-up of circulatory motion about the aerofoil.

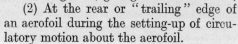

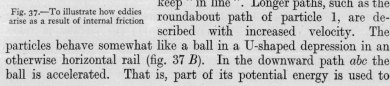

Fig. 37.—To illustrate how eddies arise as a result of internal friction

These eddies or vortices in the neighbourhood of obstacles arise as a result of *internal friction*. This may be clearly seen as follows. Fig. 37 *A* shows a very broad stream of an ideal frictionless fluid moving past a cylinder. The crosses indicate the same particles of fluid at three different times before and after they pass the cylinder. The particles of fluid keep "in line". Longer paths, such as the roundabout path of particle 1, are described with increased velocity. The particles behave somewhat like a ball in a U-shaped depression in an otherwise horizontal rail (fig. 37 *B*). In the downward path *abc* the ball is accelerated. That is, part of its potential energy is used to increase its kinetic energy. As it ascends the other side of the depression the exact reverse takes place. In the complete absence of friction the ball would regain its original level at e with its original velocity. In actual fact, however, external friction between the ball and the rail gives rise to energy losses. The velocity of the ball is less at e than at a. If the friction is great the whole kinetic energy of the ball may have been consumed by the time it reaches d. The ball then comes momentarily to rest at d, and then rolls backwards.

Internal friction has a similar effect in the case of particle 1 in fig. 37 C. In "climbing up" again to regions of high pressure it must flow past the layer of fluid which adheres to the cylinder and is therefore at rest. The particle is thus greatly retarded as a result of internal friction and is forced to turn back at the point d. This can only take

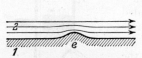

Fig. 38.—Waves arising at the
boundary of two fluids

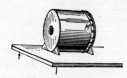

Fig. 39.—Apparatus for
producing vortex rings (1/40.)

place in the direction shown by the arrow in the figure, for this direction of rotation is forced on it by the more-rapidly moving particle 2 which is following it from outside. That is, the eddy begins by the detachment of a boundary layer which has separated itself from the advancing fluid.

So much for the formation of eddies in fluids moving past obstacles. In the last resort this is only a particular case of a more general phenomenon, namely, of eddy motion in the boundary layer between two currents of fluid with differing velocity in contact with one another. The boundary layer of two such fluids is unstable and gives rise to many types of eddy. We shall confine ourselves to three examples:

(1) In fig. 38 let 1 be a fluid at rest and 2 a fluid flowing parallel to it. Suppose that at e there is a small accidental bulge in the fluid at rest. This leads to a crowding together of the stream-lines in the moving fluid, roughly indicated in the figure by a few strokes. By Bernoulli's theorem the pressure in the region where the stream-lines are crowded together is diminished. The inevitable result of this is that the initial bulge must continually increase, that is, the boundary is unstable. The reader may imagine a flag laid in the boundary; it flutters and thus shows us the instability of the boundary layer in a crude but simple way.

(2) Fig. 39 shows a box with its rear surface formed from a stretched membrane M like that of a drum, and a sharp-edged circular hole in its front surface. The air inside the box is made visible by means of

smoke of some kind or other. If the membrane is struck a jet of coloured air shoots out of the hole. Internal friction arises in the boundary layer between the jet and the surroundings at rest, and gives rise to corrugation and rolling-up of the boundary layer, the result being the vortex ring well known to smokers. If the membrane is struck hard the ring may fly several yards into the room and may be made to blow down a house of cards, blow out a candle, and so on. The vortex ring possesses considerable momentum in its direction of motion.

The fluid eddies which we have hitherto discussed were bounded laterally by the walls of the vessel (the glass plates of the apparatus first described in fig. 5, p. 195). Fluids are often bounded above by their free surface and below by the bottom of the

Fig. 40.—Vortex sheet at the boundary between two fluids moving in opposite directions.

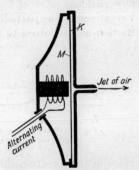

Fig. 41.—Intermittent formation of jets (2/5.)

containing vessel. In such cases the region of eddy motion extends from top to bottom.

In the interior of a fluid an eddy or vortex can never have either a beginning or an end, but always forms a more or less regularly-formed ring. The smoke-ring which we have just mentioned is a good example. It can be cut in two by a stick or other instrument, the result being two new closed vortex rings of irregular shape.

Mathematically these closed vortices are treated as vortex filaments, i.e. the radius of the core is taken as infinitesimal.

(3) The process of jet formation set up by hitting the membrane of fig. 39 lasted only a short time. A more lasting process may be set up in two ways:

(a) An excess pressure is maintained inside the box by supplying it continuously with air. Here again a vortex ring is produced at first. Subsequently the jet is separated from the air at rest by a boundary layer which must be thought of as resembling fig. 40 in section, the transition from the air at rest to the air in motion being accomplished by means of a large number of minute vortices.

(b) The air pressure inside the box is made to undergo rapid periodic alterations. For example, we may use a shallow box K (fig. 41) with a rear wall consisting of a telephone diaphragm M made to vibrate by means of an alternating current. The air is then ejected in the form of a jet, with a frequency corresponding to that of the alternating current, but is sucked in from all directions; thus the jet is an intermittent one.

If we breathe deeply we can observe the same phenomenon taking place quite slowly. In breathing *out* we can emit a puff of air strong enough to blow out a candle at a considerable distance, but we cannot " suck " it out while inhaling; when we breathe in the air flows uniformly into the nose or mouth from all sides.

Vortex sheets forming the boundary between a jet and the surrounding air are very unstable; they often break down, leaving the jet in fragments. We then have the confused mixture of fluid at rest and fluid in motion which is called *turbulence,* or, more generally, the turbulent mixture of layers of fluid with different velocities.

This turbulent flow may be demonstrated very conveniently by means of a jet of coloured water in a wide glass tube. The motion before and after turbulence has set in is shown in fig. 42 (Plate VII). Turbulence diminishes the quantity of fluid (i) passing per second for a given pressure difference $(p_1 - p_2)$. The resistance of the tube $(p_1 - p_2)/i$ is multiplied several times as a result of turbulence.

Turbulence may manifest itself by the occurrence of rushing sounds in the tubes through which fluid is flowing. When heard in the carotid arteries these sounds are a symptom of severe anæmia; normally the circulation of the blood should be free from turbulence.

Turbulent break-down of a jet is frequently brought about by trifling disturbances. " Sensitive " water-jets and flames are examples of this. We shall meet with these again in Chapter XI, § 14 (p. 265), and Chapter XII, § 6 (p. 284).

9. Reynolds' Number.

In our study of moving fluids we have been led as the result of experiment to distinguish three types of fluid motion.

(1) The laminar or creeping motion of a fluid when the effects of internal friction are predominant. This limiting case leads to the use of the stream-line apparatus as a very convenient method of investigation which saves a great deal of calculation (§ 2, p. 192).

(2) The flow of " ideal " fluid totally unaffected by friction. This limiting case may only be observed for a very short time after the beginning of the motion. It gives rise to irrotational motion without either circulation or eddies, even round sharp edges (§ 3, p. 195).

(3) The motion of ideal fluid as affected by internal friction. Internal friction leads to the formation of boundary layers, circulation, and eddies (§§ 5, 6, pp. 203, 206).

For a given configuration (pipe, slit, spiral path, &c.) the occurrence of any one of these three cases is determined by the magnitude of a certain quantity discovered by O. Reynolds in 1883. This is the quotient $lv\rho/\eta$, which is known as *Reynolds' number*. Here l is a length determining the magnitude of the body, e.g. the radius of a sphere, the diameter of a pipe, &c. (m.), v the velocity of the fluid relative to the fixed body (m./sec.), ρ the density of the fluid (kg.-mass/m.³), and η the coefficient of viscosity of the fluid (large dyne-sec./m.²). The quotient η/ρ is frequently called the *kinematic viscosity*.

Reynolds' number determines the ratio of two amounts of work done in the moving fluid, namely (1) the kinetic energy W_e and (2) the work W_f done against "internal friction".

For both amounts of work we make an assumption about dimensions; that is, we take all the lengths that occur as proportional to a length l which determines the magnitude of the body. We also leave out pure numbers occurring as factors of proportionality.

By p. 80 the kinetic energy is given by

$$W_e = \tfrac{1}{2}mv^2 = l^3\rho v^2. \quad \ldots \ldots \ldots \quad (6)$$

Using equation (7) of last chapter (p. 168) we obtain

$$W_f = F.l = \eta a \frac{v}{l}.l = \eta l^2 v \quad \ldots \ldots \quad (7)$$

for the work done against friction. From (6) and (7) we obtain the quotient

$$\frac{W_e}{W_f} = \frac{l^3\rho v^2}{\eta l^2 v} = \frac{lv\rho}{\eta}$$

for Reynolds' number. Small values of Reynolds' number mean that the work done against friction predominates, large values that the kinetic energy predominates. To the ideal frictionless fluid there corresponds the value ∞ for Reynolds' number.

In any quantitative treatment of fluid motion Reynolds' number plays a great part. Experiments on definite geometrical forms may first be carried out in linear dimensions convenient for experimental purposes, and the results obtained may subsequently be transferred to other linear dimensions. To do this it is only necessary to ensure by suitable choice of velocities and densities that the value of Reynolds' number is the same in both cases.

We give one or two numerical examples.

(1) In our stream-tube apparatus of fig. 1, Plate V, we obtained creeping motion for a value of Reynolds' number about 10. For the distance between the plates was $l = 1$ mm. $= 10^{-3}$ m., v, the velocity of the fluid, was about 1 cm./sec. $= 10^{-2}$ m./sec., and ρ, the density of the water, is 1000 kg.-mass/m³. The coefficient of viscosity is 10^{-3} large dyne-sec./m.², so that Reynolds' number

$$= \frac{10^{-3} . 10^{-2} . 10^3}{10^{-3}} = 10.$$

(2) In our stream-line apparatus with a spherical profile inserted (fig. 3, p. 193) we found that strong eddies were produced in the rear of the body for a value of Reynolds' number about 300. For the dis-

tance between the plates was $l = 1$ cm. $= 10^{-2}$ m., and the velocity v about 3 cm./sec. $= 3.10^{-2}$ m./sec., so that Reynolds' number

$$= \frac{10^{-2}.3.10^{-2}.10^3}{10^{-3}} = 300.$$

(3) Ordinary aeroplanes use values of Reynolds' number of the order of 10^7 ($v \sim 50$ m./sec., breadth of wing $l \sim 2$ m., density of air ρ at $20^\circ \sim 1.3$ kg.-mass/m.^{3}, coefficient of viscosity $1.7 . 10^{-5}$ large dyne-sec./m.^{2}.

The magnitude of this Reynolds' number has an annoying practical consequence; it prevents the study of technically important problems by means of *small models*. If the value of l for the model is small the high values of Reynolds' number attained in actual practice can only be reached by making the velocity of flow v very large. Then, however, we reach air velocities which are no longer small compared with the velocity of sound (340 m./sec.), when air may no longer be treated as an incompressible fluid. Hence for the experimental investigation of problems of practical importance it is necessary to use models of comparatively large dimensions. This is the explanation of the high expenses of institutes for aerodynamical research, several of which now exist.

(4) Turbulence in channels and tubes (§ 8, p. 215) occurs for values of Reynolds' number between 1200 and 10,000, according to the shape and nature of the wall surface.

Air passes into the nose with a velocity of about 2 m./sec. The width of the nasal canal is of the order of 0.01 m. For ordinary air the density $\rho = 1$ kg.-mass/m.^{3} and the coefficient of viscosity $\eta = 1.7 . 10^{-5}$ large dyne-sec./m.^{2}. That is, in breathing through the nose we have to reckon with a Reynolds' number of about 1000. If the internal structure of the nose is normal the flow of air takes place without turbulence. If, however, the area of its cross-section widens greatly inside, marked turbulence may set in, giving rise to an increase in frictional resistance. Abnormally wide internal nasal cavities appear to be permanently " stuffed up " (cf. p. 215). Flow phenomena in the body, such as the flow of the blood through the elastic artery-tubes, and the dangerous eddies in the veins which lead to the formation of clots, have hitherto received far too little attention from investigators. Here, of course, any advance could only be brought about by extraordinarily troublesome investigations.

This brief summary of facts concerning Reynolds' number must suffice. In this chapter on fluid motion we have confined ourselves almost entirely to a qualitative discussion of the phenomena, which enables us to go a considerable way. A quantitative treatment necessitates a great deal of mathematics, even in cases which are apparently simple. This forms the subject of an extensive but extremely specialized literature.

10. Waves at the Surface of a Fluid.

Hitherto we have only considered motions in the *interior* of fluids. Here, as throughout the whole chapter, the word *fluid* is used as a general term, including both liquids and gases. Hence in general " surface " means the boundary between two fluids of unequal density.

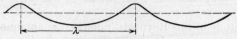

Fig. 43.—Profile of a water wave

That is, in the first place, it means the surface of a liquid in the ordinary sense of the word, and in the second place the indefinite boundary between two gases of unequal density. We are already familiar (§ 14 of last chapter, p. 183) with the boundary of two gases taking the place of a surface.

The existence of waves on water surfaces is a matter of our everyday experience. These waves have by no means the simple shape of the sine wave. Their troughs are broad and flat, and their crests high and narrow. Fig. 43 shows an instantaneous photograph of a water wave advancing from left to right. In spite of their complicated form the water waves in a wash-basin or soup-tureen will in the next chapter turn out to be a very useful means of avoiding much calculation. Here, therefore, we shall give a broad outline of the formation of waves on the surface of fluids.

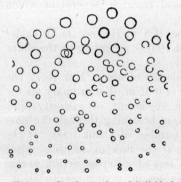

Fig. 45.—Circular motion of individual particles (orbital motion) in a progressive water wave (photographic negative with dark-ground illumination); cf. fig. 45, p. 244). The upper edge of the figure does not represent the outline of a wave but is merely due to the way in which the aluminium shavings were accidentally distributed.

For this purpose we begin by experimentally producing a train of waves which is easy to observe. For this we use a ripple tank, a long narrow lead box with glass windows at the side (about $150 \times 30 \times 5$ cm.). This is filled about half-full of water with which, as before, aluminium shavings are mixed to make the motions of the particles visible. To start the wave motion we use a block moved up and down by means of a motor. As the waves advance we see a stream-line configuration like that in fig. 44 (Plate VII), which is a time exposure of about $\frac{1}{25}$ sec. This stream-line diagram is for an observer at rest in the lecture-room and shows how the directions of the velocities are distributed.

In a wave the motion of the fluid is not steady. Hence the paths

PLATE VII

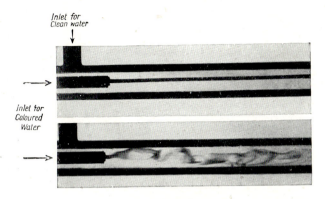

Chap. X, Fig. 42.—A jet of coloured water before and after turbulence has set in

Chap. X, Fig. 44.—Stream-lines in a progressive water-wave (photographic positive with bright-ground illumination)

Chap. XI, Fig. 17.—Apparatus for demonstrating the superposition of two simple harmonic vibrations.

Chap. XI, Fig. 35.—Projection of the vibration curves of a string with the help of a rotating lens disc.

described in the course of time by the individual particles of fluid by no means coincide with the stream-lines (cf. p. 195). The former have quite a different appearance and for moderate wave amplitudes are nearly circular. These circular paths are found both at the surface

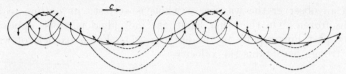

Fig. 46.—Connexion between the stream-lines and the circular motions of the individual particles in progressive water waves. By joining the small arrow-heads we obtain the profile of the wave advancing towards the right at the close of the next interval of time. The circular motions of the particles are shown only for every second arrow.

and at greater depths, but the diameters of the circles are greatest in the upper layers of the fluid.

In order to demonstrate these circular paths of the individual particles (" orbital motions ") we have merely to add a few aluminium shavings to the water. We also make the exposure of the photograph equal to a period of the wave. In this way we obtain the diagram reproduced in fig. 45.

To make observation on a screen easier we can diminish the velocity of the wave. For this purpose we put into the ripple tank two fluids of very nearly the same density in layers, e.g. salt water below, fresh water above. The block used to set up the waves is made to penetrate the boundary layer between

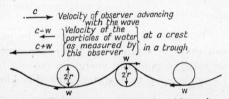

Fig. 47.—The paths of the individual particles as they appear to an observer advancing with the wave

them. The surface of the fresh water exposed to the air remains practically at rest. Along the surface between the salt water and the fresh water, on the contrary, a wave of large amplitude travels slowly towards the right.

Our experimental results serve as basis for the diagram of fig. 46, which contains the circular paths of some of the particles at the surface. Their diameter $2r$ is equal to the difference in height between wave-crests and wave-troughs.

We shall call the velocity in the circular path w, so that

$$w = \frac{2r\pi}{\tau};$$

τ, the time of a complete revolution, corresponds to the advance of the wave through a complete wave-length λ, so that $c\tau = \lambda$.

To simplify the calculations we shall assume that the surface is one of water exposed to air. We shall neglect the density and kinetic energy of the air as compared with those of the water.

Further, we shall henceforth assume that the observer is moving towards the right with a velocity c equal to that of the waves. For such an observer the wave as a whole is at rest; its outline to him seems frozen. On the other hand, the individual particles appear to hurry past him rapidly to the left (fig. 47). For a particle of water in a trough he obtains the velocity

$$v_1 = c + \frac{2r\pi}{\tau},$$

i.e. kinetic energy

$$\tfrac{1}{2}mv_1^2 = \tfrac{1}{2}m\left(c + \frac{2r\pi}{\tau}\right)^2.$$

For a particle of water in a crest he obtains the kinetic energy

$$\tfrac{1}{2}mv_2^2 = \tfrac{1}{2}m\left(c - \frac{2r\pi}{\tau}\right)^2.$$

The difference between these two kinetic energies is

$$\tfrac{1}{2}m(v_1^2 - v_2^2) = \frac{4r\pi cm}{\tau}. \quad \ldots \ldots \quad (8)$$

This gain of kinetic energy on the part of the particle in a trough can only be attained at the expense of the potential energy. The decrease in potential energy in passing from a wave-crest to a wave-trough is equal to weight × loss of height, i.e. $mg \times 2r$. We thus have

$$2mgr = \frac{4\pi rcm}{\tau} \quad \text{or} \quad c = \frac{g\tau}{2\pi}. \quad \ldots \ldots \quad (9)$$

Further, in the limiting case of small amplitudes we may neglect the diameter of the circular paths in comparison with the distance between two neighbouring wave-crests, and we may regard the outline of the wave as a sine-curve. For the sine-wave we may, as we have seen already, put

$$c\tau = \lambda; \quad \ldots \ldots \ldots \quad (10)$$

hence

$$c = \sqrt{\frac{g\lambda}{2\pi}}. \quad \ldots \ldots \ldots \quad (11)$$

The velocity of propagation of transverse waves on the surface of the water (c) depends on the wave-length (λ). There is " dispersion " (this term being used to denote any dependence of a property on a wave-length).

The above statement may frequently be verified qualitatively in ordinary experience. Short water waves are overtaken by long ones, carried on their backs for some distance, and then left behind.

In deducing equation (11) the potential energy due to surface tension was omitted as being negligible in comparison with that due to weight. This is permissible down to wave-lengths of about 5 cm.

If surface tension is not negligible the item $2\pi\sigma/\lambda\rho$ must be added to the expression under the square root sign in equation (11), where σ is the surface tension given in the table on p. 172 and ρ the density of the fluid.

Equation (11) involves the assumption that the water is " deep ", but is only applicable down to depths of $0{\cdot}5\ \lambda$.

In the opposite limiting case of infinitesimal depth h, the velocity of propagation of low waves becomes independent of λ, i.e. we have

$$c = \sqrt{gh}, \qquad \dots \qquad \dots \quad (12)$$

and there is no dispersion.

As we mentioned before, we shall find these water waves of great use in the following chapter. Here we shall merely mention three cases involving waves in boundary surfaces.

(1) *The formation of waves at the boundary between two layers of air differing in density.* Variations of density in the atmosphere are due to differences of temperature. The occurrence of these very slow progressive waves is manifested by periodic condensation of the water vapour in the form of long parallel bands of white cloud. The wave motion is to be thought of as arising in the way shown in fig. 38, p. 213.

(2) *Dead water.* The remarkable phenomenon of " dead water " is not infrequently observed near the mouths of rivers, particularly in the Scandinavian fiords. Ships moving slowly, i.e. at 4 or 5 knots, are suddenly retarded by an invisible force and sailing ships often refuse to obey the helm. Here the case described on p. 219 is realized in nature; we have a layer of fresh water superimposed on salt water. A ship of sufficient draught to reach the boundary surface sets up waves of large amplitude in the boundary surface. These are invisible to the eye, and the visible air-water surface remains practically at rest. The whole energy of this wave-motion must be supplied by the ship, hence the strong retarding effect on its motion. The circumstances are therefore similar to those associated with the resistance of bodies immersed in fluid, which results from eddies set up in the rear.

(3) The *wave resistance* of ships is chiefly due to the formation of bow and stern waves. Both of these continually carry away energy supplied by the ship backwards and sideways. By giving the ship a suitable shape it is possible to make the bow wave and stern wave partly cancel each other (by interference; cf. Chapter XII, § 12, p. 297). The shapes thus obtained for steamers and sailing ships at the waterline are quite different from the drop-shaped profile of a submarine.

ACOUSTICS

CHAPTER XI

Vibrations

1. Preliminary Remarks.

Acoustical theory originally developed in close relationship with the study of hearing and of musical problems. In the human ear, in fact, we have an extremely sensitive detector of mechanical vibrations for an astonishingly wide range of frequencies (from about 20 to 20,000 vibrations per second). The significance of the facts and relationships obtained in this way, however, extends far beyond the special subject of acoustics, or theory of sound. Hence in a modern textbook it is convenient to separate the purely mechanical problems of vibration theory from the physiological aspects of sound, and the material of this chapter and the next has been subdivided accordingly.

2. Production of Undamped Vibrations.

Hitherto we have exclusively considered the simple harmonic or sinusoidal vibrations of a simple pendulum with a linear law of force. Diagrams of this type of pendulum are given in fig. 1, p. 50, and fig. 5, p. 53. The pendulum is set vibrating by giving the bob a blow. Its motion is damped, the amplitude of vibration falling off as shown in fig. 6, p. 54. The pendulum gradually loses the energy originally given to it by the blow, chiefly as the result of unavoidable friction.

For many purposes in physics, engineering, and music, however, we require *undamped* vibrations, i.e. vibrations whose amplitude remains constant in time (fig. 6, p. 54, top). These undamped vibrations cannot be produced unless the energy losses we have just mentioned are continually replaced. The processes devised for this purpose may be briefly referred to as " automatic regulation ", and they form our next subject for discussion.

We begin with a special case which is easy to understand, namely, the automatic regulation of a gravity pendulum by means of an electromagnet. That is, at the very beginning we refuse to restrict ourselves to purely mechanical methods. We are thus enabled to present the essential features of automatic regulation in a particularly clear way. A small iron block Fe (fig. 1) is placed under the pendulum

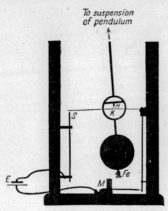

To suspension of pendulum

bob and serves as armature for the electromagnet M. For this purpose the circuit of the accumulator E must be closed by the switch S, which consists of a vibrating spring actuated at the proper instant by the pendulum itself. This is done by means of the small auxiliary pendulum H (which is made visible even in the outline diagram owing to the rod of the main pendulum being interrupted by a ring). If the amplitude of swing is large the auxiliary pendulum slips over the knob K on the spring. If the amplitude falls

Fig. 1.—Automatic regulation of a gravity pendulum

Fig. 2.—Part of the apparatus of fig. 1

to a lower limiting value, on the other hand, the pointed tip of the auxiliary pendulum catches in the middle hollow of the knob (fig. 2). Hence the pendulum in returning from right to left presses down the contact spring of the switch S. The electromagnet is excited for a brief interval of time, and the pendulum is accelerated in the direction of its motion.

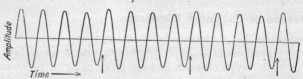

Fig. 3.—Sinusoidal vibrations of a gravity pendulum with automatic regulation, the energy being replenished after every fourth vibration

In this way the store of energy of the pendulum is repeatedly brought back to its initial value after a definite number of vibrations, and we have e.g. the vibration diagram of fig. 3. In this example the periodical replacement of energy or the periodical return to the maximum value of the amplitude takes place after every four vibrations. The interval between the successive replacements of energy may be increased or diminished as desired by altering the dimensions of the apparatus. If the energy is to be replenished after *every* vibration the construction of the apparatus may be considerably simplified as regards externals, for we then have the ordinary electric bell arrangement nowadays familiar to any school-boy (fig. 4). A pendulum with an iron rod is placed in front of the pole of an electromagnet M. The pendulum rod carries the contact spring of the switch S.

The essential feature of the working of the electric bell is frequently misunderstood. During the closing of the circuit the pendulum is accelerated by the electromagnet. This acceleration takes place not only during the quarter vibration 1—0, but also during the quarter vibration 0—1. In the path 0—1, however, the acceleration has the

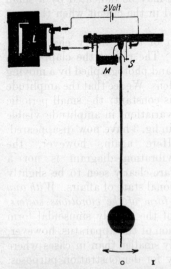

Fig. 4. — Automatic regulation of a gravity pendulum on the principle of the electric bell.

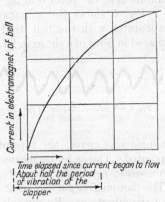

Fig. 5. — The rise of the current in the automatically-regulated electric bell circuit of fig. 4

wrong sign; it is in the direction opposite to that in which the pendulum is moving, and thus retards the pendulum and diminishes its energy. Hence it is necessary that an additional condition should be fulfilled, namely, the gain of energy in the path 1—0 must *exceed* the loss of energy in the path 0—1. It is only the *difference* of these two energies that goes to benefit the pendulum. Practically this means that the time-average of the current in the electromagnet must be smaller for the path 0—1 than for the path 1—0. That is, the current in the electromagnet must *rise* after the switch is closed, while the pendulum moves from 0 to 1 and back again, somewhat as in fig. 5. The current must not rise to a constant value immediately after the circuit is closed as it would do in a non-inductive circuit.

This slow rise of the current is ensured in practice by giving the circuit a sufficiently high *inductance*.* This is done either by using an

* See *Physical Principles of Electricity and Magnetism*, p. 168 *et seq.*

electromagnet with a great many turns or by inserting an extra coil of high inductance into the circuit. The latter arrangement is particularly well suited to lecture experiments. The pendulum used (fig. 4) is a slowly oscillating one which takes about 2 seconds to execute a complete vibration. This apparatus will not work unless the auxiliary inductance L is put into the circuit; in the absence of this coil the current in the electromagnet reaches its full value in less than $\frac{1}{100}$ second, whereas with the coil the rise of the current occupies about a second. The rise of the current may be indicated by a small electric lamp (2 volts, 1 ampere) inserted in the circuit when the pendulum is at the point 0 in fig. 4.

Fig. 6 shows the vibration diagram of an ordinary electric bell apparatus with the bell itself removed. The rod of the clapper was placed in front of a slit as in fig. 19, p. 11, and photographed by a moving

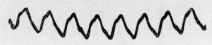

Fig. 6.—Vibration diagram of the clapper of an electric bell

lens. We see that the amplitude is constant; the small periodic variations in amplitude visible in fig. 3 have now disappeared. Here again, however, the vibration diagram is not a simple sine curve; the tops of the arcs are clearly seen to be slightly pointed. This is by no means an exceptional state of affairs. *With any automatic regulating apparatus the sine form of the vibrations suffers.* The damping is got rid of at the cost of the strictly sinusoidal form of the vibrations. By suitable construction of the apparatus, however, the deviations may be made considerably smaller than in cases where they have been exaggerated deliberately for demonstration purposes.

In the examples which follow the undamped vibrations are produced by exclusively *mechanical* means. In the two first cases a rotating shaft is used to regulate a gravity pendulum. In the first case the periodical connexion of the pendulum with its source of energy is brought about by the " sticking " or " hooking " together of two bodies at rest relative to one another.

Fig. 7 shows a side view of a gravity pendulum resembling those used in medium-sized clocks. It is clamped to a shaft about 4 mm. across by means of two padded jaws. When the shaft begins to move the pendulum is carried forwards with it. The clamps stick or hook themselves on to the shaft (statical friction). At a definite amplitude the couple due to the weight of the pendulum becomes too great for the cohesive forces and the connexion is broken. The clamps, retarded by external friction, slide freely on the shaft, and the pendulum swings back. As the pendulum again moves forward the relative velocity of the padded surface of the clamps and the surface of the shaft at a certain instant becomes equal to zero. The two bodies are at rest relative to one another, the clamps again adhere to the shaft, and the pendulum is again carried forward to the point at which it broke away. Its second vibration begins with the same amplitude as the first, and so on. Here again the vibrations are not of a good sine form; asymmetry of the vibrations can usually be detected even by the naked eye.

For long periods of time a more uniform action is obtained by the use of the anchor escapement familiar in watch - making. The driving shaft carries a balance wheel with asymmetrically-cut teeth. The pendulum rod is rigidly fixed to a bent piece of metal ending in two sharp points ("pallets"). The driving shaft is acted on by a torque produced by a spring or by a cord with weights. The shaft rotates by "jumps" of constant amount. While the pendulum is moving towards

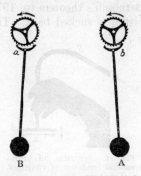

Fig. 8.—Automatic regulation of a gravity pendulum by means of an anchor escapement.

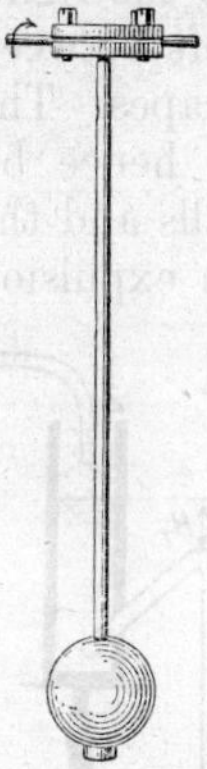

Fig. 7.—Automatic regulation of a pendulum by means of a rotating shaft and friction.

the left a tooth of the balance wheel presses on the inner surface of the pallet b in fig. 8A. The pendulum is thereby accelerated towards the left. Soon after the mean position is passed the tooth slips away from b. Immediately thereafter the pallet a engages with the wheel (fig. 8B); the pendulum then moves back towards the right.

Fig. 9.—Hydrodynamical regulation of a tuning fork

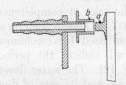

Fig. 10.—Details of the apparatus

This apparatus for automatic regulation is known as the anchor escapement and works exceedingly well* for medium frequencies (up to about 100 per second).

For higher frequencies hydrodynamical regulating apparatus is preferable. An arrangement of this kind for driving a tuning-fork is shown in fig. 9, the essential parts being shown in section in fig. 10.

* As e.g. in the balance-wheel of the stop-watch of fig. 21, Plate II, which reads to $\frac{1}{100}$ second.

A piston *a* fits into the cylinder *b* with a small amount of play, touching the cylinder nowhere. The cylinder is connected to a high-pressure air pipe. The air pressure drives the piston out of its position at rest in the cylinder and thus forces the prong of the tuning-fork to the right. When the piston leaves the cylinder a ring-shaped slit is left between the piston and the cylinder-wall, through which the air escapes. The stream-lines of the air-flow are crowded closely together; hence by Bernoulli's theorem (p. 197) the static pressure of the air falls and the piston is sucked back. This continual alternation between expulsion

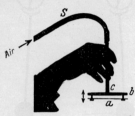

Fig. 11.—Apparatus for hydro-dynamical regulation of vibrations, e.g. of the larynx

and sucking-back of the piston may be demonstrated even more clearly by means of the apparatus sketched in fig. 11. The tube *c* ends in a disc *b*, in front of which an easily-movable plate *a* is held loosely by means of two nail-shaped rods. When air is blown into the tube *c* the plate vibrates up and down in the direction of the double arrow and emits a loud humming noise.

Fig. 12.—Relaxation oscillations

Numerous forms of automatic regulation by hydrodynamical means occur both in nature and in engineering. Some of these cases are very complicated and can be understood only superficially by a reader unacquainted with "coupled" vibrations (§ 16, p. 268). The above summary, however, will suffice for the following sections.

For the sake of completeness we shall here mention "*relaxation oscillations*". These depend on a type of automatic regulation which is but seldom used for mechanical purposes. As an illustration one example will suffice: the periodical emptying of a vessel containing water (fig. 12). The water flows in through the pipe at the top right-hand corner, and its rate of flow can be retarded at will by means of a tap. A siphon entering the left-hand side of the vessel comes into action when the water reaches a certain level. When this level is reached the air compressed in the long limb of the siphon is capable of ejecting the column of water H_2 which plugs the lower opening. Once the siphon has begun to act it speedily empties the vessel completely. In the end only the lower U-shaped end of the siphon contains water. The vessel then slowly fills again, and the process is repeated.

Relaxation oscillations doubtless play an important part in the periodic processes in organisms (e.g. the action of the heart)—the supply of liquid being slowly replenished by diffusion.

3. Representation of Non-sinusoidal Vibrations by means of Sinusoidal Vibrations.

In the production of undamped vibrations by automatic regulation we met with some vibration curves in which the deviations from the sine form were considerable. These forms, however, were very closely related to simple sinusoidal vibrations. Non-sinusoidal vibrations may either be described in a formal mathematical way in terms of simple sinusoidal vibrations (by means of a Fourier series) or built up out of simple sine vibrations by physical means.* This process is called *superposition* or *interference*.

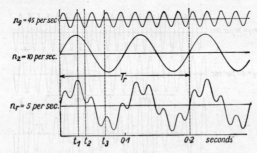

Fig. 13.—Superposition of two sinusoidal vibrations differing
widely in frequency

In both cases the component sine vibrations must have a definite frequency (n), amplitude (A), and phase (ϕ). We shall illustrate this by means of some examples.

We begin with the simplest case, that of the superposition or interference of two sinusoidal vibrations only, using their graphical representations. In the two first horizontal rows of fig. 13 we see two sine curves, one above the other. The suffixes 2 and 9 chosen to indicate these vibrations are derived from the frequencies by cancelling the common factors. Henceforward we shall always use this notation. The lower vibration has a greater amplitude than the upper, but a lower frequency $(A_2 > A_9;\ n_2 < n_9)$. Further, the figure contains four dotted vertical lines. The intercepts on these lines between the curves and the axes of abscissæ indicate the amplitudes of the two vibrations co-existing at certain instants. Amplitudes directed upwards count as positive, amplitudes directed downwards as negative. We add up these amplitudes for the various instants, t_1, t_2, &c., and plot the resultants using the lowest horizontal axis. We thus obtain a complicated curve which is not sinusoidal. In the first instance we shall formally assign it the frequency n_r. In the present case the

* For further details see p. 259.

frequencies of the two component vibrations differ considerably; $n_9 = 4.5 n_2$.

For a second example we choose frequencies which are very nearly equal ($n_7 = 7 n_6/6$ in fig. 14) and amplitudes exactly equal. We carry out the addition along the verticals just as before, and thus obtain the resultant curve n_r. This has the appearance of a sine curve with a periodically-varying amplitude. An interference curve of this kind may be called a "beats curve" *. In the example chosen the vibration comes to rest at every *minimum* of the beats, the equal amplitudes of the two component vibrations being at those instants directed in opposite directions; their difference of phase amounts to

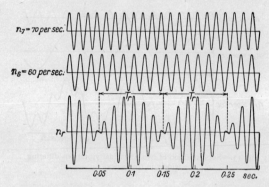

Fig. 14.—Superposition of two vibrations of almost equal frequency (beats)

180°. At the *maxima*, on the contrary, the two amplitudes have the phase difference zero and combine to give an amplitude of twice the original value.

If the two component vibrations have unequal amplitudes the beats-minima are less well-marked.

In a third case we use two vibrations of equal amplitude, but one has a frequency twice that of the other ($n_2 = 2 n_1$). This time we take their phases into account.

Case I.—In fig. 15 the two vibrations begin simultaneously with zero phase at the instant t_0. The resultant curve n_r is shown at the foot of fig. 15.

Case II.—In fig. 16 the amplitudes and frequencies remain as before, but the vibration n_2 begins at the instant t_0 with phase 90° or maximum amplitude. Although the amplitudes and frequencies are unaltered, the resultant curve n_r looks quite different from that in fig. 15. This example clearly brings out the influence of phase on the form of the resultant vibration.

* Ger., *Schwebungskurve*.

So much for these very instructive *graphical* examples, which, however, use up a great deal of time. The superposition of two simple harmonic vibrations may be demonstrated much more easily and rapidly by *mechanical* means. We start from the relationship between motion in a circle and simple harmonic motion, which we have used so frequently before. We again make a rod rotate in front of a slit and observe the successive appearances in time by means of successive images in space (by placing a rotating mirror in the path of the ray of light). The rod and slit are seen at the middle of fig. 17 (Plate VII). At first we imagine the gear wheels removed and the two upper shafts 1 and 2 each connected with a separate motor. The two ends of the rod are fixed into holes in the periphery of two circular discs I and II. Each disc may be set into rotation by its own motor at any desired rate (say n revolutions per second). To begin with we shall sup-

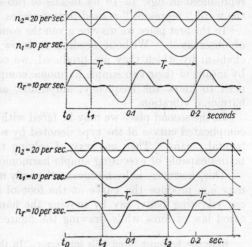

Figs. 15, 16.—The influence of phase on the form of the resultant vibration

pose that disc II is at rest and disc I rotating with frequency n_1. The rod then describes the curved surface of a cone in front of the slit. The image of the slit projected by a moving (polygonal) mirror gives a sine-wave of frequency n_1.

Similarly the disc I may be held fast and a second sinusoidal vibration of frequency n_2 obtained by rotating the disc II. If both discs are made to move at once we can get any one of the interference curves shown in figs. 13–16 at will. For in the first place we can adjust the *frequencies* of the component vibrations at will by altering the rates of revolution of the discs I and II, and in the second place we can with given frequencies vary the ratio of the amplitudes at will. For this purpose the slit is made so that it can be moved horizontally across the window. When it is close up to the disc II it gives the frequency n_2 with a large amplitude and the frequency n_1 with a small amplitude. Half-way between the discs it gives the two frequencies with the same amplitude, and so on. In the third place we can make the vibrations begin with definite phases at the instant t_0. To do this the discs I and II must be coupled together by gearing in a suitable ratio ($n_1 : n_2$) and driven by the *same* motor.* This form of the apparatus is shown in fig. 17. This simple apparatus enables us to demonstrate the interference curves of two simple harmonic vibrations differing in frequency and amplitude in a very clear way.

So much for the superposition or interference of two sinusoidal vibrations only. We can "represent" the non-sinusoidal curves reproduced in figs. 13–16 by means of *two* simple sine curves. Here the word "represent" is used in two senses.

In the first place we may be given the complicated curves as a result of observation. Without needing to know anything about the mechanism by which they are produced, we can *describe* them formally by means of their two simple harmonic components. For this we only need to know the frequency, amplitude, and phase of each simple harmonic vibration.

In the second place we may be faced with the problem of producing complicated curves of the type denoted by n_r in figs. 13–16 by experimental means. Then *one* way of solving the problem is to use *two* bodies capable of executing simple harmonic vibrations.

This, however, is by no means the only method available. For we may e.g. produce the curve at the foot of fig. 14 by means of only *one* vibrating body, say by moving the hand according to a complicated law of force while drawing the figure.

Here the beginner often falls into error. In this case also he would like to localize the individual sinusoidal vibrations required in the formal description in individual bodies separated in space and capable of executing sinusoidal vibrations. In drawing one of these curves, however, we cannot say that the finger is vibrating with frequency n_2 and the hand with frequency n_1. For such a separation in space of the two component vibrations occurs but rarely in drawing. An exception is the case of fig. 13. We attempt to draw a sine wave of large amplitude by moving along the blackboard and simultaneously making our arm move up and down in simple harmonic vibrations; our hand shakes, however, and the latter motion of high frequency is superposed on the smooth curve described as our arm moves slowly up and down.

In figs. 13–16 even the complicated non-sinusoidal curves are characterized by a definite frequency. For after every "period" of τ_r seconds a definite vibration curve is reproduced in all its details. We call $1/\tau_r$ ($= n_r$) the *fundamental frequency* of the non-sinusoidal vibration. The two component vibrations have frequencies n_a and n_β which are integral multiples of this fundamental frequency n_r. *Were it not for this integral relationship, periodic repetition of the whole vibration curve would be impossible.*

---

* It is only when the numbers are exact integers that the phases at the instant t_0 are maintained (figs. 15, 16). This cannot be attained in practice without the use of toothed wheels.

Strictly speaking, two frequencies in the ratio of $1 : \sqrt{2}$ do not give periodic repetition of one and the same vibration curve at all. The fundamental period τ_r becomes infinite or the fundamental frequency n_r becomes zero. In reality, however, every curve has an appreciable thickness, and within the degree of accuracy limited by this fact we find that a fundamental period τ_r equal to $14/n_1$ (or at least $141/n_1$) exists.

By adding further component sinusoidal vibrations we may similarly " represent " vibration curves as complicated as we like. Here the word " represent " is to be understood in the double sense explained above. The amplitudes and phases of the component vibrations must be chosen in a suitable way. Their frequencies must without exception be integral multiples of the " fundamental frequency " of

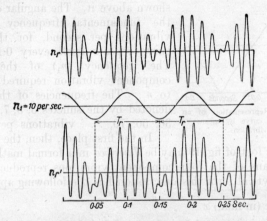

Fig. 18.—Asymmetric vibration diagram produced by superposition of a " differential vibration " (continuation of fig. 14)

the complicated curve. The fundamental frequency is accordingly the frequency of the slowest component vibration. Again, this is most readily seen from examples, of which we shall adduce three.

In fig. 18 we have again represented a beats curve obtained by combining the two component vibrations n_6 and n_7. On this beats curve we now superpose a third sine curve. This is

(a) to have a frequency equal to the *difference* of the frequencies of the two original component vibrations; i.e. $n_1 = n_7 - n_6$;

(b) to be displaced through $90°$ relative to the two first vibrations, which are of the same phase.

By this addition of the " differential vibration " we obtain an asymmetrical curve $n_{r'}$ from the beats curve, which was originally perfectly symmetrical with respect to the x-axis. It is easy to see how the amount of the asymmetry depends on the amplitude of the differential vibration used. This we chose as two-thirds the amplitude of the two other component vibrations. In analogy to electrical terminology an asymmetric curve of this kind may be referred to as a " rectified beats curve ". That is, a rectified beats curve contains a frequency

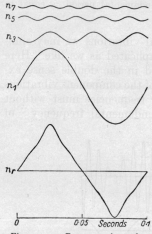

$n_{(\beta-a)}$ equal to the *difference of the frequencies of the two component vibrations* n_β, n_a. The reader should take pains to impress this very important fact on his memory.

In our second example we are to " represent " the almost-triangular curve in the lowest line of fig. 19 by means of the four different sinusoidal vibrations shown above it. The angular curve has the fundamental frequency $n_r = 10$ vibrations per second, for the whole curve repeats itself every 0·1 second. The frequency (n_1) of the slowest component vibration required is equal to n_r. The frequencies of the others, denoted by the suffixes 3, 5, 7, . . . , are 30, 50, 70, . . . vibrations per second.

Fig. 19. — Representation of an almost-triangular curve by means of four component vibrations.

In the first place, then, the vibration curve at the foot of fig. 19 may be *described* in a formal mathematical way by means of the four component vibrations reproduced above it. In analytical form this description has the following appearance:

Amplitude (in mm.) x

$$= 10 \sin 20\pi t - 1\cdot5 \sin 60\pi t + 0\cdot6 \sin 100\pi t - 0\cdot3 \sin 140\pi t.$$

In the second place, we can produce the same curve experimentally by making a ray of light fall successively on four mirrors executing simple harmonic vibrations of suitable frequency, amplitude, and phase, and then on a moving photographic plate. An experiment of this kind, however, is not worth the trouble involved. If the need arises it is much simpler to make use of a body which is vibrating but *not* sinusoidally and not with a linear law of force. In fact, we already know of a pendulum very similar to this, our electric bell. It is not in the *experimental realization* of complicated vibrations that the value of their representation by means of simple harmonic vibrations lies, but in their *description*.

In fig. 18 the period (τ_r) of the complicated vibration, and hence also its fundamental frequency $(n_r = 1/\tau_r)$, was easy to find. It is by no means necessary that this should always be so. An example where the period is not so easy to recognize is given in fig. 71, p. 312. In this connexion we may even mention fig. 13, p. 229.

4. Representation of Complicated Vibrations by Spectra.

The description of complicated vibration forms may be simplified still further by drawing, a complicated vibration being represented as a *spectrum*.

In a spectrum the *frequencies* of the component sinusoidal vibrations are marked off along the horizontal axis. The lengths of the ordinates give the *amplitudes* of each component vibration. Thus fig. 20 gives the spectrum corresponding to fig. 19. It is a *line spectrum*, the simplest way of representing the vibration of fig. 19. In one respect, to be sure, this very simple description is incomplete. *A spectrum gives no information about phase.* Now, although a knowledge of phase is indispensable for the graphical or mathematical reconstruction of the vibration curve, it is not essential for the discussion of a number of problems connected with non-sinusoidal vibrations which are more important from the physical point of view.

We shall consider three other cases of practical importance, using the spectrum mode of representation.

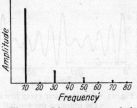

Fig. 20.—Line spectrum of the vibration represented in fig. 19 (ordinates on double scale).

Case I. — *Line spectra of damped vibrations periodically excited.* For the sake of brevity we take a numerical example. Any body capable of vibration is made to execute *undamped* sinusoidal vibrations at the rate of 400 per second. Once struck, it gives a vibration diagram consisting of a sinusoidal wave-train of unlimited length with waves of constant amplitude. Its spectrum consists of a single line at frequency 400.

The vibrating body is then damped in some way or other. Hence after a *single* excitation we now have a wave-train of diminishing amplitude and limited length (fig. 27). Above this we have the vibrations of the same body when *repeatedly* excited at regular intervals. In fig. 25 a fresh impulse occurs after every eight vibrations, in fig. 23 after every five vibrations, in fig. 21 after every two vibrations. Opposite each of these vibration curves we have the corresponding spectrum. None of these exhibit the simple spectrum of the undamped vibration, i.e. a single spectral line at frequency 400. The original frequency 400 is accompanied by a whole series of other lines. In each of the three spectra the lowest frequency is that of the series of impulses or " excitation-frequency ", which in the three given spectra, reading downwards, is 200, 80, and 50. The excitation-frequency is the fundamental frequency (n_r) of the three non-sinusoidal wave-trains. All the other spectral frequencies must be integral multiples of the excitation-frequency used in each case. Hence the spectral lines can only coincide

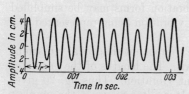

Fig. 21.—Impulse after every two vibrations
(200 impulses per second)

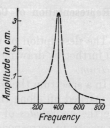

Fig. 22

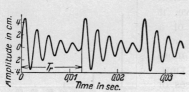

Fig. 23.—Impulse after every five vibrations
(80 impulses per second)

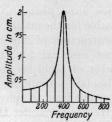

Fig. 24

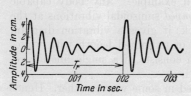

Fig. 25.—Impulse after every eight vibrations
(50 impulses per second)

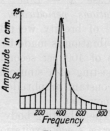

Fig. 26

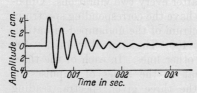

Fig. 27.—Single impulse only

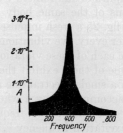

Fig. 28.—Continuous spectrum of
the damped single-impulse vibration
opposite. The product of the ordinate
A and the frequency interval Δn
gives the mean amplitude of the
vibrations in this frequency interval.

in isolated cases, seeing the three excitation-frequencies are different. But—this is important—they always fall in the same region. If the ordinate scale is suitably chosen *all three line spectra may be enclosed in the same dotted curve.*

As the frequency of excitation falls, the number of component vibrations or spectral lines required to represent the spectrum continually increases. A greater and greater number of sine vibrations is required in order that certain amplitudes may cancel one another and thus produce the wide gaps between the damped wave-trains. We thus pass in the limit to the very important Case II.

Case II.—*Continuous spectrum of a damped vibration excited only once.* In fig. 27 we have the damped wave-train resulting from a single

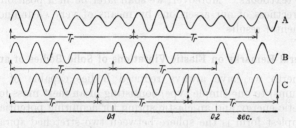

Fig. 29.—Sinusoidal vibrations, A with periodically varying intensity,
B with periodic gaps, C with periodic discontinuities of phase

impulse, and in fig. 28 its spectrum. The spectral lines are now crowded indefinitely close together and entirely fill the region enclosed by the dotted enveloping curve; this area is accordingly shown in black. The line spectrum has been replaced by a *continuous spectrum.**

Case III.—*Line spectra of periodically disturbed sinusoidal wave-trains.* There is nothing fundamentally new about this case. It is merely meant as a warning to the reader against a confusion which often arises.

We sometimes have occasion to use sinusoidal vibrations with a periodically-varying amplitude (fig. 29A), or with periodic gaps (fig. 29B), or with periodic discontinuities of phase (fig. 29C).

In all these curves it is obvious that, as always, the fundamental period τ_r and the fundamental frequency n_r are determined by the recurrence of an identical vibration curve. This is shown in all three figures. To represent these curves by a number of sinusoidal curves we have to use the frequency $n_1 = n_r$ and integral multiples of it. It is to be noted that the fundamental vibration of frequency $n_1 = n_r$ occurs with a considerable amplitude in the spectra of such curves, which are thereby distinguished from the beats curve, to which they

* In the mathematical treatment this passage to the limit means that a Fourier series is replaced by a Fourier integral.

superficially bear a strong resemblance. The beats curve is altogether an exceptional case; it contains only two component vibrations with high suffixes not greatly differing from one another. In the particular case of the beats curve the amplitudes of all the other component vibrations associated with the fundamental frequency n_r, i.e. the vibrations of frequency $n_1 = n_r$, $n_2 = 2n_r$, &c., are zero. Due consideration of these facts would have spared us a great deal of literature about tones supposed to arise from regular fluctuations of amplitude or differences of phase.

In this section we have merely discussed certain important relationships in a descriptive way. To deduce them graphically would take up too much time; a full account of them is given in all mathematical textbooks. Moreover, we shall later be in a position to prove the correctness of our assertions by means of very clear and definite experimental results.

5. General Remarks on Elastic Vibrations of Solid Bodies of any Shape.

Hitherto we have always reduced bodies capable of vibration (pendulums) to one simple type, consisting of an inert mass for acquiring kinetic energy and an elastic spring for acquiring potential energy. The simplest form is the sphere between two stretched spiral springs (fig. 1, p. 50). Henceforth we shall call this apparatus an *elementary pendulum*. This diagram sufficed to represent most of the vibrating bodies we have hitherto used, although too crudely in some cases. It will not, however, cover all the cases which occur. Very often it is impossible to localize the inert mass and the spring separately, for, of course, as we know from everyday experience, a body of *any* shape is capable of vibration. We are thus led to the problem of the elastic vibrations of any body whatever.

For the sake of simplicity we shall confine ourselves in the first instance to bodies of particularly simple geometrical form. In §§ 6–9 we shall deal with the vibrations of linear bodies, that is, of bodies whose longitudinal dimensions preponderate, such as rubber tubes, wires, spiral springs, chains, rods, &c. To deduce the proper vibrations of these linear solid bodies we may use two different methods, namely:

(1) the coupling together of a long series of elementary pendulums;

(2) the interference of two progressive elastic waves moving in opposite directions.

We shall use the first method in §§ 6, 7, and the second in §§ 8, 9.

6. Transverse Elastic Vibrations of Linear Solid Bodies.

A simple elementary pendulum is illustrated in fig. 1, p. 50. A vibration along the length of the spring will henceforth be called a *longitudinal vibration*, a vibration at right angles to the length of the spring a *transverse vibration*. We shall first consider transverse vibrations.

In figs. 30 and 31 two of these elementary pendulums are attached or " coupled " to one another. The body thus formed can vibrate in two ways. In the first case the two masses move in the same direction or vibrate " in phase "; fig. 30 shows two instantaneous photographs of these vibrations. In the second case the two masses move in opposite directions or vibrate " with a phase-difference of 180° "; fig. 31 again gives two instantaneous photographs.

The frequencies differ in the two cases. By using a stop-watch we find that the frequency is greater in the second case than in the first. That is, if two elementary pendulums are coupled together we have two modes of transverse vibration and two different frequencies of vibration.

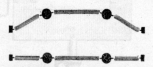

Fig. 30. — Transverse vibrations of two elementary pendulums coupled together; the two masses in phase.

Fig. 31.—Transverse vibrations of two elementary pendulums coupled together; the two masses with a phase-difference of 180°.

Similarly, in fig. 32 we have three elementary pendulums coupled together. With a chain of n elementary pendulums coupled together we have n different proper frequencies. In the limit we pass to continuous linear bodies. Thus the latter may be expected to possess a practically unlimited number of proper vibrations. We shall bring forward some experimental examples.

Fig. 33 (Plate VIII) shows a piece of rubber tubing several metres long, fastened at the top to the ceiling and below to a small sliding bar which can be moved backwards and forwards about 1 cm. in the direction of the double arrow by means of an eccentric and a powerful (say $\frac{1}{2}$-kilowatt) electric motor. By varying the rate of revolution of the motor we can excite any one of the first twelve proper vibrations of the tubing. Figs. 33 A, B, C, are time exposures of the ninth, eleventh, and twelfth proper vibrations. These are pictures of undamped vibrations or " standing waves ". We can clearly distinguish between " nodes " and " loops ". The variation in time of these proper vibrations or standing waves is best viewed " stroboscopically " (p. 12). The process can then be slowed down as much as we like. More simply still, we may illustrate it in a purely kinematical way by means of a wire bent into the form of a sine curve and having a handle at one end (fig. 34). This wire is placed in front of the projection lantern and set rotating about its longitudinal axis. The successive appearances

presented by the vibrations (often called " phases " for short) may
then be observed one after another on the screen. By turning the
handle rapidly we can conveniently obtain the appearances of fig. 33.
This apparatus, though primitive, is very useful.

We return once again to fig. 33B and imagine that a blow is struck
on the rubber tubing, which is vibrating in the plane of the paper in
its eleventh proper vibration. The tubing then begins to vibrate as a
whole in its first proper vibration or fundamental mode, and the two
standing waves occur *simultaneously*. This simultaneous occurrence of

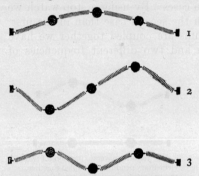

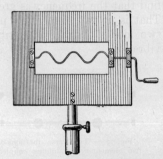

Fig. 32.—The three possible modes of transverse vibration of three elementary pendulums
coupled together: 1 fundamental or first
proper vibration, 2 second proper vibration, 3 third proper vibration.

Fig. 34.—Apparatus for illustrating
standing waves

more than one proper vibration or standing wave is used very frequently in stringed musical instruments. Fig. 35 (Plate VII) shows a
lecture experiment in which a horizontal wire is stretched tightly and
excited in the ordinary way by a violin bow so that it executes undamped vibrations.

In principle the action of the violin bow is the same as that of the automatic
regulating apparatus shown in fig. 7, p. 227. The circumference of the rotating
shaft used there may be regarded as an endless violin bow.

A slit S is placed in front of the string and at right angles to it, and
an image of the slit is projected in the usual way by means of a lens
moving horizontally.

In practice it is found convenient to replace motion of the lens in
a horizontal line by motion in a large circle, by using the " lens disc "
shown in fig. 35. The different lenses come successively into action as
the disc is rotated by turning the knob K with the thumb and forefinger. The unavoidable curvature of the time-axis which arises from
the use of the lens disc is a harmless blemish.

In this way we obtain vibration diagrams of the type shown in
fig. 36. We see that any particular point of the string, in fig. 35 the
middle point, executes vibrations in its path at right angles to the

PLATE VIII

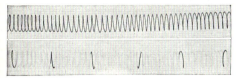

Chap. XI, Fig. 39.—Spiral spring, above at rest, below in longitudinal vibration; only the "nodes" of the vibration are clearly seen.

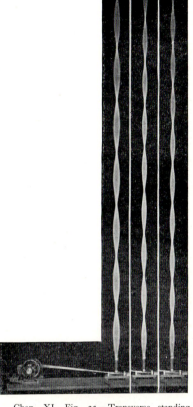

Chap. XI, Fig. 49.—Apparatus for demonstrating the "fence phenomenon". The gaps in the fence are white and the spokes of the wheel opaque. The wheel is nickel-plated, so that on the screen it will stand out well against the black fence.

Chap. XI, Fig. 33.—Transverse standing waves on a piece of rubber tubing whitened with talcum powder and placed in front of a black curtain; on the left an electric motor and eccentric. The top quarter of the picture has been cut off.

Chap. XI, Fig. 55. —To illustrate a longitudinal standing wave in a column of air (fundamental).

Chap. XI, Fig. 56.—To illustrate a longitudinal standing wave in a column of air (first overtone).

length of the string which are by no means sinusoidal. On the contrary, the vibration diagram is usually very complicated, arising from the superposition of a very large number of proper vibrations. The excitation of any proper vibration by itself can only be brought about by handling the bow in a particular way, and then only approximately. The strings of musical instruments in general give a very complicated vibration spectrum.*

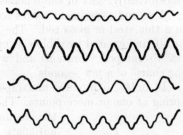

Fig. 36.—Vibration diagrams of a string, photographed by means of the lens disc.

If the velocity of rotation of the lens disc is known, the length of an individual period τ_0 for the vibration curves of fig. 36 can easily be found. Thus e.g. we find that the fundamental frequency $n_0 = 1/\tau_0$ in our lecture experiment is of the order of several hundred.

7. Longitudinal and Torsional Vibrations of Linear Solid Bodies.

At the beginning of last section we defined a *longitudinal* vibration of an elementary pendulum as a vibration of the pendulum bob along the direction of the spring. Figs. 37 and 38 represent the two possible modes of longitudinal vibration of two elementary pendulums coupled together. In fig. 37 the two pendulums are vibrating in the same

Fig. 37. — Longitudinal vibrations of two pendulums coupled together; the two masses in phase

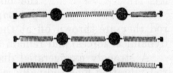

Fig. 38. — Longitudinal vibrations of two pendulums coupled together; the two masses with a phase-difference of 180°.

direction or " in phase "; in fig. 38 in opposite directions or " with a phase difference of 180° ". Proceeding to couple further elementary pendulums together, we find that a system of n pendulums has n proper vibrations. Thus in the limit we again pass to the case of a linear body with a practically unlimited number of longitudinal proper vibrations. We shall confine ourselves to two experimental examples.

We begin by producing *undamped longitudinal vibrations* in a fine spiral spring (fig. 39, Plate VIII). The vibrations are kept up by fixing one end of the spring to the clapper of an ordinary electric bell. The fundamental frequency of the clapper must coincide with one of the

---

* [The proper vibrations other than the first (or *fundamental*) are often referred to as (*upper*) *partials*, *overtones*, or *harmonics*.]

proper frequencies of the spring. The figure is a reproduction of a photograph. Only the " nodes " of the spring, which are at rest, are shown clearly. Six of these nodes are clearly visible in the figure.

A second experiment serves to demonstrate longitudinal vibrations in a thin steel or glass rod. The rod is suspended by two cords as in fig. 40 and excited by a blow at one end. This sudden excitation gives rise to damped vibrations and we hear a sound which dies away in the course of a few seconds.

In order to produce undamped vibrations we have to clamp the spring at one or more points. The clamps must be placed at nodes of the required proper vibration. To obtain automatic regulation we may use one of the methods described in § 2, p. 223, e.g. a suitable practical form of that illustrated in fig. 7, p. 227. A rotating disc faced with leather is pressed against the rod, the cohesive effect between rod and leather being possibly enhanced by damping the leather or sprinkling it with powdered resin. In the case of a rod vibrating without damping it is not difficult to discover where the nodes are. If we put paper riders on the

Fig. 40.—Longitudinal vibrations of a rod suspended by cords (length $l = 25$ cm., fundamental frequency $N = c/2l$, where c is the velocity of sound in the rod).

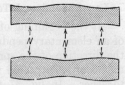

Fig. 41. — Diagrammatic representation of longitudinal vibrations in a rod.

rod they are thrown off at the vibrating loops (as a result of hydrodynamical forces) and remain at rest at the nodes.

A strongly exaggerated diagram of the state at any instant of a rod vibrating longitudinally is given in fig. 41. The rod is alternately blown out, figuratively speaking, in the region of a node N, or contracted, forming a " waist ". That is, the ring-shaped lines of nodes encircling the rod are *at rest* relative to the longitudinal direction of the rod, but their *diameter* alters *periodically* with the same frequency as the longitudinal vibration. In reality these changes in the diameter of a rod vibrating longitudinally are but trifling and can only be demonstrated as the result of refined observations.

In addition to the transverse and longitudinal vibrations of linear solid bodies there is a third type, namely, *torsional vibrations*. We rotate the bob of our elementary pendulum about the centre line of the spring and let it go. The elementary pendulum then executes torsional vibrations. Their frequency is inconveniently high, as the moment of inertia of the sphere is very small (cf. equation (8), p. 104). To bring down the frequency we replace the sphere by a dumb-bell-shaped body as in fig. 42. We may then even replace the spiral spring

by a short piece of steel wire. In spite of the larger torsional rigidity
we still obtain torsional vibrations of sufficiently low frequency. We
may couple n of these elementary pendulums together in the usual
way; we thus obtain the apparatus shown in fig. 43, which enables us
to demonstrate a whole number of torsional vibrations, including, for

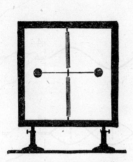

Fig. 42.—Torsional pendulum

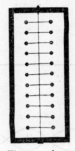

Fig. 43. — Apparatus for demonstrating torsional waves.

example, the torsional vibration of highest frequency. For this purpose
it is necessary to have some auxiliary apparatus for letting go the
even-numbered balls on the left and the odd-numbered balls on the
right at the same instant. The lowness of the frequency of these torsional oscillations makes observation considerably simpler. Here
again a passage to the limit leads us to the torsional vibrations of
strings and rods.

8. Elastic Proper Vibrations of Linear Solid Bodies as derived from the Interference of Progressive Waves.

In accordance with what we said in § 5, p. 238, we shall now explain
the production of proper vibrations or standing waves in linear solid
bodies in another way. We begin with an experiment.

In fig. 44 we see a long wire spiral fixed to the wall on the left and
held in the hand on the right. The sag of the wire due to its own
weight is not shown. We give the right-hand end of the wire a short
jerk in the direction of the vertical double arrow. We then see an
elastic disturbance run to the left along the wire with a conveniently-
observable velocity of only a few metres per second.

This finite rate of propagation of an elastic disturbance is the
essential point for us. On this fact depends the possibility of progressive elastic waves arising. For if the propagation of an elastic disturbance were absolutely instantaneous the wire would follow the
movements of the hand as a whole, like a geometrical straight line.

To demonstrate these progressive transverse waves we make our

hand vibrate vertically up and down. The waves advance snakily along
the wire. We interrupt our observations for a moment before the head
of the wave-train has reached the wall.

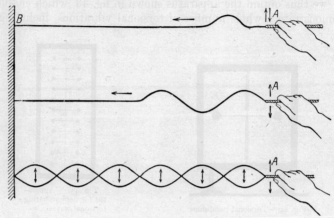

Fig. 44.—To illustrate how progressive and standing transverse waves arise

In progressive waves it always seems to the unpractised observer
that the body moves forward as a whole like a wriggling adder. In
reality, however, this is out of the question.* In progressive waves we
are merely concerned with the advance of a vibration and a psycho-

Fig. 45. — Path of
a particle in progressive
waves of large amplitude
on the surface of water.

logical interpre-
tation of what
we observe. The
student must
think this out
clearly for him-
self. In this
connexion he will find the lecture
experiment illustrated in fig. 46
helpful.

Two discs are fixed to a shaft,
and corresponding points on their

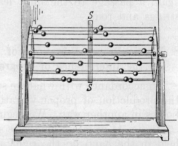

Fig. 46.—Spiral wave machine

circumferences are joined by means of thin strings so that a cylindrical
cage is formed. Light wooden balls are threaded on the strings and
arranged in a spiral. Looked at sideways the spiral has the appearance
of a dotted sine curve. By means of a screen with a vertical slit-like

* The *waves on the surface of water* which are otherwise so useful in the study of
wave motion are partly to blame for this misunderstanding. In these the water *does*
actually advance in the direction of motion of the waves if the amplitudes of the waves
are *large*. For large amplitudes the circular paths of the particles of water photographed in fig. 44, p. 218, degenerate into the curves sketched in fig. 45. Hence bodies
floating on high waves may be washed up on shore.

opening S (raised out of the way in fig. 46) we can cover all the balls but one. When we rotate the cage we see this one ball vibrating up and down in the brightly illuminated slit. If we lower the screen we may also observe the same for all the balls simultaneously by rotating the cage very slowly. We then clearly see a *difference of phase* between the individual vibrations of the balls along the row. As the rate of rotation increases, however, the appearance we see suddenly changes in a surprising way. Instead of the dotted sine curve we have a nearly black continuous wave-train advancing in a horizontal line as in fig. 47. Above a certain rate of rotation our brain makes mistakes in the *identification* of the separate individuals and their arrangement in

Fig. 47.—Instantaneous photograph of a progressive wave

a row. The process is similar to that occurring in the well-known "fence phenomenon". If we look through a garden fence the spokes of the wheels of passing vehicles appear to be curved in a peculiar way as shown in fig. 48. This phenomenon may be easily reproduced on the screen by using the apparatus of fig. 49 (Plate VIII). The eye sees the moving points of intersection of the fence uprights and the spokes of the wheel as a continuous curve.

a   *b*

Fig. 48*a*, *b*.—The "fence phenomenon" for different wheel velocities

We now resume our experiments with the long wire spiral. By moving our hand up and down we set up a wave-train of limited length. It runs up to the wall end of the wire and is reflected there. Another reflection follows when it returns to the hand, and so on. We then repeat the experiment, this time continuing to move our hand up and down. We now have two opposing wave-trains of equal frequency, namely, that starting from the hand and that reflected from the wall. At first their superposition gives a confusedly-varying motion. By cautious attempts, however, such as altering the frequency of the movements of the hand slightly, we soon obtain well-defined *standing waves*. It is merely necessary that half the wave-length of the progressive wave should be exactly equal to some integral fraction of the length of the wire.

Here, therefore, we see *a standing wave arising experimentally from the superposition or interference of two wave-trains of equal frequency moving in opposite directions.* The development of this interference as

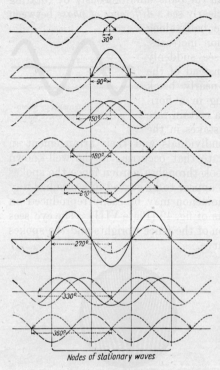

Fig. 50.—The formation of standing waves

time goes on may be followed in the graphical representation of fig. 50, which begins at the top shortly after the two opposing wave-trains have met for the first time. The waves coming from the right and left are indicated by dots and dashes respectively and their resultant by a continuous line. These instantaneous curves indicated by thick lines correspond to those known to us from fig. 33, Plate VIII.

Standing longitudinal and torsional waves may be produced in exactly the same way by the interference of *progressive* longitudinal and torsional waves.

The velocity of propagation of all these elastic waves may be calculated from the elastic constants of the bodies used. As an example we give the calculation of the velocity (c) of longitudinal waves in a solid body.

We base our calculations on a rod-shaped body (fig. 51). The elastic compression of a rod of length l and cross-section q by the linear amount Δx requires a force

$$F = E\frac{q\,\Delta x}{l};\qquad \ldots \ldots \ldots \quad (1)$$

the factor of proportionality E is called Young's modulus.

The compression by an amount Δx is to be brought about in time Δt by the impulse $F\Delta t$. During time Δt the elastic disturbance advances towards the right and affects a length of bar l equal to $c\Delta t$. Hence the impulse is given by

$$F\Delta t = \frac{E.\Delta x.q}{c}\qquad \ldots \ldots \ldots \quad (2)$$

This impulse gives the length of bar l momentum

$$mv = lq\rho\frac{\Delta x}{\Delta t} = cq\rho.\Delta x, \quad \ldots \ldots \quad (3)$$

for after the time Δt has elapsed the right-hand end of the bar has advanced through a distance Δx.

The impulse and the momentum must be equal. Combining equations (2) and (3), we obtain

$$c = \sqrt{\frac{E}{\rho}}. \quad \ldots \ldots \ldots \ldots \quad (4)$$

The velocity of longitudinal elastic waves is equal to the square root of the quotient of Young's modulus and the density.

Numerical Example.—For steel Young's modulus E is $2 \cdot 10^{12}$ dynes/cm.^{2} and the density ρ is $7 \cdot 7$ gm.-mass/cm.^{3} The velocity is therefore

$$c = \sqrt{\frac{2.10^{12}}{7 \cdot 7}} \frac{\text{cm.}}{\text{sec.}} = 5 \cdot 1 \text{ km./sec.}$$

The velocity of propagation of elastic longitudinal waves in bodies is commonly called the *velocity of sound*, for the frequencies of such waves are usually within the range of our ears. If we know the velocity of sound we may calculate e.g. the frequencies of the various longitudinal vibrations of rods. Thus for the steel rod used in fig. 40, p. 242, we find the frequency of the fundamental to be 10^4 vibrations per second.

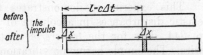

Fig. 51.—To illustrate the calculation of the velocity of sound in a bar

With a thin bar of rock salt 5 cm. long we reach a fundamental frequency of 43,000 and so on. The longitudinal vibrations of crystals (usually of quartz) are used to an increasing extent in engineering as "standards of frequency". They are more convenient than the familiar tuning-fork when high frequencies are required.

9. Longitudinal Elastic Waves in Columns of Liquid or Gas.

Here, as always, we shall discuss liquids and gases together. Our experiments will be chiefly carried out using air.

In the interior of liquids and gases (in contradistinction to the surface) no transverse or torsional vibrations are possible, but only *longitudinal* vibrations. This follows immediately from the fact that all the particles of a liquid or gas are free to move relative to one another. A particle * of liquid or gas can only drive its fellows forward in the longitudinal direction of its own motion.

* In the sense of an element of volume, not an individual molecule.

As in the case of solid bodies, we shall begin by discussing *linear* bodies, i.e. linear columns of liquid or gas bounded by a pipe.

As we found in last section, the existence of progressive and standing waves depends essentially on the fact that the velocity of propagation of elastic disturbances is *finite*. In air this velocity of propagation, or velocity of sound, amounts to about 340 m./sec.

This velocity may be measured experimentally in the lecture-room by using a pipe about 150 metres long and several centimetres across. The right-hand end is closed by a rubber membrane and a pressure gauge of small inertia is connected to the left-hand end. It is convenient to use a " sensitive flame ". Its membrane forms one wall of a shallow box included in the supply-pipe of a gas jet. By giving the membrane at the right-hand end of the pipe a short, sharp blow, the air pressure at that end of the pipe is temporarily raised and a sudden compression results. The time this compression takes to reach the left-hand end of the tube is measured by means of a stop-watch reading to hundredths of a second.

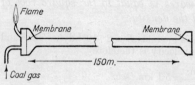

Fig. 52. — Apparatus for finding the time taken by longitudinal elastic disturbances to traverse a column of air.

This experiment demonstrates the same facts for *longitudinal* elastic disturbances in a column of air as the experiment of fig. 44, p. 244, does for the *transverse* elastic disturbances of a solid spiral wire. The simple harmonic vibration of our hand in fig. 44b may now be imagined to be replaced by a corresponding simple harmonic motion of the membrane at the right-hand end of the pipe. A longitudinal elastic wave then traverses the pipe. *The individual particles of air execute simple harmonic vibrations about their equilibrium positions*, but this time *along* the line of the pipe. Each vibrating particle begins its motion a little later than the one ahead of it in the direction of motion of the wave. Or, to put it another way, an alternating current of air flows through the pipe. For the *wave-length* of this progressive wave or alternating current of air we again have equation (10), p. 220: $\lambda = c\tau = c/n$, where n is the frequency of the vibrations of the membrane and c the velocity of sound.

In carrying out the actual experiment we insert a pressure gauge of sufficiently low inertia, say another sensitive flame, into one side of the tube about the middle. By means of a rotating mirror we then observe how the regions of increased and diminished air pressure, i.e. the wave-crests and wave-troughs, whisk past. To be sure, it is necessary to have a pipe several hundred metres long, otherwise the waves reach the other end of the pipe too soon; reflection of the wave-train then sets in, leading to the formation of standing waves.

This reflection is not confined to a closed end of a pipe, but also takes place at an open end or in general at any place where the cross-section of the pipe is altered. Thus it is particularly easy to produce *standing waves* in pipes deliberately. For a lecture experiment, for example, we need only have a cardboard tube about a metre long and several centimetres across, closed by a rubber membrane at one end. By striking or pulling at the membrane this column of air may be set vibrating, the result being a loud sound, which, however, dies rapidly away. Or one end of the tube may be made solid and a tightly-fitting cap pulled off the other. In both cases the *backward and forward*

Fig. 53. — Apparatus for the hydrodynamical detection of the alternating current of air in a pipe.

motion of the air inside the tube is easy to demonstrate. For this purpose we have to make use of a property of moving gas which depends, to be sure, on the *direction* of the flow, but not on

Fig. 54. — Stream-lines of the flow round two balls, to illustrate the hydrodynamical detection of the alternating current of air in a pipe.

its *sense*. For what we have to represent is the direction of a current of air which is continually varying in sign (an "alternating" current). This problem is solved by means of the apparatus in fig. 53.

Two small pith balls are suspended by thin threads in the interior of a pipe of square cross-section. Two windows of glass or cellon enable the balls to be observed directly or projected on a screen. The angle between the line joining the two balls and the axis of the tube may be varied. To begin with it is made a right angle. Then for a flow in either direction parallel to the long axis of the pipe the stream-line diagram must be as in fig. 16, p. 200. Between the two balls the stream-lines are crowded together. The two balls must therefore *attract* one another while the pipe is vibrating (giving out a sound). This is in fact the case.

We then set the balls so that the line joining them lies along the axis of the pipe.

A flow parallel to the line joining the centres of the balls is found to give rise to the stream-line diagram of fig. 54, which is new to us. Between the balls the stream-lines are very far apart. While the pipe is vibrating the two balls must *repel* one another. This also is easily observed.

It is this motion of the particles of air parallel to the long axis of the tube that gives rise to the characteristic distributions of pressure and density in a longitudinal standing wave. We represent these

diagrammatically by three "snapshots", in the first place for the
fundamental vibration of a pipe closed at both ends (fig. 55, Plate VIII).
By its uniform grey tint the middle picture indicates the constant pressure and constant density which originally exist all along the pipe. In
the two other pictures we have *loops* of pressure and density at *both*
ends of the pipe. In the upper picture the black loop on the left means
a *wave-crest*, a region of increased air pressure and increased air density.
The white loop on the right means a *wave-trough*, a region of reduced
air pressure and reduced air density. For the lower picture exactly
the reverse is true; in it we have low pressure at the left and high
pressure at the right.

Carefully to be distinguished from this (sinusoidal) distribution of
the pressure and the density of the air is the distribution (likewise
sinusoidal) of the velocities and of the displacements of the individual
particles of air as they swing backwards and forwards along the pipe
about their equilibrium positions. In fig. 55 the velocity and the displacement have their *nodes* at the ends of the pipe and their *loops*,
i.e. their maximum values (alternately to the right and to the left),
at the middle. That is, in these standing waves the nodes of pressure and the nodes of velocity are a quarter of a wave-length apart,
measured along the length of the pipe.

In the case of a tube executing one of the other proper vibrations
we have to imagine the pictures of fig. 55 repeated symmetrically. The
second proper vibration (in music called the first overtone), shown in
fig. 56 (Plate VIII), will serve as an example. The distributions of
pressure and of density are again represented at the moments of
greatest contrast. Here again the *loops* of pressure and of density
coincide in position with the *nodes* of the velocities and of the amplitudes of the particles. These periodic distributions of pressure may
be demonstrated for the upper partials in a column of gas very
elegantly by means of the "flame-tube" of fig. 57 (Plate IX). A
pipe about 2 metres long, supplied with coal-gas, has a series of burners
running right along its top. One end of the pipe is closed by a rubber
membrane, which is in some way or other made to execute undamped
vibrations. The frequency of these undamped vibrations must coincide with that of some proper vibration of the column of coal-gas. The
variations in height of the flames along the pipe reveal the standing
waves in the interior of the pipe in a very striking way. By suitable
variation of the frequency of the membrane a whole series of different
proper vibrations of the column of gas may be demonstrated in this
way in succession.

In practice the proper vibrations of columns of gas play a large
part in the construction of wind instruments. In these pipes undamped
vibrations are produced by means of hydrodynamical regulating
apparatus. The external appearance of the types in common use may

be taken as known. The details of their working are extremely complicated, and physics has only been able to explain the main features of
their behaviour qualitatively. In the case of the flue pipes of the organ

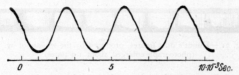

Fig. 58.—Approximately sinusoidal vibration curve of a pipe
(photograph by F. Trendelenburg)

the jet of air blowing on the sharp edge periodically breaks down into
isolated eddies. The jet of air and the column of air in the pipe form
two " coupled " vibrating systems. The same is true of the reed and
gaseous column in the reed-pipe. This complicated mechanism for
automatic regulation generally causes the vibrations of the pipe to
differ widely from the sine form.
Figs. 58, 59 represent the vibration of a pipe and its line spectrum
in a case which from the practical
point of view is a very simple
one.

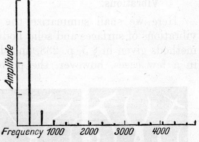

Fig. 59.—The spectrum of the vibration
represented in fig. 58

We shall confine our demonstration of the action of pipes
to two experiments. First we
" blow " a flue-pipe with water,
by means of a water-jet. Here
at last we exhibit a case of longitudinal vibration in a column of
liquid.   Secondly, we demonstrate a small whistle of *high frequency*
which we shall continually require in the course of the next chapter.
A side view and longitudinal section are shown in fig. 60 (Plate IX).
The mouth L and the sharp edge S are surfaces of revolution. The
actual space inside the whistle is only a very poor approximation
to a linear column of air.
     For subsequent use we shall determine the frequency of this little
whistle, which we shall always blow with compressed air. We do this
by measuring the length of the standing waves which the whistle sets
up in a glass tube. The glass tube is about 4 mm. across and about
15 cm. long. By moving a brass piston we can make the length of the
column of air equal to a multiple of half the wave-length of the whistle.
To demonstrate the standing waves and measure their length we use
the very elegant method of " Kundt's dust figures ".
     We scatter a light dry powder along the inside of the tube. We
then bring the little whistle (emitting its unpleasantly piercing note)

in front of the opening of the tube, and slowly move the piston along. After one or two trials the dust arranges itself in very characteristic figures in periodic succession (fig. 61). These figures arise as a result of

Fig. 61.—Kundt's dust figures, produced by the whistle of fig. 60
(photograph enlarged about 1·2 times)

hydrodynamical forces such as we observed on a larger scale with the pith balls of fig. 53. We find that the period of the dust figures is about $\frac{3}{4}$ cm. This is the distance between two nodes, or half the desired wave-length. From this wave-length of about 1·5 cm. it follows, from the familiar equation (10) on p. 220, that the little whistle has a frequency of about 23,000.

10. Proper Vibrations of Surfaces and of Solid Bodies. Thermal Vibrations.

Here we shall summarize the facts quite briefly. The proper vibrations of surfaces and solid bodies may be investigated by the two methods given in § 5, p. 238, and dealt with mathematically. Except in a few cases, however, the problems are exceedingly complicated

Fig. 62.—Chladni's figures (photographic positive)

from the mathematical point of view. In most cases of practical importance we have to depend on experiment. The two main problems which we then have to face are the determination of the various proper frequencies which occur and the discovery of the nodal lines.

The frequencies are usually determined by photographic registration of the vibration curves and their subsequent analysis into sine curves ("harmonic analysis ").

The nodal lines are usually demonstrated by utilizing the fact that if dust is strewn on the vibrating surface it heaps itself up along these lines. Fig. 62 shows the nodal lines of thin square and circular metal plates for various modes of vibration.

A glass or bell may be regarded as a bent form of plate. The vibra-

PLATE IX

Chap. XI, Fig. 57.—Longitudinal standing waves in a column of coal-gas
(Rubens' flame-tube)

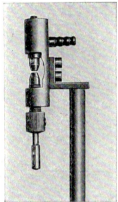
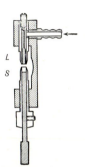

Chap. XI, Fig. 60a   Chap. XI, Fig. 60b
Whistle giving a note of high frequency (about
10,000–30,000 vibrations per second)

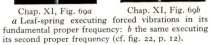

Chap. XI, Fig. 68.—Re-
sonance between the case
and the balance wheel of a
watch.

Chap. XI, Fig. 69a   Chap. XI, Fig. 69b
a Leaf-spring executing forced vibrations in its
fundamental proper frequency: b the same executing
its second proper frequency (cf. fig. 22, p. 12).

tions even of these bodies, which from the geometrical point of view are relatively simple, are complicated to an alarming degree. The simplest mode of vibration of a glass is shown in fig. 63 (in which the glass is looked at from above). The points N represent the points of intersection of four "meridional" nodal lines with the plane of the paper. Even the simplest vibration of our skulls, in whose walls our organs of hearing are enclosed, must be thought of as very similar to this.

In the realm of extremely high frequencies (up to the order of 10^{18}) all solid bodies possess a vast number of elastic vibrations, quite independently of their shape. The energy of these vibrations forms the heat content in the solid bodies or crystals (cf. p. 153). In the case of the highest frequency mentioned the individual atoms or molecules of the crystal lattice vibrate in a way roughly indicated by fig. 38, p. 241.

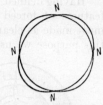

Fig. 63. — Simple vibrations of a wine-glass, as seen from above (diagrammatically).

Among the proper vibrations of hollow spaces filled with gas, those of spherical or bottle-shaped bodies with a short open neck and containing air deserve special mention. These are Helmholtz's resonators, which play an important part in acoustical experiments. They take the place of pipes of definite fundamental frequency and are much more convenient in shape. When in action they frequently exhibit the process of jet formation, apparently continuous, but in reality intermittent, which was described on p. 214. A Helmholtz resonator in action can "blow" quite hard.

The proper vibrations of large living-rooms and public halls are of considerable importance to architects. The details form the subject of a special technical literature.

11. Forced Vibrations.

If a body capable of vibration is excited in any way, whether once and for all or repeatedly by means of some means of automatic regulation, it vibrates in one or more of its *proper* frequencies. Any body capable of vibration, however, may also be made to vibrate with other frequencies not coinciding with any of its proper frequencies. In this case the body is said to execute "forced" vibrations. These forced vibrations play an exceedingly important part in all branches of physics.

In order to discuss these vibrations we must first gain a more accurate idea of the damping of a pendulum. As a result of unavoidable energy losses or intentional supplying of energy to another body, the amplitude of a pendulum excited once and for all is bound to decrease. The changes in the vibrations as time goes on are represented by curves like those in fig. 65. In most cases of simple harmonic motion these curves exhibit a simple law. The ratio of the amplitudes of two successive vibrations in the same direction remains constant right along the curve. It is called the *damping ratio* (k) and its logarithm to base e is called the *logarithmic decrement* (λ). The numerical values of the damping ratio and logarithmic decrement are given beside the curves in fig. 65.

Having defined these quantities, we shall now illustrate the essential features of forced vibrations by means of a demonstration experiment made as clear and easy to understand in detail as possible. For this purpose we use torsional vibrations of very small frequency. In the case of very slow vibrations all the details become easy to observe.

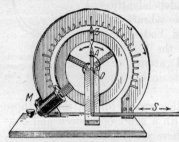

Fig. 64.—Torsional pendulum for demonstrating forced vibrations

Fig. 64 shows a torsional pendulum with a single proper frequency. Its inert mass consists of a copper wheel. The force giving rise to the couple is produced by the deformation of a coiled spring attached to the axis of rotation, the upper end (A) of the spring being displaced in the direction of the double arrow. This displacement is brought about by the lever pivoted at D and connected with the long bar S; the latter can be moved to and fro practically in simple harmonic motion with any desired frequency or amplitude by means of an eccentric and a slowly-running (geared-down) motor. In this way we can cause the shaft of the torsional pendulum to be acted on by sinusoidally-varying forces of any desired frequency. These periodic forces are meant to set up forced vibrations of the torsional pendulum. The amplitudes of the pendulum may be read off by means of the pointer Z and a scale which may be made visible to a large audience by projection on a screen.

At the lower left-hand side of the apparatus there is an auxiliary piece of mechanism M for altering the degree of damping of the torsional pendulum. This consists of a small electromagnet with its poles on either side of the rim of the wheel. The latter can vibrate backwards and forwards in the magnetic field without coming into contact with either of the poles. As the current in the electromagnet varies, this electromagnetic damping apparatus works like a brake pressed more or less hard against the rim. The sole advantages of this electrical damping apparatus over a mechanical one acting by friction are its uniformity of action and convenience of adjustment.

Before beginning the actual experiment we determine the *proper frequency* n_0 and the *damping ratio* k of the pendulum. For both purposes the pendulum is struck with the bar S at rest and we observe the point on the scale at which its motion is reversed. Using a stop-watch we find that the time of vibration τ_0 is 2·08 sec., so that the proper frequency is $1/2·08 = 0·48$ vibrations per second. The ratio of

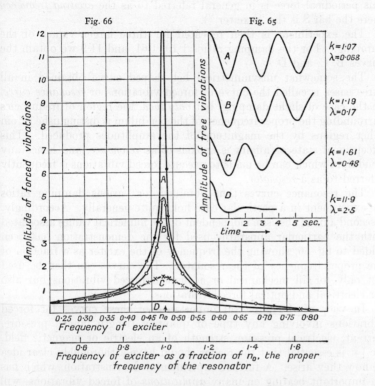

Figs. 65, 66.—The relationship between the amplitudes of forced vibrations and the damping (from experiments with the torsional pendulum of fig. 64). The couples (and hence the amplitudes of the exciter) have the same maximum value for all frequencies. For zero frequency (i.e. a constant couple, the end of the spring A always remaining in one or other of its extreme positions) the amplitude of the resonator in this example is practically 0·5 scale division. Hence the numerical values of the ordinate multiplied by two give the " magnification " of the amplitude of the resonator due to periodic excitation instead of constant (in this example, that due to a periodic couple instead of a constant couple).

two successive amplitudes in the same direction is found to have the approximately constant value of 1·07; this is the required damping ratio. In order to exhibit it graphically the successive amplitudes of vibration to left and to right at intervals of 1·04 seconds have been plotted graphically in fig. 65, and the points joined by a freehand curve.

Now comes the real experiment. We set the crank S going, determine its frequency with the stop-watch, and observe the amplitudes of the forced vibrations of the torsional pendulum. Corresponding pairs of values of the frequency of the crank and the resulting amplitudes are plotted in curve A of fig. 66. The abscissæ are the frequencies of the crank, i.e. of the periodically impressed force. The frequency of this periodical force is in general referred to as the *exciting frequency* (here the bar S is the " exciter ").

The experiment is then repeated for three larger values of the damping. For the damping ratios 1·19, 1·61, and 11·9 we obtain the curves B, C, and D.

The somewhat unsymmetrical bell-shaped curve obtained in all three cases is called the curve of forced vibrations or *resonance curve*. In the case of slight damping, but *only then*, the region of frequencies surrounding the proper frequency of the pendulum is distinguished from other regions by the magnitude of the amplitudes produced. This exceptional state of affairs is spoken of as *resonance*, and hence any pendulum which can be made to execute forced vibrations is frequently referred to as a *resonator*.

The resonance curves thus obtained for various damping ratios by experiment in a particular case hold quite generally. Accordingly, a second axis of abscissæ independent of the numerical values associated with the particular apparatus used for the demonstration has been added to fig. 66, showing the frequency of the exciter as a fraction of the proper frequency of the resonator. The curves can then be used not only for all mechanical or acoustical forced vibrations but also for electrical and optical forced vibrations.

In view of the universal importance of these curves for forced vibrations involving any type of quantity (length, angle, pressure, current, electrical pressure, strength of an electric or magnetic field, &c.) it is essential that the reader should have a thoroughly clear idea of how they arise. A further experimental demonstration, which has an important bearing on many *applications* of forced vibrations, will be helpful here. It deals with the relationship between the *frequency* of the exciter and the *difference of phase* of the amplitude of the resonator and the amplitude of the exciter. We accordingly have to observe Z, the pointer of the pendulum, and A, the end of the spring, at the same time. To make the observations easier, we increase the amplitude of the bar S and at the same time increase the damping so as to prevent the amplitude of the pendulum from becoming too great.

The results are shown in fig. 67. The abscissæ give the frequency of the exciter as a fraction of the proper frequency of the resonator. The ordinates give the difference of phase between the maximum amplitude of the pendulum and that of the exciter.

For very small frequencies the pointer Z and the end of the spring A move in the same direction, and their motion is reversed at the same instant. Their difference of phase is zero. As the frequency of the exciter increases, the maximum amplitude of the force "leads" the maximum amplitude of the pendulum or resonator more and more. In the case of resonance the difference of phase amounts to 90°; e.g. the end of the spring is moving to the right through its equilibrium position at the instant when the pendulum is turning back at its position of maximum displacement towards the left. As the frequency of the exciter increases still further, the phase-difference rises to 180°; the pointer Z and the end of the spring A then pass through their positions of equilibrium at the same instant but in opposite directions.

If the experiment is repeated with smaller damping the region of phase change moves nearer the proper frequency of the resonator (see the middle curve of fig. 67). Apart from the *slope* of the curve the fundamental phenomena remain unchanged; in particular, the difference of phase in the case of resonance remains equal to 90° even when the damping is small.

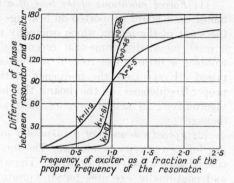

Fig. 67.—The effect of damping on the difference of phase between the exciter and the resonator

The meaning of this phase difference of 90° is easy to see; it gives rise to an acceleration of the *right sign* throughout the motion of the pendulum. For example, when the pendulum is in its position of maximum displacement to the left the exciter gives rise to a couple tending to bring the pendulum back towards the right. For at that instant the end of the spring leaves its position of equilibrium and moves towards the right. The force reaches its maximum value (the end of the spring A is furthest to the right) when the pendulum is passing through its equilibrium position, and ceases (the spring returns to its slack mean position) at the instant when the pendulum ceases to move towards the right. The same is true, with a reversal of sign, for the swing from right to left.

In the case of resonance, then, energy is continually being supplied to the pendulum throughout the whole of its motion, owing to the accelerating couple "leading" by 90°. Here the amplitude would increase without limit if it were not for the energy losses which cause damping.

When the frequencies of the resonator and exciter do not agree, the acceleration due to the couple has the wrong sign over more or less of the path. The total amount of energy supplied accordingly remains trifling.

12. Resonance and Individual Sinusoidal Vibrations.

According to what we said in last section, forced vibrations of a pendulum or resonator may attain very high amplitudes even when the periodic forces acting are very small. For this it is necessary

(a) that the pendulum should be only slightly damped;

(b) that its proper frequency should agree as closely as possible with that of the exciting force.

A whole series of lecture experiments have been devised to exhibit the often surprisingly large amplitudes which may be reached in this way. We shall confine ourselves to three examples.

(1) *Forced vibrations of the base of a machine.* We set an electric motor on a board supported on bearings at both side. The balance of the shaft of the motor, which originally is good, is disturbed by fixing a small metal disc somewhat eccentrically on the shaft. The shaft wobbles a little. The rate of revolution of the motor is gradually raised. Every time the frequency of the motor reaches one of the proper frequencies of the board the board resonates and vibrates vigorously. In actual practice such vibrations may cause serious damage.

(2) *Forced vibration of a suspended watch.* A watch hanging on a peg forms a gravity pendulum. The torsional pendulum in the watch (the balance wheel) acts as *exciter*. The whole watch acts as a *resonator*, and continually executes forced vibrations of small amplitude with the high frequency of the balance wheel, which in English watches is 5 vibrations per second, i.e. considerably greater than the proper frequency of the swinging watch. In American watches the balance wheel executes only 3 vibrations per second. With a watch of this type we can obtain *resonance* between the balance wheel and the watch. The watch is suspended as shown in fig. 68 (Plate IX) on cup bearings (to reduce the damping), and the proper frequency of the watch as a whole is made equal to that of the balance wheel by means of a small auxiliary mass. In this example of resonance the watch will continue to execute forced vibrations with an amplitude of about $\pm\,30°$.

In no case, of course, are these forced vibrations without some effect on the exciter, i.e. the balance wheel. Hence at night one should hang up one's watch so that it is *not* free to move (e.g. by leaning it against a velvet-covered support).

(3) *Forced vibrations of a leaf-spring.* When previously explaining the stroboscopic method (p. 12) we used a leaf-spring capable of executing vibrations of large amplitude (fig. 69a, Plate IX). We then utilized the *forced* vibrations of the spring, the exciter being a shaft passing vertically through the holder of the spring. The shaft was made to wobble slightly by means of a rod at the side. The damping of a leaf-spring in a metal holder is very small. Hence the resonance

curve of the spring is inconveniently pointed. In order to keep to the resonance frequency it is necessary that the rate of revolution of the motor should be adjusted to about one in a thousand and kept constant. This involves a considerable amount of trouble, which, however, is avoided by artificially increasing the damping of the spring. This is done merely by fixing the spring between rubber pads instead of metal. In a spring of this kind, moreover, forced vibrations of the second proper frequency may also be obtained; we thus obtain the appearance photographed in fig. 69b (Plate IX), showing a node.

According to these lecture experiments *resonance phenomena* obviously form a very sensitive means for *detecting* vibrations of small amplitude. Here, however, a very important point is to be noted: it is only in the case of *simple harmonic* vibrations that the form of the curve of the resonator agrees with that of the exciter. It is only in the case of simple harmonic motion that the *reproduction is true to form*. If the vibrations are not simple harmonic the use of resonance leads to distortions of the form of vibration which are usually unbearable. In order to *reproduce non-sinusoidal vibrations without distortion*, forced vibrations may only be used if studious care is taken to *avoid* resonance. This will be discussed in more detail in the next section.

In spite of this restriction, however, resonance is of invaluable service to us. It enables us to detect the individual sinusoidal vibrations which " represent " a non-sinusoidal vibration. Hitherto we have only been in a position to regard these component vibrations as a simple means for formally *describing* non-sinusoidal vibration curves. Now, however, there comes a very important advance: *according to the experiments which follow, we may henceforth treat a non-sinusoidal vibration simply as a physical mixture of independent sinusoidal vibrations*. We may speak of its " composition " and " decomposition ", or in other words, of its " synthesis " and " analysis ".

From among the numerous pieces of apparatus capable of being used in support of this hypothesis we shall choose one which is used in engineering, namely, the *vibrating reed tachometer*. It consists of a large number of leaf-springs or reeds attached to a common holder. The ends of the reeds are usually thickened to make them more visible. By suitable choice of the lengths and masses of the reeds their proper frequencies are made to form a continuous series of whole numbers. Fig. 70 shows an instrument of this type in its case; with 61 springs in two rows it covers a range of frequency from 77 vibrations per second to 108.

For lecture experiments we use a holder carrying reeds without a casing and lengthen it by an added handle as in fig. 71. We let three vibrations of differing frequency act on this bar. These are most readily produced by means of three electric motors with eccentrics. Under the simultaneous action of the three sinusoidal vibrations the holder vibrates in a very complicated way, which may be made visible by any of the usual methods, most simply by means of a mirror and optical lever. This complicated motion is transferred to the holder of

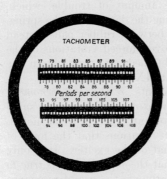

Fig. 70.—Vibrating reed tachometer

the spring. Nevertheless the tachometer merely indicates the three sinusoidal vibrations produced by the motors, none of the three being affected by the presence of the others. No new frequencies occur, so that the complicated vibration of the bar may simply be regarded as a mixture of the three sinusoidal vibrations. Only those springs with frequencies close to those of the three sinusoidal vibrations are made to execute vibrations of appreciable amplitude.

The usefulness of the tachometer, however, is by no means exhausted by the experimental demonstration of this fact.

By damping the springs sufficiently and making the resonance curves sufficiently broad, slight discrepancies between the frequency of the exciter and that of the spring may be nullified. Then the forced amplitudes of the springs are to a good degree of approximation

Fig. 71.—Silhouette of the reeds of a tachometer, side view

proportional to the amplitudes of the exciter. The tachometer turns out to be a typical "spectrometer". Apart from phase, it decomposes any arbitrarily-complicated vibration into a spectrum of simple harmonic vibrations.

In the examples just mentioned we used a crude mechanical means of transferring the vibrations to be investigated to the holder of the springs or reeds. In the laboratory and in practice the vibrations are often transferred to the springs of the tachometer by *electromagnetic* means. For this purpose an electromagnet is brought under the common holder of the springs, and a current, which varies in time according to the rhythm of the vibrations, is sent through it. An example will make this clear.

The alternating current supplied by the town mains is represented by a simple sine curve with a frequency of 50 cycles per second. Thus when passed through the electromagnet of the tachometer it excites the spring opposite the mark 50 on the scale to execute vigorous vibrations.

We then break the current quite sharply but regularly twice a second by means of a switch. The vibration curve of the alternating current then assumes the form shown diagrammatically in fig. 29B, p. 237; the fundamental frequency $n_r = 1/\tau_r$ becomes 2 vibrations per second. Its spectrum must therefore consist of integral multiples of this fundamental frequency. A whole series of these are indicated by the vibrating reed " spectrometer ", particularly the frequencies 48 and 52.

13. The Importance of Forced Vibrations in the Distortionless Reproduction of Non-sinusoidal Vibrations. Recording Apparatus.

For the mere *detection* of mechanical vibrations our sense organs suffice in the majority of cases. For example, our body can detect vibrations of the ground ($n =$ about 10 vibrations per second) with a horizontal amplitude of only $3 . 10^{-3}$ mm. If we lightly touch a vibrating body with our finger-tips we can detect vibrations with an amplitude of about $5 . 10^{-4}$ mm. when $n = 50$. Numerical details illustrating the extraordinary sensitiveness of the human ear will be given in § 17 of next chapter (p. 309). In general, however, the mere detection of vibrations is not sufficient; on the contrary, we require to have a faithful or *distortionless reproduction* of their form or a written record of them.

In any recording apparatus the vibrations under investigation are made to set some piece of mechanism (a lever, membrane, &c.) in motion. This motion, usually considerably magnified by mechanical or optical levers, is recorded on a continuously-moving strip of paper with ink or by photography. From the physical point of view the whole of this process involves *forced vibrations*. For in every case the recording system as a whole has its own series of proper vibrations. This is at the bottom of all the main difficulties met with in the recording of vibrations: to put it briefly, every recording apparatus is a resonator, in the simplest case with only *one* proper vibration. Any complicated vibration will " excite " the " resonator " to execute each of the component sinusoidal vibrations of the exciter. Each of these component vibrations imposes its *own* frequency on the resonator. It is by no means true, however, that the amplitude of the forced vibration is determined solely by the amplitude of the corresponding component of the exciting wave-train. For the resonator reacts to component vibrations of the same amplitude but of different frequency entirely according to its *own* resonance curve determined by its own frequency and damping. A vibration in the region of the proper frequency of the resonator is reproduced on much too large a scale as compared with one in a more remote region. This is the first error. The second consists in a faulty reproduction of *phase*.

The amplitude of a resonator always differs in phase from that of the exciter. The amplitude of the forced vibration lags behind the amplitude of the exciting vibration by a definite angle. This phase angle has quite different values (between 0° and 180°) for the different component vibrations of the exciting vibration. The phase angle is determined by the law illustrated in fig. 67: component vibrations of very small frequency are reproduced true to phase, but component vibrations in the resonance region of the recording apparatus have their phase changed by 90°. Such a phase difference, however, leads, even in quite simple cases where the curves merely arise from the combination of two component vibrations, to a complete change in the whole form of the curve (compare figs. 15 and 16, p. 231).

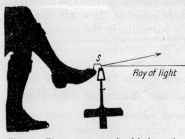

Fig. 72.—Elementary example of faulty registration of the blood pressure curve

As a warning to the reader we shall give an elementary example of a method of recording vibrations which is hopelessly wrong all through. The arrangement in fig. 72 is intended to record a man's blood-pressure curve. The " tracing lever " consists of the right leg crossed over the left knee as shown. This " resonator " is excited by the periodic expansion of the popliteal artery. The toe is connected with a small mirror S which is free to rotate. This mirror is used to reflect a ray of light to the observation screen. A rotating mirror is also inserted into the path of the ray in order to change the succession of phenomena in time into a succession of phenomena in space.

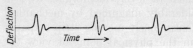

Fig. 73.—A curve " recorded " with the above apparatus

A very pretty curve is registered in this way (fig. 73). It may also be recorded permanently by photography. It has only one disadvantage: it does not resemble the actual blood-pressure curve in the slightest! The only thing that is correctly reproduced is the period of the variations of blood pressure.

Similar sins, although not always of quite such a crude kind, are in many instances committed in the course of the recording of vibrations.

According to what we have just said, to obtain *satisfactory* recording of vibrations we have to avoid two things:

(1) the enhancing of the amplitudes of component vibrations in certain regions of frequency;

(2) the altering of the phases of component vibrations relative to one another.

The first condition is comparatively easy to fulfil. By fig. 66, p. 255, we have to make the proper frequency (n_0) of the recording apparatus about equal to the highest frequency to be recorded (n_{max}) and also to see that the proper vibrations of the recording apparatus are very strongly damped. The curve of its forced vibrations must be even flatter than the curve D in fig. 66, p. 255. In this way we obtain the right amplitudes for all the frequencies between 0 and n_{max}. If, however, the phases are also to be reproduced correctly, the problem becomes considerably stiffer. The proper frequency n_0 of the recording apparatus must then be made *large* compared with *all* the frequencies n occurring in the vibration to be recorded. This we see from fig. 67, p. 257, where the change of phase is negligible only for very small values of n/n_0.

These two requirements can be satisfied by many different forms of apparatus for the low-frequency region (under 20 vibrations per second). As regards satisfactory recording apparatus for higher frequencies (up to several thousand vibrations per second) the state of affairs is very depressing. *Purely mechanical* methods have not succeeded and probably must now be regarded as hopeless. The only really satisfactory apparatus is the type of recording galvanometer known as the *oscillograph*. In the most important practical form of it, the essential part consists of a stretched loop carrying the electric current and surrounded by a magnetic field *NS*, and carrying a tiny mirror (0·5 mm.^{2} in area) for photographic registration. In the best instruments the fundamental frequency is about $2 \cdot 10^4$. In order to ensure the necessary damping the whole system is embedded in oil.* In order to use this electrical recording instrument we have first to transform the vibrations to be recorded into *variations of an electric current* without distortion of their form. The apparatus we use is in principle the same as the microphone, with which everyone is now familiar owing to its use in telephony. Its essential part is a membrane which presses a carbon contact included in the circuit more or less closely together; the resulting variation in the resistance of the circuit causes the electric current to vary in the rhythm of the vibrations of the membrane. This microphone, and even the much more convenient condenser microphone, however, introduce fresh difficulties into the problem of recording vibrations. In the first place, the *membranes* used to pick up the vibrations *themselves* execute forced vibrations. In order that the reproduction may be distortionless they must therefore be constructed in such a way as to have high proper frequencies

---

* In the telephones with attached mirrors which are frequently recommended for use as oscillographs the damping necessary for *quantitative* experiments cannot be kept very steady; they may, however, be used to demonstrate beats curves, &c., very nicely.

and strong damping. This, however, can only be attained at the cost of sensitiveness. The currents which are then produced are no longer sufficient to actuate an oscillograph. Hence the electric current must previously be *amplified without change of form*. This is done by valves resembling the well-known ones used in wireless. In the last resort these are really elaborate electrical variants and developments of simple mechanical apparatus such as is dealt with in the next section.

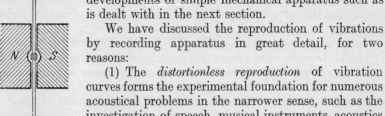

Fig. 74. — Diagram of a somewhat old-fashioned type of oscillograph.

We have discussed the reproduction of vibrations by recording apparatus in great detail, for two reasons:

(1) The *distortionless reproduction* of vibration curves forms the experimental foundation for numerous acoustical problems in the narrower sense, such as the investigation of speech, musical instruments, acoustics of rooms, &c.

(2) It is the most exacting of reproduction processes. What is true *here* may be applied, with suitable restrictions of the requirements, in the construction of other means of reproduction, such as the gramophone and the loudspeaker.

Problems similar to those occurring in connexion with recording apparatus arise in the construction of the seismographs used to record earthquakes. A seismograph for recording horizontal vibrations, for example, consists of an inverted gravity pendulum, suspended by means of suitable springs. If the earth vibrates, this pendulum is in an accelerated system of reference. As a result of forces of inertia, it is moved against the springs in the rhythm of the earth's vibrations and actuates a lever giving a large magnification (up to $5 . 10^5$). The necessary damping of the pendulum is attained by the use of air or liquid brakes.

The forces of inertia are proportional to m, the mass of the pendulum. We therefore use masses running up to several thousand kilograms. In order to have great sensitivity the extensibility of the springs is also made very small (astatic system). This, however, by equation (2), p. 52, makes n_0, the proper frequency of the seismograph, extremely small. It lies far below the smallest frequency to be recorded (n). This at first seems to be inconsistent with one of the fundamental requirements in reproduction technique, according to which the proper frequency of any recording apparatus must lie *above* the highest frequency to be recorded (see p. 263). This we deduced from the curves of forced vibrations in fig. 66, p. 255. But these curves held for the case where the amplitude of the sinusoidal vibrations of the exciter was *constant*. The forces by which the resonator was accelerated had the *same* maximum value for all frequencies. In the case of the seismograph, however, the amplitudes of the exciter, i.e. the forces of inertia, *increase* in proportion to the frequency of the exciter. The amplitudes in fig. 66 must therefore be multiplied by a factor which increases from left to right in proportion to the frequency of the exciter. The *left*-hand ends of the curves accordingly approach close to the axis of abscissæ, while the right-hand ends are practically *horizontal*. It is only in this horizontal region, i.e. in the high-frequency region, that the vibrations of the ground will be reproduced with the proper amplitudes. (The damping is made to correspond to the case D in fig. 65.)

14. The Reproduction of Vibrations by Mechanical Amplification and Reduction of Damping.

In ordinary cases of reproduction of vibrations by purely mechanical means the entire supply of energy required has to be supplied by the vibrations which are to be reproduced. If electrical means are used, the position is fundamentally different: the energy required to actuate the reproducing instrument is then supplied by a source of electrical energy, and all that the mechanical vibrations have to do is merely to *regulate* this flow of energy in their own rhythm. Here the amounts of electrical energy regulated may be considerably larger than the energy of the mechanical vibrations regulating them. In this case we have *reproduction with amplification*. In theory such amplification may be obtained by means of the microphone used to pick up the vibrations. In practice, however, wireless amplifying valves are used.

The process of reproduction with amplification is steadily growing in importance. We may mention the loud-speaker, the gramophone, the observation of the sounds made by the heart and lungs, the study of musical instruments, and so on. A *purely mechanical* explanation of this problem should make it easier for the reader to understand the main features of the process.

In this mechanical method the energy used up in the reproduction is taken not from a current of *electricity* but from a current of *water*, the latter being *regulated* by the vibrations to be reproduced. This may be realized with the primitive apparatus shown in fig. 75. A stream of water flows almost horizontally from a glass nozzle and impinges on a strongly-damped stretched membrane (e.g. a tambourine), forming a fine smooth jet. A thin jet of this kind is a very unstable body (p. 214). Minute motions of the nozzle will cause its end to break down turbulently into a large number of drops. These drops impinging on the membrane excite it to loud audible vibrations which die rapidly away. Thus a light tap on the nozzle is amplified so that it sounds like a heavy blow. In exactly the same way the vibrations of a small tuning-fork held with its stem against the nozzle (fig. 75) become audible in a very large room. As the amplitudes of the vibrations of the tuning-fork die away, the place where the jet breaks down moves farther and farther away from the nozzle. The sound emanating from the membrane meanwhile becomes softer. As a final experiment we hold a watch against the nozzle; its ticking becomes audible throughout the lecture-hall.

Amplifying valves are also used in radio engineering in very large sizes for producing *undamped electrical vibrations.* Our mechanical amplifier may similarly be made to produce undamped mechanical vibrations. We have only to bring about "back coupling" between the body capable of vibrating, in this case the membrane, and the glass nozzle, i.e. the vibrations of the membrane have to be transferred

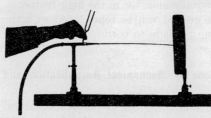

Fig. 75.—A jet of water acting as an amplifier

to the nozzle by a mechanical connexion. The membrane then "regulates" the break-up of the water-jet in the rhythm of its own frequency. All that is necessary is to lay a metal rod on the nozzle and the membrane as in fig. 76. Undamped vibrations audible at a distance are at once set up. Their frequency may be altered at will; all we have to do is to give the membrane another proper frequency by altering its mechanical tension.

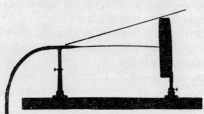

Fig. 76.—A water-jet made to produce undamped vibrations by automatic regulation ("back coupling")

The only means of attaining large amplitudes available in the older methods of reproduction was *resonance.* This, however, is not reliable for vibrations other than the purely sinusoidal ones, for it *distorts* the vibrations. In modern methods of reproduction large amplitudes are obtained by *amplification.* Resonance is only used in cases where it is reliable; but even in these another very effective device is mostly used, reduction of damping.

The amplitudes obtainable from a resonator with any particular exciting vibration are the greater, the smaller the losses of energy between successive amplitudes. Despite the most careful construction these energy losses due to friction, to the heat developed by a current, &c., can never be reduced below a certain minimum. But these unavoidable losses can to a great extent be replaced at will by a periodic supply of energy. This supply must be regulated by the resonator itself, otherwise the phase and frequency will not be right. With an automatic auxiliary apparatus of this kind any body may be made to vibrate with only very slight damping. Once excited, it will only come to rest after a large number of vibrations; it has a damping ratio scarcely differing from unity, and hence a very pointed resonance curve (fig. 66, p. 255).

The methods for amplification and reduction of damping described in this section may be realized in practice in a great variety of ways. The chief problem is the avoidance of too great distortion of the vibrations to be reproduced. In this respect electrical valve apparatus has shown itself vastly superior to all other types of apparatus, and plays a very important part in radio engineering.

15. Non-linear Combination of Sinusoidal Vibrations.   Differential Vibrations.

According to sections 12 and 13 (p. 258) the representation of non-sinusoidal vibrations by sinusoidal component vibrations is of much greater value than that of a formal description. Physically, a non-sinusoidal vibration behaves like a *mixture* of separate sinusoidal vibrations acting independently. In a resonator excited by the vibration, the amplitudes of the forced vibrations due to the various components are merely combined additively.

Physically, this addition of the individual amplitudes is necessarily associated with a *linear* law of force of the pendulum executing the forced vibrations. If the law of force is not linear, the resonator exhibits vibration curves which are distorted to one side. For example, under the action of two sinusoidal vibrations of frequency n_1 and n_2 it executes vibrations like those shown in the familiar graph of fig. 18, p. 233. *In the resonator a third and different sinusoidal vibration is set up.* Its frequency is equal to the difference $(n_1 - n_2)$ of the frequencies of the two original sinusoidal vibrations. We obtain a *differential vibration*, sometimes also referred to as an " objective differential tone ".

As a rule the one-sided distortion does not strictly follow the very simple diagram on which fig. 18, p. 233, is based. Other so-called "combination tones" then occur as well as the differential tone. Their frequencies are given by the formula $n_k = an_1 \pm bn_2$ (where a and b are small integers).

In the case of purely *mechanical* vibrations these differential vibrations occur only very occasionally. At the ordinary amplitudes, fortunately, any law of force may without much loss of accuracy be replaced by a linear one (p. 53). When *electrical* vibrations are used, unintentional differential vibrations occur very commonly, e.g. in the ordinary carbon microphone.

Differential vibrations may be *deliberately* produced by any kind of *rectifying device*. That is, the " beats curve " must be made to assume the one-sided distorted form of fig. 18, p. 233.

For a purely *mechanical* method of rectification we may use e.g. the intermittent formation of jets exhibited by Helmholtz's resonators (p. 253). *Electrical* rectifiers, however, are much more convenient.

For a lecture experiment we send two simple harmonic alternating currents of frequency 50 and 70 simultaneously through the electromagnet of a tachometer (fig. 70). The tachometer indicates both of them. We then insert into the common circuit a crystal detector such as is used in wireless. The frequency 20 (among others) appears immediately.

These differential vibrations obtained by electrical means are utilized in many ways. On them, for example, is based an elegant method for measuring the amplitudes of individual sinusoidal components of complicated vibration curves. The component vibration to be investigated is made to " beat " in a definite frequency n with an auxiliary sinusoidal vibration of variable but known frequency. The " beats curve " is then rectified and the amplitude of the resulting differential tone measured in some way. From the practical point of view this has the great advantage of involving the measurement only of amplitudes of one and the same frequency n, that of the differential tone; hence the name of " analysis by transformation of frequency ".

16. Two Coupled Pendulums and their Forced Vibrations.

Hitherto we have only mentioned the coupling together of two pendulums quite briefly. In fig. 30, p. 239, we hooked two elementary pendulums together. Considering the matter more strictly, we have to distinguish between three different ways in which pendulums may be coupled together:

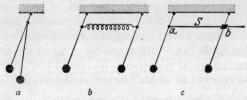

Fig. 77.—*a*, Acceleration coupling; *b*, force coupling; *c*, frictional coupling

(1) Acceleration coupling (fig. 77*a*): the one pendulum is suspended from the other. It is in an accelerated system of reference, and is therefore subject to forces of inertia.

(2) Force coupling (fig. 77*b*): the two pendulums are connected by an elastic spring.

(3) Frictional coupling (fig. 77*c*): part of the one pendulum, e.g. the bar S, which is free to rotate about *a*, rubs against a part of the other pendulum, say inside the socket *b*, which is free to rotate.

In all three cases each pendulum is again to have the same proper frequency. After coupling, the system in all three cases possesses the two proper frequencies with which we are already familiar. The lower,

n_1, is obtained when the pendulum bobs move in the same direction, the higher, n_2, when they move in the opposite direction (p. 241).

Now comes a new observation: we begin by moving only *one* of the two pendulums (no. 1) out of its position of equilibrium, and then let it go (fig. 78). Then a somewhat surprising thing happens. Pendulum no. 1 gradually gives up its whole energy to pendulum no. 2, which was previously at rest, and causes it to execute vibrations of large amplitude. Pendulum no. 1 is thereby itself brought to rest. The same process then begins again with the rôles reversed.

We may describe this process in two ways: (1) as *beats* of the two superposed frequencies n_1 and n_2, (2) as *forced vibrations with resonance*.

Pendulum no. 1, which is originally released at a point of reversal, acting as exciter, "leads" pendulum no. 2, acting as resonator, by 90°. Throughout its whole path it gives pendulum no. 2 an accleration of the proper sign, and is thereby itself retarded by the equal reaction which occurs. We have forced vibrations, together with a strong reaction of the resonator on the exciter.

Fig. 78.—Two coupled
pendulums

We shall mention three further examples of coupled vibrations.

(1) An electric bell-push hangs from a chandelier. The bell-push and the chandelier have the same proper frequency. If the chandelier is given a small, almost imperceptible blow, the bell-push begins to execute vibrations of large amplitude.

(2) A ball suspended by a spiral spring simultaneously represents two bodies capable of vibration: if the length of the spring is constant, it acts as a gravity pendulum, being deflected laterally to one side or the other, whereas if the axis of the spring remains vertical it acts as a spring pendulum. If the frequencies of the two motions are the same, the two modes of vibration recur alternately as a result of the coupling.

(3) A strongly damped leaf-spring is placed as a small rider on a tuning-fork, as in fig. 79 (Plate X). The damping of the spring is brought about in the usual way by fixing it in rubber. The spring and the fork each have their own frequency.

To begin with, let us prevent the spring from vibrating by touching it with the finger. The tuning-fork is excited and the vibrations allowed to die away, which they do very slowly, say in a minute. The vibrations may be made visible at a distance by means of the mirror Sp. We then repeat the experiment without interfering with the spring. Once excited, the tuning-fork comes to rest in barely a second. The energy of vibration transferred to the spring is wasted in the form of heat in the rubber padding. Instead of the lasting vibrations of the previous case the tuning-fork executes only a few vibrations. If the dimensions are suitably chosen the energy may even be annihilated by the time the first vibration minimum is reached.

So much for the *free* vibrations of two pendulums coupled together. In engineering, *forced* vibrations of two coupled pendulums play an important part. We shall confine ourselves to a single example, the prevention of rolling in ships.

Imagine that the tuning-fork in fig. 79 (Plate X) represents a steamer and the spring a strongly-damped pendulum built into the ship. Further, imagine the single blow on the tuning-fork replaced by the periodic impacts of the waves. We then have the essential features of the matter. In actual practice the strongly-damped pendulum takes the form of a column of water in a U-tube.

Fig. 80.—Model of a rolling-tank

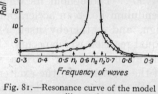

Fig. 81.—Resonance curve of the model rolling-tank

Fig. 80 shows a model " rolling-tank " of this kind attached to a suspended board with the profile of a ship section. The two limbs of the tube are connected by a tube containing air controlled by the clip H. When the clip is closed the column of water cannot oscillate. The board, i.e. the model ship, if once tilted through 40°, will complete say 20 vibrations. If the clip is opened the column of water is free to oscillate and may also be suitably damped. This time the model after being tilted through 40° comes to rest after only two or three vibrations.

In order to discuss the process more accurately we must obtain the whole curve of the forced vibrations of the coupled system, i.e. the ship and the rolling-tank, for differing degrees of damping in the rolling-tank. In order to do this we set the model on a see-saw (to imitate the action of the waves) and connect the latter by a crank to the eccentric of a slowly-running motor. If not coupled together the ship and the rolling-tank would attain their maximum amplitudes at the same frequency n_0. As a result of the coupling (chiefly frictional) they form a system with two proper frequencies n_1 and n_2. According to the construction of the tank the resonance curves exhibit two maxima of differing height. The frequencies of the waves encountered in actual practice at sea must lie in the range of frequency between 0 and n_2.

CHAPTER XII

Waves and Radiation

1. Preliminary Remarks

All vibrating bodies are subject to damping. Energy is lost between every pair of successive extreme positions. Hitherto we have ascribed these losses to the inevitable external and internal friction. This, however, is not the whole truth. In the majority of cases considerable losses of energy arise from the *radiation* of progressive waves. Three types of progressive waves are known to us:

(1) waves on the surface of liquids (Chapter X, § 10, p. 218);

(2) transverse, longitudinal, and torsional waves in solid bodies (Chapter XI, §§ 6–8, p. 238);

(3) longitudinal elastic waves in liquids and gases (Chapter XI, § 9, p. 247).

In all three cases we have hitherto confined our discussion to the propagation of these waves in *linear* bodies (wires, rods, pipes, &c.). We shall now abandon this restriction and consider the propagation of waves in all directions simultaneously. The material then falls naturally under the three following headings:

(1) How do progressive waves propagate themselves in all directions? In particular, why do we speak of the radiation of waves?

(2) How are good radiators or emitters of sound constructed?

(3) How are good detectors or receivers of sound constructed?

We shall begin by considering the propagation of waves on the surface of a liquid (water) in § 2.

2. The Propagation of Waves on the Surface of Water.

As we saw in Chapter X, § 10 (p. 218), waves on the surface of water have a complicated shape very far removed from the sinusoidal form. In addition, their velocity of propagation depends on the wavelength used, i.e. the water waves exhibit "dispersion". So far the phenomena are by no means simple. These waves, however, are propagated in a *plane*. This makes them easy to represent graphically. Their velocity of propagation is small. This makes observation easier.

For the production of water waves we use the ripple-tank and accessory apparatus shown partly in silhouette and partly in section in fig. 1. The tank has the profile of a shallow dish. Its gently-sloping "banks" allow the oncoming waves to die away, and prevent undesirable reflections. The cone of light from an arc lamp below projects an image of the waves on to the screen.

For some experiments the wave-trains used must be of limited length, i.e. must consist of only a few crests and troughs (as in fig. 10, Plate XI). In such a case the waves are excited by a single dipping of a pencil point or something similar into the surface of the water. For the majority of experiments, however, we require wave-trains of un-

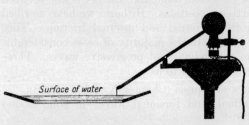

Surface of water

Fig. 1.—Apparatus for projecting water waves; in the upper right-hand corner an electric motor and eccentric

limited length. These are produced by making a small body move continually up and down in simple harmonic motion. It is convenient to use a metal rod at the end of a lever moved by an eccentric. The depth to which the rod is immersed varies by a few millimetres from the mean position and it executes about 12 vibrations per second. Then the actual wave-length is about 2 cm., which is sufficiently magnified on the screen for observation. Now for the experiments.

When the rod is vibrating we see the waves progressing outwards in the form of continually-widening circles and traversing the whole surface of the water. The centre of the waves appears as a *point source of radiation* (fig. 2, Plate X).

By using intermittent illumination we may slow down the velocity of these waves as it appears to our eyes as much as we like (p. 12). With a suitable frequency of illumination their direction of motion may even be reversed. The waves then appear to move towards the central point in concentric circles of continually-diminishing diameter. We speak of convergent waves running together to an *image*. So much for the free propagation of waves.

In the following experiments we place obstacles (of sheet lead) in the paths of the waves. We shall write down our observations in two parallel columns.

| The waves are made to pass through a slit. | The waves are interrupted by a disc-shaped obstacle. |

PLATE X

Chap. XI, Fig. 79.—Tuning-fork with a strongly damped spring attached to it (Max Wien).

Chap. XII, Fig. 2.—Waves in the surface of water (figs. 2–14 are from photographic positives).

Chap. XII, Fig. 3 *

Chap. XII, Fig. 4

* Note the reflection of the waves at the side of the slit; it also comes out very clearly in a number of the following illustrations.

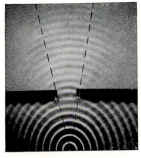

Chap. XII, Fig. 5

Chap. XII, Fig. 6

The breadth of the slit is equal to the breadth of the obstacle. Both are *large compared with the wave-length*.

A cone or bundle of concentric waves is selected from the wave-train (fig. 3, Plate X).

A cone of shadow * arises behind the obstacle (fig. 4, Plate X).

To a good degree of approximation the wave-trains are bounded by the straight lines or " rays " subsequently added to the figures. If these rays are produced backwards they meet in the " point source of radiation ".

More careful observation shows that the individual wave-crests and troughs do not suddenly break off at the boundaries marked by dotted lines, but, on the contrary, extend in low and rapidly decreasing amplitudes

beyond the boundaries of the cone of radiation to the left and to the right.

into the cone of shadow from the left and from the right.

The waves are " diffracted " beyond the region limited by the geometrical rays.

For our subsequent observations we shall now make

the slit

the obstacle

narrower. The breadth is now to be only about *three times the wave-length*. The diffraction is clearly seen. In figs. 5, 6 (Plate X) it extends considerably beyond the rays subsequently added to the figures.

In fig. 5 the vertical angle of the cone of waves is greatly widened. Its boundaries are blurred.

In fig. 6 the region of shadow is largely filled with diffracted waves. These chiefly appear at a considerable distance from the obstacle.

As the breadth of the slit or obstacle approaches the order of magnitude of a wave-length, the geometrical ray construction therefore becomes only a very moderate approximation.

If the breadth of the slit or obstacle is reduced still further it loses all meaning.

In fig. 7 (Plate XI) the breadth of the slit

In fig. 8 (Plate XI) the breadth of the obstacle

has been made about *equal to the wave-length*.

---

* A shadow being the simplest " representation " or " image " of the obstacle.

The waves fill a region subtending an angle of nearly 90°.

A shadow no longer exists. The presence of the obstacle is betrayed only by slight disturbances of the waves in its immediate neighbourhood.*

Passing to the limit we make

the breadth of the slit

the breadth of the obstacle

small compared with the wave-length. Here we use

a wave-train of indefinite length, as before.

a very short wave-train consisting of only one or two crests and troughs.

Observation yields very important results:

The slit in fig. 9 (Plate XI)

The obstacle in fig. 10 (Plate XI)

becomes the starting-point of a wave-train which is propagated in the form of

semicircles.

complete circles.

That is, both arise as limiting cases of diffraction. In this limiting case we speak of waves arising as a result of scattering, or "scattered waves". The names "elementary waves" and "wavelets" are also used. Their amplitude decreases with

the breadth of the slit.

the breadth of the obstacle.

They exist, however, no matter how small the dimensions of the slit or obstacle. If the intensity of the primary waves is sufficient they can be demonstrated in all cases. Even the minutest bodies betray their presence by scattering waves.†

Summarizing these facts, we may say that *the propagation of waves and their lateral restriction by obstacles may be represented by means of simple geometrical rays*, subject to the necessary proviso that *the geometrical dimensions* (breadth of the slit or obstacle) *must be large compared with the wave-length used.*

The physical meaning of this geometrical representation of the propagation of the wave ("geometrical optics") will be illustrated by four examples. In all of these a geometrical ray construction is given for the boundary of the wave phenomena.

(1) In fig. 11 (Plate XI) waves fall obliquely on a smooth plane

---

* Here reproduction processes reach the limit of their powers.

† Ultramicroscopic detection of minute particles.

PLATE XI

Chap. XII, Fig. 7

Chap. XII, Fig. 8

Chap. XII, Fig. 9

Chap. XII, Fig. 10

Chap. XII, Fig. 11.—Reflection
of circular waves at a mirror

Chap. XII, Fig. 12.—Shallow-
water lens

obstacle. The mechanical faults in its surface (scratches, hollows, &c.) are small compared with the wave-length. The bundle of rays is re-flected as by a mirror. In front of the *mirror* on the left we see the intersection or interference of the direct and reflected trains of waves. At the top of the figure behind the mirror we see the cone of shadow thrown by the mirror, with its borders blurred as a result of dif-fraction.

(2) In shallow water waves advance less rapidly than in deep water (equation (12), p. 221). We make use of this fact to construct a " shallow-water lens ". We place a lens-shaped body in the ripple-tank. Between its surface and that of the water there is only a space of about 2 mm. The " lens " is supported on both sides by a screen (fig. 12, Plate XI). The waves are retarded most in passing the thick middle of the lens and less as the edges are approached, owing to the decrease in thickness. As a result of this retardation the sign of the curvature of the waves is reversed. Behind the lens they run together concentrically to the " point-image " B (which in reality is a region of finite diameter) and only subsequently diverge.

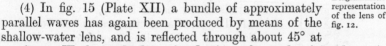

(3) Fig. 14 (Plate XII) was obtained by deeper im-mersion of the lens. After passing it the wave-trains are to a first approximation parallel straight lines; the image has moved away to an " infinite " distance from the lens. The source of radiation is at the " focus " of the lens.

Fig. 13.—
Diagrammatic representation of the lens of fig. 12.

(4) In fig. 15 (Plate XII) a bundle of approximately parallel waves has again been produced by means of the shallow-water lens, and is reflected through about 45° at a mirror. We have the law of reflection: the angle of incidence is equal to the angle of reflection.

In many cases a bundle of parallel waves may be satisfactorily represented by a single line, the central line of the bundle (as in fig. 15). In view of this commonly-employed mode of representation a bundle of parallel waves is often briefly referred to as a " beam " or " ray ". Thus we speak of sound rays and light rays instead of bundles of parallel sound waves or light waves. The terms sound ray and light ray are convenient and are frequently used, but for the sake of clear-ness we shall apply the word " ray " only to the geometrical lines in the diagrams.

3. The Fresnel-Huyghens Principle.

All the experimental results of § 2 may be summarized and ex-plained to a first (but nevertheless remarkably accurate) degree of approximation by a formal geometrical principle known as the Fresnel-

Huyghens principle. This principle applies with equal importance to all wave phenomena in physics. Here we shall explain it with reference to *one* case only, but in great detail. The case we consider is that of the limitation of a train of waves by a slit (fig. 16).

The experimental starting-point of the Fresnel-Huyghens principle is the existence of the wavelets in figs. 9 and 10 (Plate XI). The

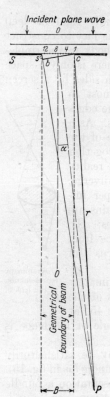

slit is imagined to be subdivided into a large number (N) of similar segments 1, 2, 3, &c. Each of these segments is regarded as the starting-point of a wavelet, no. 1, no. 2, &c. All these N wavelets intersect or are superposed at any arbitrary point of observation P behind the plane of the slit. The amplitudes of the wavelets combine to form the total amplitude actually observed at P. In the addition the important point is the difference of phase between the various wavelets; the greatest difference of phase ($\Delta\lambda$) is equal to the difference of the distances of the point of observation P from the left-hand and right-hand edges of the slit.

For the sake of clearness we shall carry out the addition process for a number of different points of observation, not by calculation but by graphical means. To simplify the problem we make three assumptions, none of which affect the essentials of the matter in any way:

(1) The centre of the waves, or source of radiation, is to be a long way from the slit and to lie on its axis of symmetry OO. This has the result of making the wave-crests (the black lines at the top of fig. 16) practically straight lines. All points of a wave-crest reach the plane of the slit at the same instant or in the same phase.

Fig. 16.—Limitation of plane waves by a slit (Fraunhofer's diffraction; $\angle bcS = \alpha$).

(2) All the points of observation P are to be a long way from the slit in a plane parallel to the slit. In this limiting case the arc cb described with centre P and radius r may be regarded as practically a straight line. For the *greatest* difference of phase occurring between two wavelets ($\Delta\lambda$) we have the geometrical relationship

$$s = \Delta\lambda = B \sin\alpha,$$

where B is the breadth of the slit. Further, in this limiting case we may regard the difference of phase of any two neighbouring wavelets ($d\lambda$) as the same, and may put

$$N\, d\lambda = \Delta\lambda = B \sin\alpha,$$

or
$$d\lambda = \frac{B \sin\alpha}{N}.$$

(3) We make N, the number of elements of the slit, equal to 12. With only 12 wavelets we obtain quite a sufficient degree of accuracy.

We now carry out the graphical addition of the twelve component amplitudes for several points of observation P. For the point P_0 on the axis of symmetry of the slit we have

$$s = 0, \quad a = 0, \quad \sin\alpha = 0, \quad d\lambda = 0.$$

All the twelve amplitude vectors are accordingly combined as in the auxiliary figure 0, as there are *no* differences of phase. Their sum or resultant R_0 is drawn below and plotted in fig. 17 at the point with abscissa $\sin\alpha = 0$.

For the next point P_1 we choose $s = \lambda/3$; then $\sin\alpha = \lambda/3B$ and the difference of phase between any two neighbouring wavelets is $d\lambda = \lambda/36$, or in angular measure $d\phi = 120°/12 = 10°$.

The amplitudes of the twelve wavelets combine according to the auxiliary figure 1, and as their resultant we obtain the line R_1, which is plotted in fig. 17 at the point with abscissa $\sin\alpha = \lambda/3B$.

We continue in the same way. For the point P_2 we choose

$$s = 2\lambda/3, \quad \text{i.e. } \sin\alpha = 2\lambda/3B, \quad d\lambda = \lambda/18, \quad d\phi = 20°.$$

The auxiliary figure 2 gives us the line R_2.

For the next point we choose

$$s = \lambda, \quad \text{i.e. } \sin\alpha = \lambda/B, \quad d\lambda = \lambda/12, \quad d\phi = 30°.$$

The amplitudes of the twelve wavelets combine to form a closed polygon as in the auxiliary figure 3. Their resultant is zero. In fig. 17 we accordingly have to insert a point on the axis of abscissæ at $\sin\alpha = \lambda/B$.

Finally we put

$$s = 3\lambda/2, \quad \text{i.e. } \sin\alpha = 3\lambda/2B, \quad d\lambda = \lambda/8, \quad d\phi = 45°.$$

The graphical addition is shown in the auxiliary figure 4. The amplitudes of the first eight wavelets form a closed octagon; their resultant is zero. The remaining amplitudes give half an octagon, i.e. the resultant R_4.

For $s = 2\lambda$ or $d\phi = 60°$ the amplitudes of both the first six wavelets and the last six give the resultant zero, and the point in fig. 17 for abscissa $\sin\alpha = 2\lambda/B$ again lies on the axis of abscissæ.

This will suffice. We can now complete fig. 17 without further discussion by making the curve symmetrical. The formal procedure of the Fresnel-Huyghens construction leads to the fluctuations of amplitude which are of such great importance in connexion with all kinds of wave motion. It accounts for the actual state of affairs, which is sketched quite roughly in fig. 18 and may be described in words as follows:

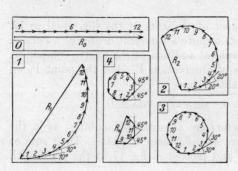

(1) The train of waves extends on either side beyond the lines (shown dotted) given by the geometrical ray construction: it is diffracted. It vanishes only at an angle of deviation α whose sine is equal to λ/B, i.e.

$$\sin \alpha_{\min} = \frac{\lambda}{B}. \quad (1)$$

(2) The geometrical ray construction is a satisfactory approximation only for *small* values of the quotient λ/B.

(3) In the region where diffraction takes place other wave-trains occur at certain angles, in addition to the main wave-train.

The two first points are completely borne out by our experimental results for waves on the surface of water (Plate X).

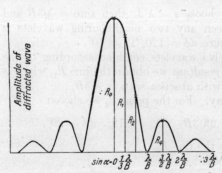

Fig. 17.—The amplitude-crests which arise when a plane train of waves is cut off by a slit, with the auxiliary figures required in the construction (above). The intensity or energy of the waves is proportional to the square of the amplitude. Hence for a comparison with experiment (e.g. fig. 32) we have to square the ordinates of the amplitude-crests.

The third point, the existence of subsidiary wave-trains in the region of diffraction, may, as we shall see in § 9, p. 290, be demonstrated in a very impressive way with sound waves. Water waves are not well adapted for this purpose, owing to certain of their peculiarities which we shall discuss in the next section. We shall therefore break off and resume our line of discussion in § 9, p. 290.

In figs. 16, 17 we dealt with the limiting case of " Fraunhofer's diffraction ": both the source of radiation and the plane of observation lie far away (at an " infinite " distance) from the slit. The incident wave crests are practically straight lines. With the water waves, on the contrary, we had the more general case of "Fresnel's diffraction": the incident wave-crests were arcs of circles with an appreciable curvature. This makes the graphical combination of the amplitudes of the elementary waves in the discussion based on the Fresnel-Huyghens principle a little more difficult, but does not alter the underlying principles in the least.

According to the Fresnel-Huyghens principle we combine the amplitudes of the wavelets while taking account of their mutual differences of phase. This somewhat lengthy process is unavoidable in dealing with diffraction phenomena. If the quotient λ/B is small, however, diffraction may be neglected and the wave-train may be regarded as bounded by the geometrical rays. In this limiting case we

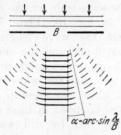

Fig. 18.— Rough diagram of the diffraction of waves by a slit.

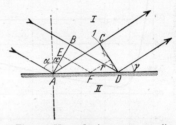

Fig. 19.—How reflection occurs, according to Huyghens' principle. The train of waves is bounded at the sides by rays represented by arrows.

may confine ourselves to the original Huyghens principle, according to which the resultant wave is constructed as the common tangent or envelope of the wavelets. We shall illustrate this for the case of "reflection". A wave-train bounded by parallel lines is to be thrown back from the plane boundary surface of two media (a "mirror"), (fig. 19).

AB is a wave-crest before reflection, CD a wave-crest after reflection. Each point of the boundary surfaces is regarded as the starting-point of a wavelet. Three of these are drawn in the figure; the fourth, starting from D, has not yet attained a finite radius r. For every wavelet we have the equation (see the dotted lines in fig. 19)

$$EF + r = BD.$$

As the result of this Huyghens construction, we have the law of reflection: *the angle of incidence is equal to the angle of reflection.* The original form of Huyghens' principle will prove useful to us in § 10, p. 292, in explaining the diffraction grating.

4. Dispersion of Waves on the Surface of Water: Group Velocity.

In this section we conclude our account of waves on the surface of a liquid (e.g. water). The velocity of waves on the surface of a liquid depends on the wave-length; there is "dispersion". For sufficiently low waves of practically sinusoidal form the velocity may be calculated from the formulæ on pp. 220, 221. Fig. 20 gives numerical values for the velocity of water waves with wave-lengths such as are used for lecture experiments.

This dispersion of waves on the surface of water has a very important and surprising result: it forces us to distinguish between *phase velocity* and *group velocity*.

In order to illustrate these two ideas we begin, as always, with

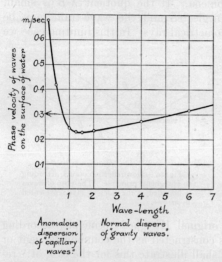

Anomalous dispersion of "capillary waves."

Normal dispers of "gravity waves."

Fig. 20—Velocity of low, practically sinusoidal waves on the surface of water, for the wave-lengths used in lecture experiments.

observation. By dipping a rod into the ripple-tank (fig. 1, p. 272), we set up a finite wave-train (cf. the wide arc in fig. 10 (Plate XI), where the wave-length is about 3 cm. and the amplitude one or two millimetres). We see that the wave-train as a whole, the "group" of waves bounded in front and in the rear by still water, advances only about half as fast as the individual wave-crests within the group. We can also see the reason for the slow advance of the group, though as yet without fully understanding it: the wave-crests die away at the front of the group and new wave-crests are simultaneously formed at the back. Thus we at once see that there are two different velocities, (1) the *group velocity*, the velocity of the beginning or end of the wave-train; (2) the velocity of the individual wave-crest, which in general we call the *phase velocity*.

In the strictest sense the term phase velocity only applies to a sinusoidal wave. It means the velocity of a definite phase, e.g. of a wave-crest, in a sinusoidal wave-train of *infinite* length and constant amplitude.

The profile of our short train of water waves, however, is by no means that of a simple sine curve, but resembles the curve of fig. 27, p. 236, i.e. the spectrum corresponding to it is continuous (fig. 28, p. 236). The finite train of waves is to be regarded as the result of the superposition of an "infinite" number of sinusoidal components (spectrum lines), and in the case of dispersion each has its own phase velocity. These individual phase velocities, phase velocities in the strict sense of the word, cannot be picked out by observation of the wave-train, and are therefore inaccessible to measurement. The most important of these components, however, belong to only a narrow spectral region, corresponding to the sharp maximum of the spectrum in fig. 28, p. 236.

For this narrow region the phase velocity is approximately constant. Its average value is equal to the wave-crest velocity which we observe, or, in other words, the middle part of our finite wave-train may, to a good approximation, be replaced by a length of ordinary sinusoidal wave of constant amplitude. The observed velocity of the wave-crests is the phase velocity of this substituted sine-wave.

How does the difference between the phase velocity and the group velocity arise? The answer is that in measuring the wave velocity we must keep some definite reference point in view, the beginning or end of the train of waves, the place where the crest is lowest or highest, &c. When there is dispersion the different sinusoidal components of the wave-train advance at differing rates. Hence the profile of the wave-train, which arises from the combination of the individual sinusoidal components, is subject to continual change: the reference point we selected does not remain fixed,

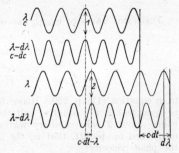

Fig. 21.—To illustrate group velocity

but is displaced relative to the individual components as they advance with their phase velocities. We see this clearly from the following example, which illustrates the quantitative treatment of group velocity. For this we need only consider the simplest wave-group, a "beats curve" consisting of only two sinusoidal components (see the bottom of fig. 14, p. 230).

The two components are sketched diagrammatically in fig. 21. In the beats curve itself it is impossible to recognize these two sinusoidal components as individuals, and their phase velocity therefore cannot be measured. All that can be measured is the group velocity. As a reference point it is convenient to choose the wave-crest which rises highest above the general level. This point is distinguished from other points of the beats curve by the fact that there the two component waves are in the same phase. It is marked in fig. 21 by the arrow 1.

In fig. 21 this mark moves considerably more slowly to the *right* than the two individual waves λ and $(\lambda - d\lambda)$. For in fig. 21 the *smaller* wave has also the *smaller* velocity,* i.e. only $(c - dc)$. It is accordingly

---

* Corresponding to normal dispersion in optics.

overtaken by the larger wave λ, at the rate of $d\lambda$ in time $dt = d\lambda/c$. As a result of this the point where the two waves are in phase moves through a whole wave-length λ *backwards to the left* in the train of waves which is advancing towards the *right*. The two component waves in fig. 21 advance to the right through distances $c\,dt$ and $(c\,dt - d\lambda)$ respectively in time dt. The reference point or place where the two waves are in phase, however, only retreats through the much smaller distance $(c\,dt - \lambda)$ in the same time dt to the double arrow 2. (In fig. 21 it happens that the intervals $(c\,dt - \lambda)$ and $d\lambda$ have come out almost equal.) That is, the velocity of the reference point, or group velocity, is given by

$$U = \frac{c\,dt - \lambda}{dt} = c - \lambda\frac{dc}{d\lambda}. \quad \ldots \quad (2)$$

For water waves of length exceeding about 5 cm. we have

$$\frac{dc}{d\lambda} = \frac{1}{2c}\frac{g}{2\pi}$$

or $\qquad\qquad \lambda\frac{dc}{d\lambda} = \frac{c}{2}$

from equation (11), p. 220. Hence by equation (2) above, the group velocity U is equal to $c/2$, i.e. equal to half the phase velocity. Again, for wave-lengths below about 2 cm. (capillary waves) we obtain $U = 3c/2$ from the equation mentioned on p. 221, that is, the group velocity is 50 per cent greater than the phase velocity.

The distinction between phase velocity and group velocity is by no means confined to water waves, but plays an important part in many wave phenomena in optics and electricity. The production of progressive waves of strictly sinusoidal form, in fact, is always impossible. Every train of waves has a finite length, i.e. in the last resort has the form shown at the bottom of fig. 6, p. 54, and in fig. 27, p. 236. To any such wave-train there corresponds a narrow continuous spectrum consisting of a large number of closely neighbouring frequencies. Hence in any case where dispersion exists it is impossible to measure anything but the *group velocity*. *Measurement of phase velocity is only possible when the wave velocity is entirely independent of the wave-length.*

Water waves occupy a privileged position merely because they are easy to observe. Thus their characteristic usefulness as an experimental aid to the study of wave-motion in general extends to the question of the distinction between phase velocity and group velocity.

5. Longitudinal Elastic Waves in Air: Sound Waves.

Among all progressive elastic waves the most important for us are longitudinal waves in *air*. We have already given a full discussion of linear waves confined in pipes (Chapter XI, § 9, p. 247). When the limitations of the pipe are removed these longitudinal waves are propagated symmetrically into space from a centre which is more or less a "point". A section in a meridional plane at a definite instant shows a distribution of air pressure and density as in fig. 22 (Plate XII). In the heavily shaded regions the pressure and density are *greater* than in the air at rest. Following our usual graphical representation, we call the former regions *wave-crests*. Similarly the high lights represent *wave-troughs*. In these the pressure and density of the air are *less* than in the air at rest. If the amplitudes are large the density maxima of progressive sound waves may be made visible by the shadows they cast in instantaneous photographs ("Schlieren" method with dark

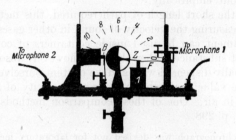

Fig. 23.—The Behm chronograph

ground). Pictures of this kind are shown on Plate XIII. Details of the magnitudes of the pressure changes which actually occur in sound waves will be found in § 17, p. 309.

The whole distribution shown in the instantaneous picture of fig. 22 advances symmetrically outwards with a velocity of about 340 metres (about 1100 feet) per second. This, the velocity of sound, is measured by finding the time required to traverse a known distance. Of the large number of methods used we shall demonstrate only one experimentally. It requires a length of path of only a few metres, the beginning and the end of which are occupied by microphones. As the sound wave passes, these microphones actuate a type of chronograph (fig. 23). This little instrument forms the only really distinctive feature of the experiment. Its essential part is an incomplete disc of aluminium, the outline of which appears in the figure. It is free to rotate with very little friction about a horizontal axis. Its iron pointer Z is held down in the equilibrium position by the electromagnet 1. In this position the pointer holds the spring F down so that it is curved slightly downward. The current of the electromagnet is meanwhile flowing through a contact formed by the spring and the iron core of the magnet. The sound wave approaches from the right and excites microphone 1 to execute forced vibrations. At the first minimum of the current three things happen:

(1) The iron pointer and the spring are let go.

(2) Owing to the above-mentioned spring contact being broken the circuit of the electromagnet 1 is broken once and for all.

(3) The spring at once straightens itself and thereby gives the disc angular momentum.

This angular momentum makes the disc rotate with constant angular velocity. After a few hundredths of a second the sound wave reaches the second microphone. At its first current minimum the latter releases the electromagnetic brake B and stops the rotating disc. Here again the measurement of time is reduced in quite an obvious way to uniform rotation. The graduation of the scale is best carried out empirically.

Owing to the short length of path required, this method may also be used for measuring the velocity of sound in other gases or in liquids, for the instrument can be used inside containers of convenient size. In the case of substances other than air, however, absolute measurements are usually dispensed with and we content ourselves with a comparison of the value there obtained for the velocity of sound and its known value in air. One of these comparison methods will be mentioned in § 13, p. 298.

The Behm chronograph was devised not for laboratory measurements but for a technical purpose, namely the acoustic method of depth sounding.

A sound signal is produced at the surface of the sea. It travels to the bottom and is reflected from there as an echo. The total time taken by the sound is measured and the length traversed (twice the depth of the water) is then calculated from the value of the velocity of sound in water (about 1500 m. per second). The interval of time may be measured in a large variety of ways, one being by the use of the Behm chronograph.

6. Bundles of Parallel Sound Waves in Air.* Diffraction of Sound Waves.

Waves on the surface of water give us a very clear picture of the way in which progressive waves are propagated. The processes which there take place *in one plane* may be represented in large measure by means of the Fresnel-Huyghens principle. This formal construction must with suitable modifications remain valid for the propagation of elastic waves *in space*. In the construction it cannot make any difference whether a sine curve represents the profile of a water wave or a distribution of air pressure. *In the case of the propagation of waves in space the ratio of* λ, *the wave-length, and B, the size of the openings and obstacles used, must again be the decisive factor*. Hence we can go ahead more rapidly with the experimental observation of the propagation of sound waves in air than we could previously with water waves. As source of radiation we use the small whistle shown in fig. 60, Plate IX.

* See p. 275 above.

PLATE XII

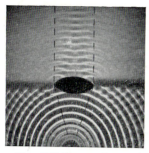

Chap. XII, Fig. 14.—Source of radiation at the focus of the shallow-water lens. The two dotted lines indicate the boundaries of the bundle of parallel rays.

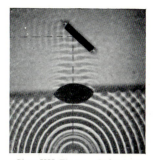

Chap. XII, Fig. 15.—Reflection of an almost parallel bundle of rays. The dotted line indicates the central ray of the bundle.

Chap. XII, Fig. 22.—Section of a spherical wave in air (sound wave)

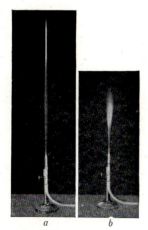

Chap. XII, Fig. 25.—a Sensitive flame, b the same flame under the influence of short sound waves

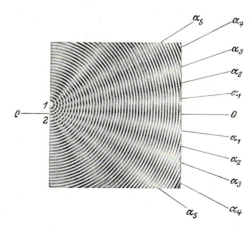

Chap. XII, Fig. 46.—Interfering wave-trains in the double slit interferometer (model experiment)

The length of the waves produced by it in a pipe was found to be about 1·5 cm. That is, it is certainly of the same order of magnitude as the wave-length of the water waves in the ripple-tank. We may take over these geometrical dimensions without more ado.

With these considerations in mind we proceed to set up a " sound reflector " (fig. 24), which is intended to supply a train of plane sound waves bounded by parallel lines (bundle of parallel waves). For this we use the parabolic mirror of a large motor-lamp placed on the optical bench B which may readily be tilted and rotated. Using a small electric lamp, we determine the focus of this mirror in the ordinary way; the image of the filament must appear sharply on a distant ("infinitely" distant) wall. By means of a suitable stop we may then replace the electric lamp by the little whistle and bring the latter exactly to the place formerly occupied by the filament. In all the following experiments we shall again adjust the path of the beam of sound by this optical method.

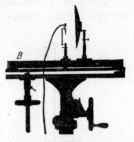

Fig. 24.—A sound reflector; whistle as in fig. 60, Plate IX.

To detect the sound waves we shall meanwhile use a *sensitive flame*. This is a long coal-gas flame with the velocity of the issuing jet suitably adjusted (fig. 25a, Plate XII). The flat, quietly burning flame is very sensitive; it reacts to mechanical disturbances in very much the same way as the sensitive jet of water in fig. 75, p. 266. The smooth filament breaks down turbulently * into a short, unsteady, roaring flame (fig. 25b, Plate XII).

We use this sensitive flame as a detector to demonstrate the sharply-bounded bundle of parallel sound waves emitted by the reflector. The sound may impinge on the flame directly or by way of a mirror (a smooth wooden or metal plate fixed so as to rotate about a vertical axis). We find that the law of reflection (angle of incidence equal to angle of reflection) is strictly satisfied. Obstacles throw *shadows*, provided they are large compared with the wave-length used (about 1·5 cm.). The waves bend round small bodies, as may be shown by many kinds of object, one or two fingers, the whole hand, &c.

This diffraction of sound waves has a very important bearing on their use in ordinary human intercourse. The sound-waves used in speech, for example, mostly have wave-lengths of one or two decimetres. Hence they are diffracted round obstacles of ordinary size between speaker and listener. In addition, the nuisance that would be caused in ordinary life by the shadows cast by large obstacles is prevented by the frequent reflection of the sound waves at the walls of the room (see also § 17, p. 309).

* Even when the gas is not lighted.

Sound shadows may also be demonstrated very prettily without the help of any instruments at all. If we rub our right thumb and forefinger together at a distance of about 20 cm. from the right ear we hear a sound of high pitch somewhat like that of our whistle. If we then close the right ear with our left hand we find we hear absolutely nothing, for the left ear is entirely within the sound shadow.

7. The Sound Radiometer.   Subjects to be studied in the next few sections.

The results obtained by using the sensitive flame are not accurate enough for the remainder of our experiments on sound waves. This defect is avoided by the use of another form of sound detector, the "sound radiometer". To discuss acoustics without this instrument would be something like discussing electricity without a galvanometer.

Externally the sound radiometer consists of a metal plate fixed to the arm of a sensitive torsion balance. When sound waves are reflected at this metal plate the plate is acted on by a force in the direction of the incident train of waves. The incident wave exerts a " pressure " on the obstacle. This is an empirical fact.

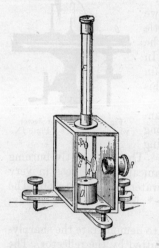

Fig. 26.—Sound radiometer (represented by R in fig. 28 and subsequent figures).

From this experimental fact we draw an important conclusion: the linear law of force can only be a first approximation, although a very good one, even for sound waves in air. For with a linear law of force the increase in the air pressure at the wave-crests would have to be exactly equal to the decrease of the air pressure at the wave-troughs. Unless there were an additional quadratic term (cf. equation (3), p. 53) a radiation pressure could never arise from the sound waves. Quantitatively we find that for air the radiation pressure P is equal to $1 \cdot 2I$, where I is the intensity of sound waves defined on p. 303 below.

A sound radiometer sufficiently sensitive for lecture experiments is shown in fig. 26; a diagrammatic sketch is given in fig. 28. The torsion balance is enclosed in a rectangular case with glass windows. A is an aluminium plate, G a counterpoise. The torsion axis consists of a thin bronze filament F. Sp is a mirror forming part of an optical lever. The sound waves enter through the short side-tube R. They are usually conducted to the instrument by means of a concave mirror (cf. fig. 28). The plate A is placed approximately at its focus. This sound radiometer is very convenient to handle.

Waves of all kinds are of tremendous importance in modern physics. Thus for example the whole of atomic and molecular physics is at present developing into "wave mechanics". To a large extent the same formal treatment applies to all wave phenomena. The whole of this chapter is intended in the first place to give a clear and simple account of this formal basis of wave propagation. Water waves, owing to their special simplicity, have been of considerable service to us in this respect. Above all they have made the meaning of the geometrical ray construction (geometrical optics) clear to us. Short sound waves will prove even more useful to us. Using the radiometer as a quantitative detector, we shall by their means be enabled to extend our knowledge of wave propagation very considerably. A great number of instructive experiments may be carried out with these short waves, but we shall confine ourselves to really important facts; these will form the subject of the next three sections.

8. Refraction, Reflection, and Scattering of Sound Waves by Layers of Air differing in Density.

(1) *Refraction of sound waves by a prism.* Waves, no matter of what kind, almost always advance with different (phase) velocities in different substances or media. Hence the direction of the waves is altered as they cross the boundary between two media. This phenomenon is referred to in general as *refraction*. According to Huyghens' principle (p. 275), refraction is to be thought of as occurring in the manner shown in fig. 27.

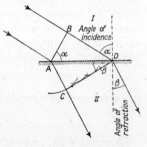

Fig. 27.—How refraction arises, according to Huyghens' principle

A plane wave-train (bundle of parallel waves, represented here by two boundary rays) falls obliquely on the boundary surface of two media I and II. AB is a wave-crest in the first medium, CD a wave-crest in the second medium. CD is the resultant or envelope of the wavelets shown emanating from the boundary surface. The distances BD and AC are described in the same interval of time. They are therefore in the ratio of the velocities of the waves in the two media:

$$\frac{BD}{AC} = \frac{v_{\mathrm{I}}}{v_{\mathrm{II}}}.$$

From fig. 27 we also have the geometrical relationship

$$\frac{BD}{AC} = \frac{\sin \alpha}{\sin \beta},$$

and for the constant quotient of the two wave velocities, v_1/v_{II}, we put n. We thus obtain the *law of refraction*,

$$\frac{\sin a}{\sin \beta} = n, \quad \ldots \ldots \ldots \quad (3)$$

or, in words, *the quotient of the sine of the angle of incidence and the sine of the angle of refraction is equal to a constant*, which is called the *index of refraction* or *refractive index*. The refractive index is equal to the ratio of the wave velocities in the two adjacent media.

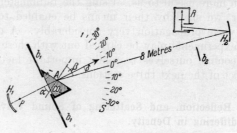

Fig. 28.—Refraction of a bundle of parallel sound waves by a prism full of CO_2 (acute angle of prism equal to a)

In order to refract sound waves we use a plane boundary between carbon dioxide and air. The velocity of sound in carbon dioxide is about 269 metres per second at room temperature (it may be determined e.g. by the method shown in fig. 52, p. 248). By calculation we accordingly find that the index of refraction for the passage of waves from air to carbon dioxide is

$$n_1 = 340/269 = 1 \cdot 26$$

and from carbon dioxide to air

$$n_2 = 269/340 = 0 \cdot 79.$$

A suitable experimental arrangement is shown in fig. 28. A sound reflector (fig. 24) is again used to supply an approximately parallel bundle of waves, represented in the figure by the dotted central line (see p. 275). This bundle of plane sound waves passes through the hollow prism filled with carbon dioxide. Its " transparent " walls are made of silk material.* The sound emitter, H_1P, the hollow prism, and the large screen b_1 are all mounted on the same optical bench. The bench is free to rotate about the vertical axis A, and the angle through which it is turned may be read off on the scale S. The sound waves are detected by means of the sound radiometer R, which is placed at a distance of about 8–10 metres. To begin with we use the hollow prism filled with air, so that the rays go through without

* Paper, cellophane, and gutta-percha membranes are practically " opaque ".

deviation. After filling the prism with carbon dioxide we find that there is a deviation δ of 9·8°. Only the second surface is concerned in this refraction, for with normally incident waves the angle of incidence α is zero, hence by the law of refraction (equation (3)) the angle of reflection β is also zero, i.e. with normally incident waves there is no change of direction.

The angles of incidence (α) and of reflection (β) are shown for the second surface: in the prism used $\alpha = 30°$. From the figure we see that $\beta = \alpha + \delta$, i.e. $\beta = 39·8°$. Hence

$$n = \sin 30°/\sin 39·8° = 0·5/0·64 = 0·78.$$

The value thus obtained experimentally is in good agreement with the value calculated above from the velocities in the two gases.

Fig. 29.—The gas burner used to produce a vertical column of hot air in fig. 30.

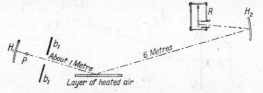

Fig. 30.—Reflection of a ray of sound from a hot layer of air

(2) *Reflection of sound waves at the boundary of two layers of air of unequal density.* By means of a comb-like gas burner we produce a tolerably plane vertical wall of hot air of low density (fig. 30). This reflects the bundle of parallel waves very clearly, although not quite so accurately as a wooden or metallic mirror.

(3) *Scattering of sound waves in air varying in density from place to place.* We direct the bundle of parallel waves thrown by the sound reflector straight at the radiometer, and then interpose in the path of the waves hot flame gases from a gas burner moved irregularly here and there. Or we let gaseous carbon dioxide stream out of the rose of a watering-can in the path of the waves. In both cases the waves are scattered irregularly in all directions as a result of reflection and refraction. The radiometer gives only a small deflection. Of the original sharply-defined bundle of waves nothing is to be seen; it has been completely destroyed by the "turbulent" medium.

The bearing of these three experiments on the propagation of sound waves through the atmosphere is obvious. The second experiment explains the "atmospheric echo" or echo in the absence of any solid wall. The two other experiments enable us to understand the great variations of intensity of distant sources of sound. According to the third experiment, the sound on its way to the ear may be scattered as a result of marked irregularities in temperature and density. On the other hand, reflections and refractions of the sound rays may favour the sound reaching our ears from great distances (sound mirage).

9. Limitation of a Train of Sound Waves by a Slit.*

In § 3, p. 275, we learned how to represent the propagation of waves according to the Fresnel-Huyghens principle, but did not complete the first example. This example was concerned with the limitation of a plane train of waves by a slit. Here we start directly from the geometrical construction given previously (fig. 17, p. 278). In its main features the result of this construction agrees with what

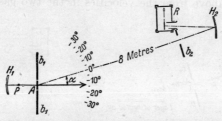

Fig. 31.—Limitation of a bundle of plane sound waves by a slit (Fraunhofer diffraction); R the sound radiometer with receiving mirror H_2.

Fig. 32. — The slit b_1 of breadth B used in fig. 31.

we found experimentally for water waves. For large values of the quotient wave-length/breadth of slit it gives strong diffraction, i.e. the effects of the wave-train persist for a considerable distance beyond the boundaries indicated by the geometrical rays.

This construction according to the Fresnel-Huyghens principle, however, gave us still more, namely, the maxima and minima in the region of diffraction. These finer details cannot be satisfactorily reproduced with water waves. An excellent demonstration, however, may be carried out with short sound waves, using a sound radiometer. A primitive type of apparatus is shown in fig. 31. In the bundle of parallel waves emitted by our sound reflector we place a screen with a rectangular hole 11·5 cm. across. The whole apparatus may be rotated about a vertical axis A passing through the centre of the slit. The angle (a) through which it is rotated is read off on the scale. The wavelength of the note emitted by the whistle is 1·45 cm. (cf. Chapter XI, § 9, p. 247); all the other particulars may be seen from the figure.

Fig. 33 gives the radiometer deflections for various values of a. The plotting of this curve (sound-peak, wave-peak) point by point, however, takes up some time. For a rapidly-performed lecture experiment it is more convenient to confine ourselves to alternate deter-

* Continued from p. 278.

mination of the individual maxima and minima. The first minima lie at the angles $\alpha_{min} = 7\cdot2°$ from the middle line. By equation (1), p. 278, we then find that $\lambda = 1\cdot44$ cm., in excellent agreement with previous measurements by means of Kundt's dust figures.

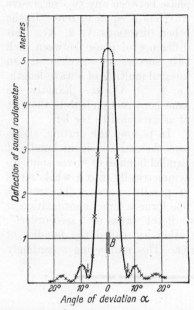

Fig. 33. — Fraunhofer's diffraction curve (sound-peak) for wave-length 1.45 cm. and the slit shown in fig. 31. The shaded region B indicates the geometrical boundaries of the ray.

In fig. 34 the same experimental results are represented in polar co-ordinates. The radius vector r denotes the magnitude of the radiometer deflection or the intensity of the sound. This method of representation is preferred in engineering circles.

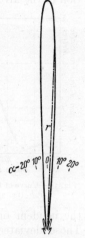

Fig. 34. — The Fraunhofer diffraction curve of fig. 32 plotted in polar co-ordinates.

Finally, fig. 35 gives the corresponding results for another slit of breadth 5 cm., giving a correspondingly stronger diffraction effect. The angle α_{min} for the first minimum turns out to be 17°. Hence $\lambda = 1\cdot46$ cm., i.e. a result again in good agreement with the value just given.

These impressive experiments are among the fundamental experiments of wave theory. The same is true of those which follow in §§ 10–12.

10. Fraunhofer's Diffraction Grating with Many Equidistant Slits.

A train of plane waves bounded by parallel lines (bundle of parallel waves) is to impinge on the grating at right angles. The essential phenomena may be obtained even from the simple graphical construction in fig. 36. Each of the five narrow slits in the grating becomes the starting-point for a train of wavelets. The wave-crests have been drawn in the figure and the intervening valleys left blank. At the wave-crests of the neighbouring curved wavelets *common tangents* may be drawn at right angles to the directions indicated by the arrows, 0, 1, 2, ... Along these lines the crests of the wavelets again combine to form a plane or uncurved wave-train. In the direction of the arrow 0 the difference of phase between any two successive

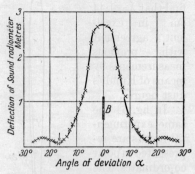

wavelets is equal to zero. In the other directions 1, 2, &c., this difference of phase between each two successive wavelets is an integral multiple of a wave-length, i.e. λ, 2λ, 3λ, &c. Exactly the same construction is to be thought of as extended to the left.

In passing the grating, therefore, the single incident bundle of parallel bundles of waves splits up symmetrically into a whole series of parallel waves. The middle one forms a direct continuation of

Fig. 35.—Curve of Fraunhofer diffraction for a slit b_1 (fig. 32) only 5 cm. across

the incident one; it is called the bundle of waves of " zero order ". Those deviated to one side or the other are called the bundles of waves of first, second, . . . , nth order. This numbering accordingly

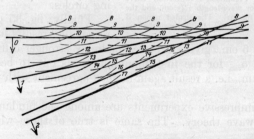

Fig. 36.—Fraunhofer's diffraction grating, explained by means of Huyghens wavelets

expresses the differences of phase between neighbouring wavelets.*

From fig. 36 we see at once that the angles of deviation of the bundles of parallel waves of the various orders are

$$\sin a_1 = \lambda/d, \quad \sin a_2 = 2\lambda/d, \quad \sin a_n = n\lambda/d, \quad . \quad . \quad (4)$$

* Here we have only used Huyghens' principle. By combining the wavelets according to the Fresnel-Huyghens principle we obtain in addition $N-2$ low subsidiary maxima between each order when the grating has N slits. This combination may be carried out straightforwardly by the graphical method illustrated on p. 278. These subsidiary maxima are too small to be detected in the measurements recorded in fig. 39.

where d is the "grating constant", i.e. the distance between the middle points of adjacent slit openings. That is, the angles of deviation for a given grating are determined by λ, the wave-length of the incident

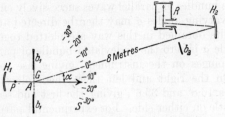

Fig. 38. — The Fraunhofer diffraction grating used in fig. 37.

Fig. 37.—Diffraction grating for sound waves

bundle of rays, alone. Hence the diffraction grating acts as a *spectrometer*. It will decompose a mixture of sinusoidal waves incident simultaneously into components separated in space. We imagine that at a great distance from the grating and parallel to its plane there is a plane of observation. In this plane the various wave-lengths arranged

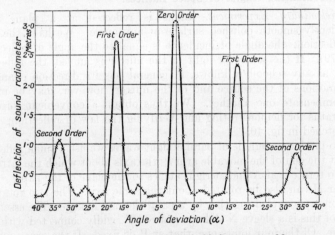

Fig. 39.—A diffraction spectrum obtained with the apparatus of fig. 37

in the order of their magnitude or frequency appear beside one another as a *spectrum*. The shortest waves appear least deviated, the longest ones most deviated. In the region of large deviation, however, the spectra of bundles of waves of different order may overlap. We may trace out these spectra for sound waves by using the sound radiometer. The arrangement of the experiment, which again is quite primitive, is shown in fig. 37. In fig. 36 we began with a bundle of parallel waves incident at right angles. A bundle of parallel waves of this kind is produced by means of the sound reflector. In the path of the waves we put a grating of wooden laths. It has seven slits and a grating constant of 5 cm. (fig. 38). The sound reflector and the grating may be rotated simultaneously about a vertical axis A passing through the plane of the grating (fig. 37). The bundles of parallel waves successively emerging from the grating at varying angles α may then be directed at the sound radiometer. Results obtained in this way are plotted together in fig. 39. When the angle α is zero the undeviated bundle of parallel waves of zero order impinges on the instrument, giving rise to the maximum deflection. On the right and left there then follow two further maxima at angles 16·8° and 33·6°, giving the first and second order spectra of the whistle on either side. For experimental purposes it is quite sufficient to determine the individual sharp maxima and flat minima alternately. The experiments are ridiculously easy to carry out. This spectrum, then, consists practically of only *one* spectral line. Its wave-length, by equation (4), p. 292, is found by calculation to be 1·45 cm. In this case, therefore, the whistle has produced practically only *one* sinusoidal wave.

11. The Glancing Angles of Space Lattices.

Preliminary Experiment.—Light is reflected at a plane and sufficiently smooth polished surface at any angle. We have the law of reflection: the angle of incidence is equal to the angle of reflection (p. 279). If the source of light or the mirror is moved, the reflected light, as any schoolboy knows, in general has its direction changed. The motions of the source and of the mirror, however, may be made to compensate one another. We thus obtain a convenient form of apparatus for demonstrating and verifying the law of reflection; this is sketched in fig. 40.

To produce a bundle of parallel light waves we use a small reflector HP at the end of the movable arm *r*; the axis about which the latter is free to rotate is at A and at right angles to the plane of the paper. The mirror M can also be rotated about the same axis. The two rotatory motions are coupled together by a parallelogram linkage. The essential part of this is a sleeve N sliding on a rod F rigidly connected with the mirror. Of the four joints one, that at B, is fixed. If the arm *r* moves through the angle α the mirror rotates through an angle α/2. Hence the reflected bundle of light waves S remains fixed, seeing that the law of reflection (angle of incidence equal to angle of reflection) holds. We now carry out the experiment with sound waves. We use our ordinary sound reflector (fig. 24, p. 285) and a smooth board as mirror. We obtain a strong reflection for *any* angle of incidence α.

Now comes a second experiment: we replace the mirror M by a plane grating, e.g. the lattice of thin wires and wooden balls shown in fig. 41. We again obtain a *definite reflection* for *any* angle of incidence.

With this wide-meshed netting the intensity of the reflected waves is naturally much less than with the opaque mirror of fig. 40.

According to Huyghens' principle this reflection arises in exactly the same way as is shown in fig. 19, p. 279. In the geometrical con-

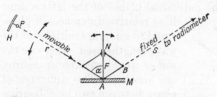

Fig. 40.—To demonstrate the law of reflection with a constant direction of the reflected wave-train

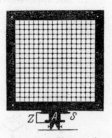

Fig. 41. — Plane grating (Z is the pointer for the angle scale S).

struction every point of the lattice is regarded as the starting-point of a wavelet.

According to the strict Fresnel-Huyghens principle laterally-deviated wave-trains of 1st, 2nd, ... order occur on either side of the regularly reflected wave-train of "zero order". In the present case we may neglect these details of diffraction in view of their very small intensity.

Now for the main experiment: we replace the two-dimensional lattice by a three-dimensional one. We put together at least three or

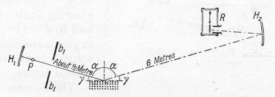

Fig. 42.—Glancing angle of a space lattice, demonstrated with a bundle of parallel sound waves

four equidistant plane gratings as shown in fig. 42. All the points of the lattice are at corners of cubical elementary cells with an edge of 3 cm. (this lattice constant is thus only a little greater than the wavelength of the sound waves used (1·45 cm.)).

The experiment yields an extremely important result. The space lattice also reflects the bundle of parallel sound waves according to the law of reflection. But this reflection occurs at a few sharply defined angles of incidence only, in our case at $a_1 = 61°$ and $a_2 = 76°$; or in other words, the space lattice exhibits one or two sharply-defined *glancing angles* (the latter being the name given to the complements of the angles of incidence, i.e. the glancing angles are $\gamma_1 = 29°$, $\gamma_2 = 14°$).

The occurrence of these glancing angles is explained by the geometrical construction in fig. 43. We do not even need to bring in the Huyghens wavelets.

In this approximation we again inevitably neglect the small effects mentioned in the paragraph in small type above.

The waves reflected from the individual planes of the lattice must

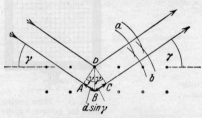

have relative differences of phase equal to integral multiples of the wave-lengths used, otherwise the waves would weaken or destroy one another. This difference of phase between two neighbouring layers is in fig. 43 equal to the paths ABC. AB and BC are both equal to $d \sin \gamma$. Thus the glancing angle is given by

Fig. 43.—To illustrate how the phenomenon of the glancing angle arises

$$n\lambda = 2d \sin\gamma, \quad \ldots \ldots \ldots \quad (4a)$$

where n is a small integer. That is, the glancing angle γ, the wavelength λ, and the distance d between successive lattice planes are connected by a very simple relationship. If the wave-length is known, the unknown distance between successive lattice planes may be determined by finding the

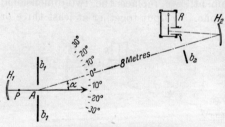

glancing angle experimentally. The values of the glancing angle given above lead to $d = 2\cdot99$ cm., 3 cm., in good agreement with the actual construction of our cubical lattice.

These glancing angles which we have demonstrated here with sound waves

Fig. 44.—Double slit interferometer (Thomas Young, 1802)

have become of extreme importance in the investigation of crystal lattices. The waves used are those of X-rays ($\lambda \sim 10^{-8}$ cm.).

Our space lattices may of course also be used to exhibit Laue diagrams for sound waves.

12. Interferometers.

Interferometer is a general name for a type of measuring instrument, varying widely in form, which is used for all cases of accurate measurement in any way connected with the magnitude of wavelengths. The underlying principle is simple: a train of waves is split into two component wave-trains, which are made to traverse

different paths and thus acquire a difference of phase. The two wave-
trains are then brought together and give rise to interference. We
shall describe two typical types of interferometer for sound waves.

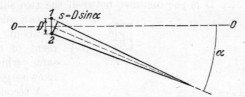

Fig. 45. —
Double slit for
the interfero-
meter shown in
fig. 44.

Fig. 47.—To illustrate the principle of the double slit
interferometer

(1) Fig. 45 shows a double slit. Together with the sound reflector
H_1P this is fixed to an optical bench, which is free to rotate about a
vertical axis A in the plane of the slit (fig. 44). The width of the slit
(about 1 cm.) is of the same order of magnitude as the wave-length of
the whistle. Expanding trains of waves accordingly emerge from both
slits (cf. e.g. fig. 7, Plate XI). These wave-trains intersect as in the
model experiment of fig. 46 (Plate XII), where black lines denote wave-
crests and white line wave-troughs.

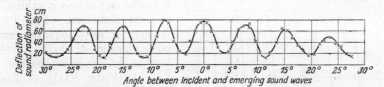

Fig. 48.—Results of measurements with the double slit interferometer of fig. 44

We see that the wave-motion is maintained, i.e. there is a periodic
succession of black and white lines, along the axis of symmetry. On
either side of the axis of symmetry, on the other hand, there is a
system of " interference bands ". At the angles a_1, a_3, a_5 the alternation of black crests and white troughs characteristic of the wave
fails. There is no wave-motion there. This is a result of the difference
in phase (s) of the two waves, the meaning of which is easily seen from
fig. 47. At the angles where the interference bands arise it amounts
to an odd number of half wave-lengths, i.e. $\lambda/2$, $3\lambda/2$, $5\lambda/2$, &c. At
these places the crests of the one train coincide with the troughs of
the other.

As the interferometer is turned about the axis A, the deflections of
the radiometer must vary periodically between their maximum value
and (practically) zero. This is actually the case. The results of a series
of observations are plotted in fig. 48. For example, we find the third

minimum or the third interference band at the angle $a_5 = 19.2°$. According to the geometrical construction in fig. 47 we must have

$$\sin a_5 = 5\lambda/2D,$$

where D is the distance between the two slits (about 11 cm.). This

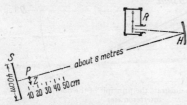

Fig. 49.—Interferometer with moving mirror

gives a calculated value for λ equal to about 1·45 cm. The measurement of the wave-length of the same whistle (fig. 39, p. 293) likewise gave the value 1·45 cm.

A second type of interferometer is illustrated in fig. 49. The one wave-train (no. 1) proceeds directly from the whistle P. The second (no. 2) arises from reflection of the first at the plane mirror (sheet metal) S. Here it is particularly easy to see the difference of phase of the two waves: it is given by twice the distance between the whistle and the mirror. For differences of phase equal to an even number of half wave-lengths we get the configuration of fig. 50 (Plate XIII) on either side of the axis of symmetry (shown by the arrow). The wave-motion is maintained, black crests and white troughs following one another periodically. If the radiometer

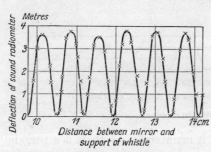

Fig. 52.—A series of measurements with the interferometer of fig. 49

is placed on the axis it exhibits a large deflection. For differences of phase equal to an odd number of half wave-lengths fig. 50 is replaced by fig. 51 (Plate XIII). The periodic succession of crests and troughs fails, for the crests of the one wave-train coincide with the troughs of the other. The radiometer deflection accordingly reverts practically to zero. In fig. 52 the radiometer deflections are plotted for various positions of the mirror. The successive minima correspond to positions of the mirror 0·72 cm. apart, hence $\lambda/2 = 0.72$ cm. or $\lambda = 1.44$ cm., again in good agreement with the values obtained previously.

13. Sound Emitters or Radiators.

In the ripple-tank the mechanism of the propagation of waves was easy to follow. The body dipping into the water displaced the latter rhythmically with the frequency of its vertical oscillations. This experiment may be adapted to the radiation of longitudinal elastic waves in three dimensions in air, water, &c., by altering the volume of a sphere in the rhythm of sinusoidal oscillations. We thus obtain an "ideal" emitter or radiator of sound. All points on its surface vibrate in the same phase, and completely symmetrical spherical waves are produced. Hitherto it has not been possible to realize this ideal emitter in practice, although many attempts represent very good approximations to it. In the first rank are thick-walled vessels with a vibrating membranous wall. The membrane is best actuated by electromagnetic means from inside. On this principle it has been possible to produce submarine signals with membranes of about 50 cm. in diameter, giving powers up to $\frac{1}{2}$ kilowatt. The expression "membrane", however, must not be taken too literally: in the case mentioned it is a steel plate about 2 cm. thick.

We shall demonstrate an emitter for use in air, which is constructed in a similar way but is of much more modest dimensions. This is an ordinary motor-horn with the funnel removed. It makes a loud noise, which, however, is quite bearable in a lecture-room. The amplitude of vibration of the alternating current of air emerging from it may be conveniently demonstrated by means of hydrodynamical forces. At a distance of a few decimetres from our source of sound we suspend a cardboard disc about the size of a penny in such a way that it is free to rotate. It carries a mirror from which light may be reflected, and is protected from draughts by means of a small gauze cage. Its plane is inclined at an angle of about 45° to the line joining it to the radiating membrane. The alternating current of air flows round the disc as indicated in fig. 14, p. 199. The disc is acted on by a couple and tends to set itself at right angles to the direction of the current. This so-called Rayleigh disc may be used as a measuring instrument. It then gives the *velocity* v_0 of the air particles as they move to and fro in sinusoidal vibrations in the direction of the sound (cf. p. 304).

In the *simplest* type of vibration the membranes of a sound emitter vibrate in the same phase along the whole surface and exhibit no nodes except at the edge. In addition we shall assume as a rough approximation that the amplitude is constant over the whole surface. Then physically the conditions are very similar to those where a wave emerges all in the same phase from a slit as in fig. 16, p. 276. That is, in certain circumstances we may confine the advance of the waves to a cone in space similar to that shown in fig. 34, p. 291. For this the diameter of the membrane must be an exact multiple of the wave-length emitted.

The open ends of short thick vibrating columns of air are also tolerable radiators of sound. On the other hand the *strings* used in so many ways in music are quite poor radiators.

In fig. 53 the black disc represents the cross-section of a string standing at right angles to the plane of the paper. Let the string begin by vibrating downwards in the direction of the arrow. Roughly speaking, it thereby " displaces " the air on its under side and gives rise to a wave-train beginning with a crest. At the same time the string, again roughly speaking, leaves an empty space on its upper side and there gives rise to a wave-train beginning with a trough. In any direction the two waves have a phase difference of practically 180°, and almost entirely cancel each other owing to interference. Hence the string is a poor radiator.

Fig. 53.—
To illustrate the emission of sound from a string.

Almost exactly the same considerations apply to the *tuning-fork*. As the two prongs approach one another a wave beginning with a crest is set up in the space intervening. Simultaneously waves beginning with a trough are set up on the outer sides. These wave-trains again interfere with one another and practically cancel each other almost everywhere owing to their having a difference of phase of almost 180°. In the radiation which remains, to be sure, a variation with direction may be noted. For the breadth of a tuning-fork, in contradistinction to that of a string, may no longer be neglected in comparison with the length of the waves emitted. A note of frequency 2000 (where the ear is most sensitive) has a wave-length of only 17 cm. in air. A tuning-fork of this frequency, however, has a total width of about 5 cm., each prong being about 2 cm. across. Hence if the vibrating fork is rotated about its longitudinal axis, its intensity of radiation exhibits a very definite variation with the direction relative to the common plane of the prongs.

For practical use, therefore, the vibrations of strings and tuning-forks must be immediately transferred to good radiators. For this purpose the string or tuning-fork is connected by suitable mechanical means to some body which is a good radiator. The latter is thus made to execute forced vibrations. In certain circumstances the special phenomenon of resonance may be used to produce large amplitudes. The radiator is then made with slight damping, and its proper frequency is made equal to that of the tuning-fork or string. To illustrate what we have said we bring forward the following examples:

(1) In fig. 54a a piece of ordinary string is held by a hand on the right and allowed to slip through two fingers on the left. The string is thus made to vibrate like the string of a musical instrument, but it radiates practically no sound at all. We then fix the right-hand end to a good radiator, say a shallow metal or cardboard box (fig. 54b). The vibrations may now be heard at a distance.

(2) A vibrating tuning-fork sounds softly when held between the fingers, loudly when the stem is held against the table. The stem of the tuning-fork vibrates up and down simultaneously with the prongs and excites the top of the table, which is a good radiator, to execute forced vibrations.

(3) We make use of the special case of resonance. We bring the prong of a tuning-fork near a cylindrical glass jar open at the top. The column of air in the jar will act as a good radiator. The proper frequency of the column of air depends on its length. We can shorten it as much as we like by pouring in water. When the frequencies of the column of air and of the tuning-fork are approximately equal, a sound which is audible at a distance resounds from the apparatus.

For practical purposes the columns of air are contained in four-cornered wooden boxes open on one side, the so-called resonating boxes. We often hear the expression " the vibrations are enhanced

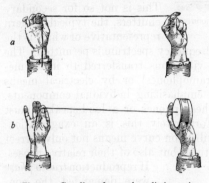

Fig. 54.—Coupling of a poorly radiating string to a good radiator (a stretched membrane)

Fig. 55.—Improvement of the radiation from a tuning-fork by means of two side walls.

by resonance ". This mode of expression is quite erroneous. The comparatively good radiating properties of the box are the only essential feature. The resonance is only an auxiliary means for transferring the vibrations from one body to the other. This may be illustrated by a very striking experiment. One prong of a tuning-fork is placed as in fig. 55 in the slit between two side walls which are not very small compared with the wave-length. The tuning-fork is audible at a considerable distance. For the interference between the waves from the inner and outer sides of the prong is now considerably diminished and the fork is thereby made into a tolerably good radiator.

In musical instruments, e.g. the violin, the relationships are extremely complicated. The strings and the body of the violin form a complicated coupled system (p. 268). The body itself has a whole series of proper frequencies. When it is set in forced vibration certain definite frequencies occurring in the vibrations of the string are therefore specially reinforced. Figs. 56 and 57 show the vibration curve of a note on the violin and the corresponding spectrum. The belly, moreover, is stiffened on the inner side by the sound-post. The belly and back regarded as membranes are by no means small compared with all the wave-lengths used in music. The radiation of sound is thereby markedly reinforced in certain directions. Many details of the action of the violin still remain to be cleared up.

The mention of the violin problem brings us to the distinction, important in practice, between *primary* and *secondary emitters of sound*. Primary emitters have to produce vibrations of definite spectral composition. We fully approve of every primary emitter, say a musical instrument, having an individual form of spectrum, or, physiologically speaking, a definite quality of tone. This is not so for secondary emitters, the typical modern representative of which is the loud-speaker. Here no choice of frequency spectrum is permitted. The secondary emitter must radiate vibrations transferred to it by mechanical means (as in many gramophones) or by electrical means (as in the loud-speaker) without emphasizing individual components. It is often said that a pair of head-phones or a loud-speaker must be free from " distortion ". Fortunately this is an exaggeration. Distortionless reproduction of a vibration curve means not only correct reproduction of the various amplitudes but also of their relative phases.

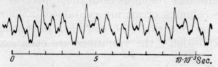

Fig. 56.—Vibration curve of a note of a violin (recorded by H. Backhaus)

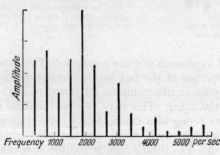

Fig. 57.—Line spectrum of the violin note reproduced in fig. 56

If reproduction *true to phase* were demanded the construction of a loud-speaker would simply be an insoluble problem. Even in spite of the minute mass of the oscillograph loop it has only been satisfactorily solved up to frequencies of a few thousand, and even then with a considerable loss of sensitiveness. Here, however, a fundamental property of the ear comes to our assistance; the ear does not attach the slightest importance to accurate reproduction of phase (p. 314). What to our ear seems to be distortionless reproduction is merely a reproduction of the amplitudes of the components in the correct ratio. With the development of broadcasting, great advances have been made in the construction of loud-speakers intended to fulfil this requirement. In these a cone-shaped paper membrane plays an important part. The external appearance of the commonest types is familiar to users of wireless. An entirely satisfactory solution of the problem, however, has yet to be obtained.

The position is similar in the case of the purely mechanical reproduction of vibrations in the sound-box of the gramophone. A very useful type of record which has recently become commercially available has recorded on it sine curves of all frequencies from 100 to 6000 in continuous succession. Here the product of needle deflection and frequency (see equation (6), p. 304) has been kept constant. A record of this kind should give sound waves of constant intensity provided the reproduction of amplitudes by the gramophone is not distorted. It should therefore enable us to demonstrate the spectral sensitiveness of the ear, which is a maximum at frequency 2000. The sound should seem loudest at frequency 2000. This is far from being the case. We obtain several regions of frequency where the intensity is great, corresponding to the proper frequencies of the mica diaphragm and possibly of the horn. In gramophone reproduction also we seem unlikely to progress farther without the use of electrical devices, and these are apt to lead to results of uneven merit.

These remarks about sound emitters, whether primary or secondary, are in no way intended as exhaustive. They are merely meant to illustrate the main problem. Only one point remains to be mentioned: not only in the case of secondary emitters but also in the case of primary, electrical devices for producing mechanical vibrations are continually gaining in importance. In addition to the oldest methods described on p. 225, valuable new ones have recently been discovered. We shall content ourselves by mentioning them by name:

(1) Alternating-current generators with sinusoidal curves up to frequencies of 10^5 (*Physical Principles of Electricity and Magnetism*, p. 148).

(2) Oscillatory electric circuits automatically regulated by valves, with a wide range of frequency (*ibid.*, p. 304).

(3) Differential vibrations of such circuits.

14. Energy of the Acoustic Field: Acoustic Hardness and Acoustic Resistance.

The energy radiated by sound emitters is propagated in space with the velocity of sound (c). The region filled with sound energy may be called an *acoustic field*. We define the energy of vibration contained in one cubic centimetre as the *energy-density* of this field. This may be measured e.g. in watt-seconds per cubic metre. In the case of plane or only slightly curved sound waves the idea of the *acoustic intensity* I may then be defined at once. This is the energy of vibration contained in a thin cylinder on a base of unit area and of length c units, where c is the velocity of light:

$$I = \delta c. \qquad \ldots \ldots \ldots \quad (5)$$

This definition of intensity may be applied both to progressive and to standing waves. In the case of *progressive* waves another suggestive physical meaning may also be attached to the sound intensity I. Imagine the base of a cylinder of section 1 sq. m. set at right angles to the direction of propagation of the waves. Then the whole acoustical energy contained in a cylinder of length c m. passes across it in one second. That is, for progressive waves the acoustic intensity I means the acoustic energy falling in one second on unit area placed at right angles to the direction of advance of the waves. It is therefore given by the rate of emission of acoustic energy from unit area of the wave surface and may be measured e.g. in watts per square metre.

The energy of vibration in the acoustic field is additively composed of the energies of vibration of all the individual particles of air moving to and fro in the direction of propagation of the sound. It may therefore be expressed by means of relationships already known to us. The total mass of air particles contained in a cube of unit volume is numerically equal to ρ, the density of the air. Suppose that the wave-length is large compared with the side of the cube. Then practically all the particles of air pass their equilibrium positions at the same instant. At this instant they have their maximum velocities (v_0) or else they all have zero velocity. Then the increase of pressure as compared with air at rest reaches its maximum value (ΔP_0).

In the first case the whole energy in the cube of unit volume is in the form of kinetic energy, and the acoustic energy-density is

$$\delta = \tfrac{1}{2}\rho v_0{}^2. \qquad \qquad (6)$$

In the second case only potential energy is contained in the cube of unit volume and the acoustic energy-density is given by

$$\delta = \tfrac{1}{2}\frac{(\Delta P_0)^2}{c^2\rho}. \qquad \qquad (7)$$

In every sinusoidal vibration the maximum velocity v_0 and the maximum deflection x_0 are connected by the equation

$$v_0 = \omega x_0$$

(equation (7), p. 82), where ω is the number of vibrations in 2π seconds, i.e. $\omega = 2\pi n$. We thus obtain a third expression for the acoustic energy-density, this time containing the frequency:

$$\delta = \tfrac{1}{2}\,\rho \omega^2 x_0{}^2.$$

All three quantities defining the vibrations of the air, namely, the maximum value of the velocity v_0, of the change of pressure ΔP_0, and of the deflection x_0, are accessible to direct measurement.

(1) The *velocity* v_0 is measured by means of hydrodynamical forces. We may for example use the attraction of two spheres (fig. 53, p. 249), or the rotation of a Rayleigh disc (fig. 14, p. 199). In both cases the calibration may be carried out with a direct current of air of known

velocity, for the forces are independent of frequency. The calibration may also be carried out by calculation.

(2) The *change of pressure* ΔP_0 may be measured by means of the forced vibrations of a manometer membrane screened on one side.

Fig. 58 shows the resonance curves with differing degrees of damping: it is a diagrammatic repetition of fig. 66, p. 255. This time the essential point for us is the course of the curves in the region of low frequencies, i.e. frequencies which are small compared with the proper frequency (n_0) of the body executing forced vibrations. In the low-frequency region the amplitude is practically independent of the damping and only slightly dependent on the frequency of the exciting vibration. Starting from this fact we give the membrane measuring the pressure a proper frequency which is *large* compared with the frequency of the acoustic field. We then observe the deflections or amplitudes of the membrane, which is best done by means of a mirror and optical lever, and then calibrate the membrane subsequently with air at a given pressure at rest, i.e. with zero frequency.

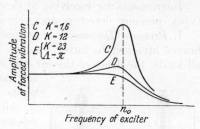

Fig. 58.—Amplitudes of forced vibrations for differing degrees of damping

(3) The *maximum deflection* x_0 is measured by bringing minute spherical particles of dust into the acoustic field and measuring their oscillations under a microscope. The small spheres are carried along by the internal friction of the gas (Chapter IX, § 9, p. 166). They have nearly as great an amplitude as the particles of air surrounding them. This method, however, is only applicable to sounds of high intensity.

Numerical values obtained by these methods follow in § 17, p. 309.

From the above three equations for the acoustic energy-density we deduce two auxiliary concepts which are frequently used, especially in technical literature:

From equations (6) and (7) we obtain the quotient

$$\frac{\Delta P_0}{v_0} = c\rho, \quad \ldots \ldots \ldots \ldots \quad (8)$$

which is called the *acoustic resistance* (or *radiation resistance*).

Hence, since $v_0 = \omega x_0$, we have

$$\frac{\Delta P_0}{x_0} = c\rho\omega, \quad \ldots \ldots \ldots \quad (9)$$

which is called the *acoustic hardness*.

The acoustic hardness, then, is the quotient of the pressure and the corresponding deflection. In very " hard " substances even minute deflections give rise to very large increases in pressure. The acoustic hardness of water, for example, amounts to about $9 . 10^8$ millibars per centimetre $= 900$ atmospheres per centimetre for frequency 1000. On the other hand, the value for air is only $2 \cdot 8 . 10^5$ millibars per centimetre or 0·28 atmosphere per centimetre.

15. Sound Receivers or Detectors.

Instruments for receiving or detecting sound include two limiting types, pressure detectors and velocity detectors.

I. *Pressure Detectors.*—Most pressure detectors consist of membranes attached to boxes, funnels, &c. Examples: microphones of all kinds, the drum of the ear, the recording disc in the older (purely mechanical) method of making gramophone records.

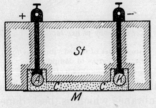

Fig. 59.—The Reisz microphone

All pressure detectors execute *forced vibrations* in the acoustic field. Their amplitudes are independent of the orientation in the acoustic field, for air pressure is a quantity independent of direction. This is shown by any aneroid barometer in a living-room. A barometer of this kind is in the last resort merely a pressure detector of longitudinal waves in the air, except that in the case of variations of atmospheric pressure the oscillations are almost always of very low frequency.

From the practical point of view the microphone now surpasses all other types of pressure detector in importance. Here the demands made on the instrument have been enormously increased by broadcasting. Nowadays good microphones are expected to reproduce sound waves as far as possible without " distortion " in the sense explained on p. 303. It is necessary that the original ratios of the amplitudes should be preserved over the wide range of frequency from about 100 to at least 10,000 vibrations per second. Here again, as with all forced vibrations, this condition can only be fulfilled at the cost of considerable loss in sensitiveness. For broadcasting the Reisz microphone (fig. 59) has been much used. M is the receiving membrane, of thin mica or similar material, bounded at the sides by the thick marble block St. The space behind the membrane is filled with powdered carbon. A and K are the terminals, also made of carbon. In contradistinction to all other carbon microphones, the current here flows not at right angles to the surface of the membrane, but parallel to it.

II. *Velocity Detectors.*—In velocity detectors the variations of velocity in the alternating current of air are used to produce forced vibrations. This is best made clear by an experimental example.

In fig. 60 (Plate XIII) a fine glass fibre about 8 mm. long is placed at right angles to the progressive sound waves, to act as a small spring (its movements being projected on to a screen). Periodic variations of air pressure have no effect on this fibre whatever. On the other hand, the alternating current of air carries the fibre along in the direction of the vibrating particles of air as a result of internal friction, and thus causes it to execute forced vibrations (the source of sound being a reed-pipe a short distance away). This fibre is a typical velocity detector. At the same time it exhibits an important property characteristic of *velocity detectors*: the amplitude *depends on the orientation* in the acoustic field. If the fibre is placed parallel to the direction of propagation of the waves it remains at rest.

Velocity detectors may be used as " direction finders ". Imagine two fibres symmetrically placed with respect to the longitudinal axis of a moving body. If the body is moving straight towards the source of sound the two receivers will " speak " with equal loudness. Deviations from the straight course manifest themselves by inequality of the forced amplitudes.

Pressure detectors and velocity detectors, as we mentioned above, represent limiting cases. Every transference of momentum by pressure involves the presence of a wall which does not yield appreciably as the pressure arises. The forced vibrations of the wall must be small compared with the movements of the vibrating particles of air or water. That is, the wall or membrane must have a greater acoustic hardness than the medium. Hence good approximations to a purely pressure detector may be made in the case of air, but only poor ones for sound waves in water. Instruments which are pressure detectors in air may even become velocity detectors under water.

16. Special Cases of the Propagation of Sound.

We have taken the *linear* law of force as a basis for all vibration phenomena. Practically, as we have repeatedly said, this means a limitation to " small " amplitudes. We dealt with the special phenomena associated with a non-linear law of force only in § 15 of Chapter XI (p. 267). These consisted in the occurrence of " differential vibrations ".

Similarly in discussing the processes of wave propagation we have always considered the limiting case of the linear law of force. In the case of waves also we practically confined ourselves to " small " amplitudes. For large amplitudes special phenomena again occur in the propagation of waves.

Sound waves of abnormally large amplitude, in which the pressure at the centre may rise to thousands of atmospheres, are produced by the detonation of explosives. Large variations of pressure may also be produced by means of the electric spark. In neither case have we to do with trains of waves consisting of a large number of crests and troughs. We now have, figuratively speaking, only one really steep crest, followed by a shallow and somewhat wavy trough. Owing to its great density the wave-crest may be photographed. The side illumination of extremely short duration required for an instantaneous photograph of this kind is always obtained from an electric spark.

Fig. 61 (Plate XIII) shows a photograph of the explosion wave * of an electric spark obtained in this way. On the right the wave has impinged on a sieve and has been partly reflected at it. The figure also serves as another illustration of Huyghens' principle. Both the wave reflected at the sieve and the wave penetrating through it appear as the envelope of a number of wavelets. These sound waves involving abnormally high pressure variations have a greater velocity than ordinary sound waves. This may be shown e.g. by means of explosion-waves of two *simultaneous* sparks of unequal intensity. In fig. 62 (Plate XIII) the stronger spark is on the left. At the instant recorded by the photograph its explosion wave has covered a path almost a third longer than that of the feebler spark. The velocity of the explosion-wave on the left must therefore have reached nearly 500 metres per second. The same is true of the muzzle waves of modern fire-arms. The shadow of a muzzle wave is shown at SS in fig. 63 (Plate XIII). At the instant of the photograph the bullet has just overtaken the muzzle wave. A wave like the bow wave of a ship starts from the nose of the bullet. Such bow waves arise in the case of all bodies which move faster than sound. A familiar example is the crack of a whip. Of course the pressure variations due to sparks and explosions decrease as the distance from the source increases. The velocity of sound then falls to its normal value.

The nose-wave of a bullet is an example of the emission of sound by a body moving very rapidly. Hitherto we have tacitly assumed that our instruments for producing and detecting sound were at rest. With moving sources and receivers the *Doppler effect* occurs. Diminution of the distance between the bodies during the emission of the sound raises the frequency observed by the receiver. Increase of the distance has the reverse effect.

In a quantitative discussion the case of the moving source must be separated from that of the moving receiver. If the source of sound is moving with velocity v, the observer at rest obtains the frequency

$$n' = \frac{n}{1 \mp v/c}, \quad \cdots \cdots \quad (10)$$

the minus sign applying to a diminution of distance.

* [Sometimes called *onde de choc*.]

PLATE XIII

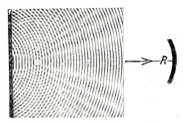

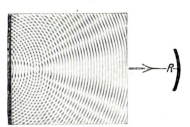

Chap. XII, Fig. 50.—To illustrate the waves produced by the interferometer with moving mirror; both waves in phase in the direction of observation.

Chap. XII, Fig. 51.—Like fig. 50, except that there is a phase difference of 180° between the two waves in the direction of motion.

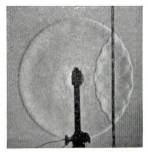

Chap. XII, Fig. 61.—Reflection of an explosion-wave at a sieve (this illustration and the two which follow were taken by the "Schlieren" method by C. Crantz; see fig. 42, p. 183).

Chap. XII, Fig. 60.—A fine glass fibre as a detector of motion (actual thickness only 0·028 mm.)

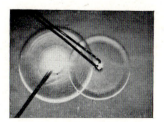

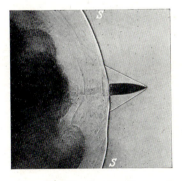

Chap. XII, Fig. 62.—Variation of the velocity of sound with the intensity

Chap. XII, Fig. 63.—Muzzle-wave of a rifle and nose-wave of the bullet (the black cloud consists of gases from the powder).

The moving receiver or observer obtains the frequency

$$n'' = n (1 \pm v/c), \quad \ldots \ldots \quad (11)$$

the plus sign applying to a diminution of distance.

This Doppler effect may be demonstrated by moving a whistle rapidly round in a circle.

17. The Sense of Hearing.

The sense of hearing and our organs of hearing are really subjects for physiological and psychological research rather than for physical. Nevertheless, the following brief summary of those facts which are most important in physics will be found useful, just as in optics it is necessary to know the properties of the eye, at least in main outline.

(1) The ear reacts to mechanical vibrations over the wide range of frequency from about 20 to 20,000 vibrations per second. That is, the ear can deal with a spectral region of about 10 octaves ($2^{10} = 1024$). The exact limits have been the subject of considerable controversy; it is certain that the upper limit falls with increasing age.

(2) The spectral distribution of the sensitiveness of the ear is

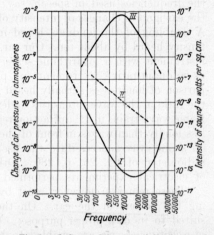

Fig. 64.—I. Curve of the spectral distribution of the sensitiveness of the ear (threshold of audibility); II. Curve of equal intensity for normal speech; III. Curve of onset of painful sensation (threshold of feeling). The ordinate at the right gives the intensity of the sound waves (§ 14, p. 303) in watts per square centimetre.

illustrated in fig. 64. The ordinates of curve I are the intensities which are just heard ("threshold of audibility"). These are measured by methods depending on the same principle as those described on p. 303. There is agreement about the general run of the curve of sensitiveness, and also about the position of maximum sensitiveness (minimum threshold of audibility) in the neighbourhood of 2000 vibrations per second. There is some uncertainty about the absolute values, but this is of practically no importance.

(3) With a rise in intensity of the waves the spectral distribution of sensitiveness is altered. With the intensity of ordinary conversation it follows a rectilinear path as a first approximation (curve II in fig. 64).

(4) If the intensity rises still higher the sensation of hearing is

replaced by a painful sensation. Curve III in fig. 64 gives the intensity required to produce this painful feeling in the various regions of the spectrum. Curves I and III enclose the *sensation area*, the totality of intensities of various frequencies which lead to the sensation of hearing.

(5) The deflection of the sound radiometer (fig. 26, p. 286) is directly proportional to the intensity of the sound. The radiometer belongs to the convenient type of instrument which has a linear scale (p. 91). The loudness heard by our ear, on the other hand, is not even approximately proportional to the intensity of the waves. In the region of the frequencies used in speech, say 10^2 to 10^4 vibrations per second, the ear ignores changes of intensity of less than from 10 to 20 per cent. A doubling of the loudness heard means a multiplication of the intensity heard by about ten. Our ear, roughly speaking, is a measuring instrument with a logarithmic scale. Hence the ear is simply useless as a quantitative instrument for physical measurements. A comparison of loudnesses actually heard gives us quite a distorted idea of the ratio of the wave intensities under comparison. Thus, for example, feeble wave-trains in a " diffraction region " are heard in the ear scarcely less loudly than waves in front of the obstacle. Physical observations on sound waves necessitate the use of a physical detector, say the radiometer of fig. 26, p. 286. Otherwise we remain at the level of electrical experiments performed without measuring instruments.

These remarks, however, must by no means be taken as disparaging the powers of the ear. On the contrary, the ear is excellently suited to its own proper purpose. Owing to its " logarithmic " scale of sensitiveness we are enabled to hear almost as well in regions of feebly diffracted, reflected, or scattered sound waves as when the sound waves are free to advance unhindered by obstacles. Further, overstrain of the ear is prevented. In the region of greatest sensitiveness the ear can deal with intensities differing in the ratio of $1:10^{14}$, which is a most remarkable feat.

(6) To sinusoidal vibrations the ear reacts with the sensation of a " pure tone ". Every tone has a definite pitch. Pitch is a subjective quality and as such is inaccessible to physical measurement. Nevertheless we speak in a general way of the frequency of a tone. This mode of expression, though convenient, is a lax one. What we invariably mean is the pitch corresponding to a sine wave of the given frequency.

(7) In the most favourable region of frequency the ear can distinguish between two frequencies differing by only 0·3 per cent. That is, the ear has a separating power $n/\Delta n$ of about 300, corresponding to the optical performance of an equilateral prism with a side of about 1 cm.

(8) To non-sinusoidal vibrations the ear reacts with the sensation

of a compound tone. A compound tone is entirely *independent* of *phase differences* between the individual sinusoidal components. This is G. S. Ohm's fundamental discovery.

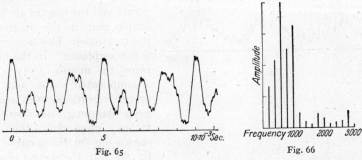

Fig. 65

Fig. 66

Vowel in a man's voice and its line spectrum, fundamental frequency 200 (figs. 65–71 are due to F. Trendelenburg)

(9) To every musical tone there corresponds a line spectrum of definite structure, distinguished by the ratios of the frequencies and amplitudes of its spectral lines. The absolute value of the fundamental frequency is of no importance. Two sinusoidal vibrations of approximately equal intensity with frequencies in the ratio $1:2$ always give the interval we call an " octave ", and so on.

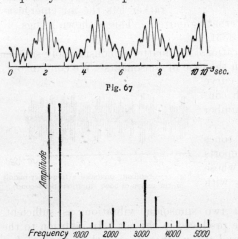

Fig. 67

Figs. 67, 68.—Vowel *i* in the upper register of a woman's voice, together with spectrum (fundamental frequency about 350).

(10) In the compound tones known as vowels, on the other hand, the structure of the line spectrum varies considerably with the frequency of the fundamental or the corresponding type of voice (bass, tenor, &c.). But for the principal lines we always obtain approximately the same *absolute value* of the frequency. The general scheme of a vowel is given in figs. 22–26, p. 236. The damped proper vibrations of the mouth cavity are excited in rapid succession by jets of air from the larynx. The fundamental frequency, and hence the type of voice, depends on the frequency of these jets. A change in the fundamental frequency alters the positions of the

spectral lines, but they remain in the same region of frequency (figs. 22, 24, 26, p. 236).*

Figs. 65–70 show some very well recorded vowels with their line

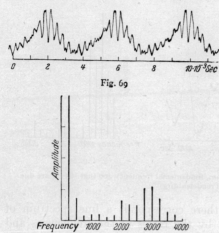

Fig. 69

Figs. 69, 70.—Vowel *i* in the lower register of a woman's voice, together with spectrum (fundamental about 250).

spectra. Both the vibration curves and the spectra are less simple than in the diagrams just mentioned. This, however, is not surprising. In the first place, the mouth cavity is of a complicated shape, and its proper vibrations, even apart from damping, are by no means sinusoidal; this is shown e.g. in the *a*-curve of fig. 65. In the second place, the larynx does not excite it once and for all; the duration of the jet of air is not small compared with the duration of the proper vibration of the mouth cavity. The larynx rather lets the air emerge as a more or less sinusoidal alternating current. This is shown in figs. 67 and 69 (note also fig. 18, p. 233).

In spite of these complications, the matter is considerably simpler in the case of vowels than in the case of consonants. We content ourselves with reproducing an *S*-curve without its spectrum, which consists of an extraordinary number of lines (fig. 71).

(11) The compound tones which we call noises or reports are produced by vibrations whose curves vary very irregularly with time.

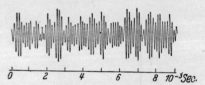

Fig. 71.—Sibilant; unvoiced *s* (frequency mainly in the region of 6000 vibrations per second)

(12) If the intensity of two sinusoidal vibrations is sufficient, "differential tones" may be produced in the ear. In addition to the two tones of frequency n_2 and n_1, we then hear a third tone of frequency $n_2 - n_1$. Differential tones † may be demonstrated very well with organ pipes. Sometimes other "combination tones" are heard, such as the "summation tone" $(n_1 + n_2)$, or the tone $(2n_1 - n_2)$.

---

* The production of vowels is a typical example of a process of "frequency multiplication" much used in practical applications.

† The term "subjective", which is often prefixed, is unnecessary: tones are always sensations and therefore always subjective.

(13) The time required to excite the ear and the time required for sound to die away in the ear are very uncertain; they appear to be of the order of a few hundredths of a second.

(14) With the two ears the direction of approaching sound waves can be detected. This is most successful with musical notes or noises in which characteristic peculiarities recur or which begin sharply. The essential point is the interval of time which elapses between the excitation of the left ear and that of the right by the same portion of the sound-wave curve. In the case of high frequencies, differences of intensity due to the shadow thrown by the head occur in addition.

This power of "directional hearing" is well adapted to the measurement of small differences of time (Δt). For example we may lead the first time signal to the left ear and the second to the right. If $\Delta t < 3.10^{-5}$ (corresponding to a difference of path of the sound of one centimetre) the source of sound is localized in the central plane of the body. The impression of greatest displacement to one side is obtained when $\Delta t = 60.10^{-5}$ sec.

18. The Ear.

The most essential part of the organs of hearing is the "internal ear", the snail-shaped labyrinth which is built into the temporal bone. The mechanical waves are led to it by two paths, (1) by way of the ear-drum and the adjoining ossicles of the middle ear, (2) by way of the soft and bony parts of the head. The first path is not essential: hearing is possible even in the absence of an ear-drum or ossicles. These parts merely have the following purpose: the body is largely composed of water and has about the same acoustic resistance as water, i.e. about 3000 times that of air. Hence sound waves in air impinging on the surface of the head are strongly reflected. These reflection losses are considerably diminished by the membrane of the ear-drum, which is acoustically soft. The ossicles act like a system of levers, reducing the large amplitudes of the "soft" membrane to the small amplitudes of the "hard" labyrinth. Ear-drum and ossicles would accordingly be quite useless to mammals living entirely in water (dolphins and whales). None of these animals, in fact, have an external ear, external ear-passage, ear-drum, and ossicles all being reduced to the merest vestiges.

The actual mode of action of the internal ear has not yet been settled by direct observation. Meanwhile we merely know the powers of the internal ear, and have to attempt to imitate them with a mechanical model. This model resembles the tachometer (fig. 71, p. 260). Imagine about 100 springs for the region of frequency 250–500, 200 for the region 500–1000, 300 for the region 1000–2000, and so on. The main point about this model, which goes back to Helmholtz, is that it explains the fact which is most important for the physicist, i.e. Ohm's discovery that the phase of vibrations has no effect whatever.

Owing to phase being ignored, the brain has merely to learn *one* spectrum for each note. For the interval of the octave it would in our model be somewhat as sketched in fig. 72. If phase had to be taken into account the brain would have to "learn by heart" a whole series of different arrangements even for this simple sound: (1) that depicted at the bottom of fig. 15, p. 231, (2) that depicted at the bottom of fig. 16, p. 231, and (3) several intermediate forms not shown in the figures. This is for the perfectly simple sound of the octave, which consists of only two sinusoidal waves.

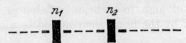

n_1 n_2

Fig. 72. — Diagram of the line spectrum of an octave, as shown by a tachometer. The damping of the springs is taken into account: the springs on either side of the frequencies n_1 and n_2 also vibrate with noticeable amplitudes.

In the case of ordinary sounds (such as words) composed of a large number of sine waves, we should have a fantastic number of differently-shaped curves even for one and the same compound tone. In retaining our vocabulary, indeed, our brain already performs an enormous task even in registering every word only in the form of a line spectrum. If phase had to be taken into account, however, the brain would have to retain thousands upon thousands of widely varying wave curves for every single word, a task absolutely inconceivable in its magnitude.

Further, the Helmholtz model of the ear explains why differential tones play such a trivial part. Differential tones imply the existence of differential vibrations. If these are not present in air they can arise only in the body. As we saw on p. 233, differential tones arise from one-sided distortion of the wave-form. This again involves the occurrence of deviations from the linear law of force. In all substances, however, these deviations are very trifling if the amplitudes of the waves are small. In general, therefore, the amplitudes of the differential tones can only be small.*

These two examples of the usefulness of the model may suffice.

In the snail-shaped cochlea of the labyrinth is concealed Corti's organ, which is described in detail in all textbooks of anatomy and physiology. In its structure Corti's organ displays in many respects a striking resemblance to the tachometer of the physicist. Perhaps it is really a tachometer of minute dimensions (its total length is about 34 mm., the length of the "springs" running from 0·04 to 0·5 mm.). The smallness of these dimensions would indeed lead to insuperable quantitative difficulties (as regards frequency, damping, &c.) if the materials concerned were those used in present-day engineering. But here again in the last resort the living material no doubt proves itself superior to the raw materials at present known to engineers.

---

* Owing to the logarithmic scale of sensitiveness of the ear we are apt to over-estimate the amplitudes of differential vibrations.

MISCELLANEOUS EXAMPLES

The problems marked * require the use of the calculus

1. A swinging pendulum is observed through a disc with four slits, which is revolving 5 times per second. For what rate of vibration does the pendulum appear at rest to the observer? (p. 10.)

2. A man in a boat wishes to row in a straight line from A to B, where A, B are on opposite sides of an estuary of width 2000 ft., and B is 1000 ft. farther inland than A. He rows through still water at the same rate as that of the incoming tide. Find the direction in which he must point the boat. (p. 17.)

3. If g is 32 ft./sec.², calculate the distance traversed in the fourth second of its fall by a body falling freely from rest. (p. 21.)

4. A parachutist after taking-off falls 50 m. without friction with an acceleration $g = 9\cdot81$ m./sec.². When the parachute opens his downward motion is subjected to a retardation of 2 m./sec.². He reaches the ground with a velocity of 3 m./sec. How long is the parachutist in the air, and at what height did he take off? (p. 21.)

5. A particle on a smooth horizontal table is acted on by forces 10 lb. wt. in the direction S. 25° E., 6 lb. wt. in the direction S. 50° W., and 4 lb. wt. in the direction N. What additional force acting in the direction N.E. must be applied to make the resultant act along the E.-W. line? Find the magnitude of this resultant and state the direction in which it acts. (p. 29.)

6. A body of weight W is to be prevented from sliding down a rough plane of inclination 40° by a force making an angle θ with the vertical. Find the least force necessary if the coefficient of friction (i.e. the ratio of the force of friction and the normal reaction) is 0·3. By varying θ find the direction of the smallest force which will hold the body at rest on the plane. (p. 29.)

7. A mass of 40 lb. has its velocity increased from 50 to 60 ft. per second while passing over 20 yards. Find the acceleration and the force producing it. (pp. 22, 33.)

8. A mass of 500 gm. rests on a smooth table 100 cm. from the edge, and a string passing from it over the edge of the table has a mass of 20 gm. fastened to the other end. If the system starts from rest in this position, calculate its

velocity and its kinetic energy when the larger mass reaches the edge of the table. Describe the subsequent motion ($g = 981$ cm./sec.^{2}). (p. 33.)

9. What is the radial acceleration of a man in latitude $51° 30'$ (say in London)? The radius of the earth is 6378 km. (p. 41.)

10. Find the maximum values of the velocity and acceleration of a body moving with simple harmonic motion if the maximum displacement from the mean position is 4 in. and the time of a complete oscillation is one second. (p. 51.)

11. A particle starts from rest at a point P and moves towards a point O where OP $= 10$ ft. If its velocity in passing through O is 20 ft./sec., draw its distance-velocity graph (1) if its acceleration is constant, (2) if its motion is simple harmonic about the centre O. Determine in each case the time taken to travel from P to O. (pp. 21, 51.)

12. A stone of mass m is whirled round horizontally by a cord of length l with angular velocity ω. What will the angular velocity of the stone become if the cord is suddenly shortened by an amount l_1? (p. 55.)

13. If the moon's gravitational acceleration is $\frac{1}{10}$ that of the earth, what would the period of the terrestrial seconds pendulum be on the moon? What is the mass of the moon if its radius is r? (pp. 66, 70.)

14. A motor-car weighing 1 metric ton is ascending a slope of 1 in 25, 1 km. long. In addition to gravity it has to overcome resistances amounting to 1% of its weight. What amount of work is done in climbing the slope? At what rate is the engine working at the top of the incline if the motor-car is then travelling at 60 km. per hour? (p. 75.)

15*. The mechanical equivalent of 1 kg. coal is $7000 . 420$ kg.-force-m. How deep may the coal lie if the work which has to be done against gravity in raising it to the surface is not to exceed the mechanical equivalent of the coal? (pp. 70, 75.)

16. A block of wood weighing 10 lb. is pulled with a force of 8 lb. wt. horizontally along a table with a flat and uniform surface. If the coefficient of friction (see Ex. 6) between the block and the table is 0·25 and is constant at all speeds, calculate the distance that the block moves in 4 sec. from rest ($g = 32$ ft./sec.^{2}). (p. 76.)

17. Two identical cylindrical vessels with their bases at the same level contain water to heights h, H cm. respectively, the area of either base being a sq. cm. Find in gm.-cm. the work done by gravity in equalizing the levels when the two vessels are connected. (p. 77.)

18. A stream of water of 1·2 kg. per minute impinges horizontally against a vertical plate. Calculate the velocity of the water if the force it exerts against the plate is 20 gm. wt., supposing the water (a) to be stopped dead, (b) to rebound with equal velocity from the plate. (p. 83.)

19. A man of mass 60 kg. is standing in a truck of mass 120 kg. and has with

him ten kilogram weights. With what velocity must he throw out one weight in order that he may attain a velocity of 0·3 km. per hour? (p. 85.)

20. A body of mass m slides without friction on an inclined plane, starting at height h with zero velocity. When at a height h_1 it receives an inelastic impulse in the direction opposite to that of its motion, of magnitude A large-dyne-seconds, which causes it to move up the plane again. What height does the body reach on the inclined plane and with what velocity does it then descend? (pp. 21, 88.)

21. A box full of sand is suspended by a cord, forming a ballistic pendulum of mass M, and receives an inelastic blow from a smaller mass m moving with velocity v. Through what height does the box rise? (p. 93.)

22. A bullet weighing 30 gm., moving horizontally with a velocity of 300 m. per second, is fired into a large block of wood weighing 3 kg., which is suspended by light strings. With what velocity does the block begin to move after impact? How is the energy originally possessed by the bullet expended? (p. 93.)

23. A mass of 30 lb. is placed at one end of a uniform lever weighing 10 lb. (a) What force must be applied at the other end of the lever in order that it may balance about a fulcrum one-fifth of its length from the 30 lb. mass? (b) What must the position of the fulcrum be if a mass of 50 lb. is placed at the other end, and what is the reaction of the fulcrum? (p. 99.)

24. What amount of work is required to set a brass sphere 1 m. in diameter rotating at 20 revolutions per minute, if the specific gravity of the sphere is 8·5? (p. 102.)

25. A body weighing 500 gm. makes small oscillations under gravity about a horizontal axis distant 20 cm. from its centre of gravity. The length of the equivalent simple pendulum is 50 cm. Find the moment of inertia of the body about its axis of suspension and about a parallel axis through its centre of gravity. (p. 107.)

26. For a uniform thin rod of length a and mass m, swinging about a point of suspension at one end, the moment of inertia is $\frac{1}{3}ma^2$. Find the length of the equivalent simple pendulum, and the moment of inertia of the rod about a parallel axis through its centre. (p. 107.)

27. At what point in its length must the rod in the preceding question be suspended to give a minimum time of vibration, and at what point must it be suspended to give the same time of vibration as if suspended at one end? (p. 107.)

28. A man of mass m on a symmetrically constructed frictionless rotating table of moment of inertia I (initially at rest) moves radially from the centre to a distance r, then walks once round in a circle of radius r, and finally returns to the centre. Through what angle does the rotating table revolve? (p. 110.)

29. A pencil 20 cm. long is held in a horizontal position and allowed to fall under gravity. After falling through 1 m. it strikes the edge of a table with one end. How does it then fall? What angular velocity will it acquire? (p. 110.)

30. A gun is fixed to a fly-wheel making 10 revolutions per second. The weight of the bullet is 4 gm. and its muzzle velocity 100 m./sec. With what force does the bullet press sideways against the barrel just before leaving it? (p. 136.)

31. What is the magnitude of the horizontal component of the Coriolis acceleration of a railway train running at 60 km. per hour along the parallel of latitude passing through London? (p. 147.)

32. A gyrostat with a horizontal axis is balanced on a sharp point at its centre of gravity, as in fig. 36, Plate IV. A horizontal plane passing through the axis of the gyrostat is rigidly attached to the system. For an observer in this plane the gyrostat is at rest, in spite of the couple acting on it. Why? (p. 147.)

33. The two cylinders of a hydraulic press are 4 cm. and 75 cm. in diameter, and a force of 10 gm. wt. is applied to the piston of the smaller cylinder. What weight must be placed on that of the larger one to maintain equilibrium? (p. 160.)

34. A hydraulic press has a small piston of cross-section a_1 and a large piston of cross-section a_2. A man exerts pressure on the small piston by applying a force of 20 kg.-force at the end of the long arm of a lever, the lever ratio being 1 : 10. If the larger piston is to raise a mass of 1000 kg., what must a_2 be? (p. 160.)

35. Find the pressure in cm. of mercury (density 13·6) at the depth of 1 mile in sea-water of density 1·026, including an atmospheric pressure of 76 cm. of mercury. (p. 162.)

36. What is the upward thrust on a solid leaden sphere of diameter r metres in water? (p. 163.)

37. A cylindrical buoy made of iron (specific gravity 7·8), 1 m. in diameter and 2 m. long, is kept submerged in water by an anchor and a chain. If the thickness of the iron is 0·5 cm., what is the tension on the chain? (p. 163.)

38. The pressure indicated by a siphon barometer whose vacuum is defective is 750 mm., and when mercury is poured into the open branch until the vacuum is reduced to half its former volume, the pressure indicated is 740 mm. Deduce the true atmospheric pressure. (p. 178.)

39. A U-tube with one limb short and the other long (of lengths h_1, h_2 metres respectively) contains water, which initially stands at the same height in both limbs. The air in the long limb is then replaced by hydrogen (through a side tube above the level of the water). What is the resulting difference of level of the water? (p. 182.)

40. A balloon of capacity 1000 cu. m. is filled with hydrogen. If the mass of the balloon and basket together is 200 kg., what is the force tending to raise the balloon from the ground? The densities of air and hydrogen are 1·3 gm. per litre and ·09 gm. per litre respectively. (p. 186.)

41. A rubber balloon of volume 10 litres is filled with hydrogen and held in mid-air by a string attached to a spring balance vertically underneath it. The

balance indicates a weight of 7·0 gm. when the height of the barometer is 76 cm. What will the reading be when the barometer stands at 74 cm., if the temperature remains unchanged? (pp. 178, 186.)

42. Two spheres of radius r_1 and r_2 and density ρ_1 and ρ_2 respectively fall in paraffin with constant velocity v_1 and v_2 respectively. Find the viscosity and density of the paraffin. (p. 193.)

43. A cylindrical tank is filled to a height h with water. Water is to flow out of a circular opening at the ground-level at the rate of 1 c.c. per second. What must the radius of the opening be? (p. 197.)

44. The lowest and highest notes of the normal human voice have about 80 and 800 vibrations respectively per second. Find their wave-lengths if the velocity of sound is 1100 ft. per second. (p. 219.)

45. Investigate mathematically the case of superposition of vibrations illustrated in fig. 14, Chap. XI, and find the interval between successive maxima of the beats. (p. 230.)

46. Notes of 225 and 336 vibrations, each containing the first two overtones, are sounded together. Show that two of the overtones will give rise to beats at the rate of 3 per second. (p. 230.)

47. A rod 8 ft. long, vibrating longitudinally in its fundamental mode, gives a note of 800 vibrations per second. Find the velocity of sound in the rod. (p. 242.)

48. Find the velocity of sound in glass of density 2·6 for which the value of Young's modulus is $6·5 . 10^{11}$ dynes per square centimetre. (p. 247.)

49. The dust-heaps in a Kundt's tube filled with air and excited by a whistle are 1·2 cm. apart. Find the wave-length and frequency of the note emitted by the whistle. (p. 252.)

50.* A particle of unit mass is restricted to move in a straight line under the influence of an elastic restoring force. The body is also acted on by a periodic force $a \sin \omega t$. What is the amplitude of its forced vibrations? (p. 253.)

51.* The particle of Ex. 50 is acted on by a force of friction proportional to its velocity, in addition to the periodic force. What is the amplitude of motion of the particle? (p. 253.)

52.* In water two light-waves of wave-lengths λ, $\lambda + d\lambda$ have optical refractive indices n, $n + dn$ respectively. What is the group velocity with which the train of waves formed by these two wave-lengths is propagated in water? (p. 282.)

53. The interval of time between a sound and its echo from the sea-bed is ·15 sec. If the velocity of sound in sea-water is 4800 ft. per sec., what is the depth in fathoms? (p. 284.)

54. What must the refractive angle (α) of the prism in fig. 28, Chap. XII (p. 288) be in order that an incident sound-wave may be unable to pass beyond the prism surface A, i.e. in order that total reflection may occur? (p. 288.)

55. A sound-wave of wave-length $\lambda = 589$ μμ is incident on a space lattice like that in fig. 43, Chap. XII (p. 296). The lattice constant d is 0·005 cm. At what angles does regular reflection take place? (p. 296.)

56. A plane sound-wave falls on a screen with four small openings situated at the corners of a square of side a cm. Where do the interference maxima nearest the axis of symmetry of the square lie? (p. 296.)

57. A plane sound-wave of wave-length λ is incident on a grating with very narrow slits. The distance between the interference bands is to be 10 cm. at a distance of a cm. from the grating. Find the grating constant d, i.e. the distance between two successive slits. (p. 297.)

58. Water-waves from a very distant source are incident on a pair of slits at a distance a apart. The diameter of the slits is small compared with the wave-length. Investigate the interference phenomena in a plane parallel to the line joining the slits and situated at a distance y from the slits which is large compared with a. (p. 297.)

59. What is the length in feet of the shortest narrow tube (1) stopped at one end, (2) open at both ends, which will resonate to a tuning-fork of frequency 520? (pp. 250, 301.)

60. A railway engine is approaching an observer at rest at the rate of 20 m. per second. The whistle emits a sound of frequency 500. What is the wave-length measured by the observer? (p. 308.)

ANSWERS TO EXAMPLES

1. Period of pendulum $= \frac{1}{20}$ sec.

2. Downstream at an angle of 53° 8′ with the bank.

3. 112 ft.   **4.** 17·4 sec.: 642·4 m.

5. 12·62 lb. wt.: 8·55 lb. wt. in direction W → E.

6. ·4130 W/(·9588 sin θ — ·4130 cos θ): a force at an angle of 23° 18′ with the horizontal.

7. 9·17 ft./sec.^{2}: 11·46 lb. wt.   **8.** 86·9 cm./sec.: 2000g ergs.

9. 2·11 . 10^{-2} m./sec.^{2} **10.** $2\pi/3$ ft./sec.: $4\pi^2/3$ ft./sec.^{2}

11. 1 sec.: $\pi/4$ sec.   **12.** New angular velocity $\omega_1 = 2ml^2\omega/ml_1^2$.

13. $\sqrt{10}$ sec.: mass of moon $= gr^2/10\gamma$.

14. 0·14 kilowatt-hours: 8·18 kilowatts.   **15.** 2440 km.

16. 176 ft.   **17.** $a(H - h)^2/4$.   **18.** 981 cm./sec.: 490·5 cm./sec.

19. 15·75 m./sec.^{2}

20. $h_2 = \{\sqrt{2g(h - h_1)} + A/m\}^2/2g$: $v = \sqrt{(2gh_2)}$.

21. $h = m^2v^2/2g(M + m)^2$.   **22.** 119·3 cm./sec.

23. 3·75 lb. wt.: $\frac{7}{18}$ of the length of the rod from the 50 lb. wt.: 90 lb. wt.

24. 0·15 watt-second.   **25.** 5 . 10^5 gm.-cm.^{2}: 3 . 10^5 gm.-cm.^{2}

26. $2a/3$: $ma^2/12$.

27. At either of the two points distant $a/2\sqrt{3}$ from the centre: at either of the two points distant $a/6$ from the centre.

28. $mr^2/(I + mr^2)$.

29. The centre of gravity of the pencil falls vertically. By Steiner's theorem the moment of inertia I_A is $ml^2/3$: then $2I_A\omega = mvl$, hence $\omega = \frac{3}{2}\sqrt{\{2g(h_1 - h_2)\}}/l$, or about 33 revolutions per second.

30. The bullet is acted on by a Coriolis force $2mv\omega = 5$ kg.-force at right angles to its path.

31. $2\omega v \sin\varphi = 4·28$ m./hr.^{2}

32. The observer's system of reference is rotating. In his system a moving body is acted on by Coriolis forces $2\omega v \sin\varphi$. The acting mechanical couple is therefore subject to the Coriolis forces in the opposite direction. They give rise to a couple which just balances the acting couple.

33. 35·2 kg.   **34.** $a_2 = 5a_1$.   **35.** 11,217 cm. of mercury.

36. 500 $\pi r^3/3$ kg.-force.   **37.** 1270 kg.-force.   **38.** 760 mm.

39. $h_1s_1 - h_2s_2$, where s_1, s_2 are the specific gravities of hydrogen and air respectively.

40. 1010 kg.-force.   **41.** 7·35 gm.

42. $2gr_1{}^2r_2{}^2(\rho_2 - \rho_1)/9(r_1{}^2v_2 - r_2{}^2v_1)$: $(r_1{}^2v_2\rho_1 - r_2{}^2v_1\rho_2)/(r_1{}^2v_2 - r_2{}^2v_1)$.

43. $(1/\pi\sqrt{2gh})^{\frac{1}{2}}$.   **44.** 13·75 ft., 1·375 ft.

45. The vibrations may be represented by $x_1 = a_1\sin\omega_1 t$, $x_2 = a_1\sin\omega_2 t$, where ω_2 is nearly equal to ω_1: $2\pi/(\omega_1 - \omega_2)$.

46. The beating overtones have frequencies $2 \times 336 = 672$ and $3 \times 225 = 675$.

47. 12,800 ft. per second.   **48.** 5000 m. per second.

49. 2·4 cm.: 14,170.

50. The differential equation of motion is of the form $d^2x/dt^2 + n^2x = a\sin\omega t$: the solution gives $a/(n^2 - \omega^2)$ for the amplitude of the forced vibrations. In the case of resonance ($n = \omega$) the amplitude becomes infinite.

51. The differential equation is of the form $d^2x/dt^2 + 2kdx/dt + n^2x = a\sin\omega t$: the required amplitude is $a/\{(n^2 - \omega^2)^2 + 4k^2\omega^2\}^{\frac{1}{2}}$.

52. The group velocity is $c - \lambda dc/d\lambda = c(1 + \lambda dn/nd\lambda)$.

53. 60 fathoms.

54. The angle for which total reflection occurs is given by $\sin\alpha = 1/n$.

55. $\sin\gamma = n \times \cdot00589$.

56. The nearest interference maxima lie on the diagonals.

57. $d = a\lambda/10$.

58. Distance between successive interference bands is $y\lambda/a$.

59. 55/104 ft.: 55/52 ft.

60. $n' = n/(1 - v/c) = 530$ approx.: $\lambda = 63$ cm.

INDEX

INDEX